MICROBIAL SURFACTANTS

Volume 3: Applications in Environmental Reclamation and Bioremediation

Books Published in Industrial Biotechnology Series

1. Silkworm Biofactory: Silk to Biology
   Enoch Y Park, Katsumi Maenaka (eds.), 2019
2. Probiotics, the Natural Microbiota in Living Organisms: Fundamentals and Applications
   Hesham El Enshasy, Shang Tian Yang (eds.), 2021
3. Microbial Surfactants—Volume 1: Production and Applications
   Riyazali Zafarali Sayyed, Hesham Ali El Enshasy, Hameeda Bee (eds.), 2021
4. Microbial Surfactants—Volume 2: Applications in Food and Agriculture
   R.Z. Sayyed and Hesham Ali El Enshasy (eds.), In Press
5. Microbial Surfactants—Volume 3: Applications in Environmental Reclamation and Bioremediation
   R.Z. Sayyed (ed.), In Press

Series: Industrial Biotechnology

MICROBIAL SURFACTANTS

Volume 3: Applications in Environmental Reclamation and Bioremediation

Editor

R.Z. Sayyed

Head, Department of Microbiology
PSGVP Mandal's Arts Science & Commerce College
Shahada, India

CRC Press
Taylor & Francis Group
Boca Raton London New York

CRC Press is an imprint of the
Taylor & Francis Group, an **informa** business
A SCIENCE PUBLISHERS BOOK

First edition published 2022
by CRC Press
6000 Broken Sound Parkway NW, Suite 300, Boca Raton, FL 33487-2742

and by CRC Press
2 Park Square, Milton Park, Abingdon, Oxon, OX14 4RN

CRC Press is an imprint of Taylor & Francis Group, LLC

Library of Congress Cataloging-in-Publication Data (applied for)

ISBN: 978-1-032-19635-0 (hbk)
ISBN: 978-1-032-19637-4 (pbk)
ISBN: 978-1-003-26016-5 (ebk)

Doi: 10.1201/9781003260165

Typeset in Times New Roman
by Radiant Productions

Preface to the Series

Industrial biotechnology has a deep impact on our lives, and is the focus of attention of academia, industry and governmental agencies and become one of the main pillars of knowledge based economy. The enormous growth of biotechnology industries has been driven by our increased knowledge and developments in physics, chemistry, biology, and engineering. Therefore, the growth of this industry in any part of the world can be directly related to the overall development in that region.

The interdisciplinary ***Industrial Biotechnology*** book series will comprise a number of edited volumes that review the recent trends in research and emerging technologies in the field. Each volume will covers specific class of bioproduct or particular biofactory in modern industrial biotechnology and will be written by internationally recognized experts of high reputation.

The main objective of this work is to provide up to date knowledge of the recent developments in this field based on the published works or technology developed in recent years. This book series is designed to serve as comprehensive reference and to be one of the main sources of information about cutting-edge technologies in the field of industrial biotechnology. Therefore, this series can serve as one of the major professional references for students, researchers, lecturers, and policy makers. I am grateful to all readers and we hope they will benefit from reading this new book series.

Series Editor

Prof. Dr. rer. Nat. Hesham A. El-Enshasy

Johor Bahru, Malaysia

Preface

Biosurfactants are surface active agents that are synthesized by various microorganisms. Biosurfactants possess similar properties that reduce the surface and interfacial tensions by mechanisms similar to those of synthetic surfactants. Biosurfactants exhibit numerous benefits over synthetic surfactants such as high biodegradability, low toxicity, low irritancy and compatibility with the human skin and ecology. Moreover, they possesses special characteristics like selectivity and specific activity at extreme temperatures, pH and salinity among other conditions. It is found that a number of microorganisms like bacteria, yeasts, and fungi show the ability to produce different classes of bio surfactants on the basis of their chemical composition and microbial origin. Since the last few decades, Biosurfactants are coming up as 'Green products' with an increasing consumer demand. Off all natural and safe products they show substantial potential in respect of implantation of sustainable industrial processes such as the use of cheap and renewable resources. In this context, biosurfactants have remarkable growth in the global market due to their promising and diverse applications, biosurfactants are used in environmental applications, phytoremediation, bioremediation, in petroleum, detergent and laundry industries.

The present book highlights the various aspects of biosurfactants including production, characterization, molecular biology, and applications of biosurfactants in various fields such Environmental, phytoremediation, bioremediation, in petroleum, detergent and laundry industries. This volume is reader friendly and easy-to-understand way, with well-illustrated diagrams, protocols, figures, and recent data. As such, this book will prove useful for students, researchers, teachers, and entrepreneurs in the area of biosurfactants and its allied fields.

Contents

1

Efficient Substrates for Microbial Synthesis of Biosurfactants

P Saranraj,[1,*] *R Z Sayyed,*[2] *Karrar Jasim Hamzah,*[3] *Neethu Asokan,*[1] *P Sivasakthivelan*[4] and *Abdel Rahman Mohammad Al-Tawaha*[5]

1. Introduction

Microbial surfactants or biosurfactants are surface-active molecules derived from a large number of microorganisms. These microbially produced surface-active compounds possess the ability to reduce the surface and interfacial tensions between two immiscible fluid phases. They are found in nature in a wide variety of chemical structures including glycolipids, lipopeptides and lipoproteins, fatty acids, neutral lipids, phospholipids, polymeric and particulate lipids. Biosurfactants are different from synthetic surfactants in being non-toxic, more effective and environment-friendly. Contrary to the chemical surfactants that are generally produced from petroleum feedstock, the microbial surfactants can be produced by using a wide variety of cheap agro-based raw materials. The features that make them commercially superior to their chemically synthesized counterparts are their stability at extremes of temperatures, pH and salinity. These properties are desirable in various industrial

[1] Department of Microbiology, Sacred Heart College (Autonomous), Tirupattur – 635 601, Tamil Nadu, India.

[2] Department of Microbiology, PSGVP Mandal's Arts, Science and Commerce College, Shahada – 425 409, Maharashtra, India.

[3] Department of Internal and Preventive Veterinary Medicine, College of Veterinary Medicine, AL-Qasim Green University, Babylon, Iraq.

[4] Department of Agricultural Microbiology, Faculty of Agriculture, Annamalai University, Annamalai Nagar – 608 002, Tamil Nadu, India.

[5] Department of Biological Sciences, Al - Hussein Bin Talal University, Maan, Jordan.

* Corresponding author: microsaranraj@gmail.com

processes such as food processing, pharmaceutical formulations, enhanced oil recovery and environmental bioremediation. Apart from the classical applications (Desai and Banat 1997), biosurfactants have also been reported to possess antibacterial, antifungal, antitumor, antimycoplasmic and antiviral properties (Cameotra and Makkar 2004, Singh and Cameotra 2004). Due to an increasing concern over the emergence of various multi-drug resistant pathogens, these molecules have emerged as potential drug molecules (Das et al. 2008). In spite of having such clear cut advantages, these molecules have not been commercialized extensively due to lower yields at the cellular level. This low level of production roots back to the genetics of these producer strains and thus, to increase the productivity that is essential to use recombinant and mutant hyper producing varieties of microorganisms. Although significant increase in the production was obtained by the optimization of the growth medium and environmental conditions (Sen 1997, Sen and Swaminathan 1997, Sen and Swaminathan 2004), the real breakthrough in their production enhancement can be achieved only by using hyper-producing recombinant and mutant varieties, as these have been reported to increase the yield manifolds. The development and use of these hyper-producers however demands a deep insight into their genetics (Mukherjee et al. 2006).

Microorganisms such as yeast, bacteria or fungi can produce biosurfactant-surface-active compounds using different substrates such as oils, glycerol, alkanes, sugars and wastes (Prasad et al. 2015, Shekhar et al. 2015). Biosurfactants are biodegradable, making them an attractive alternative to chemically synthesized surfactants which are normally petroleum-based and environmentally hazardous (Usman et al. 2016). High production costs pose a major obstacle to the widespread usage of biosurfactants (Nitschke and Sousa Silva 2018). Strategies to make biosurfactants commercially competitive include accessing agro-industrial wastes as cheap feedstock, developing overproducing robust wild-type or engineered strains, optimizing fermentation and downstream processes and combining production of biosurfactants with other biomolecules such as enzymes or bioplastics (Henkel et al. 2012, Nitschke and Sousa Silva 2018).

2. Biosurfactant Producing Microorganisms and their Industrial Uses

Microorganisms use a set of carbon sources and energy for growth. The combination of carbon sources with insoluble substrates facilitates the intracellular diffusion and production of different substances (Marchant et al. 2014). Microorganisms (yeasts, bacteria and some filamentous fungi) are capable of producing biosurfactants with different molecular structures and surface activities (Campos et al. 2013). In recent decades, there has been an increase in scientific interest regarding the isolation of microorganisms that produce tensioactive molecules with good surfactant characteristics, such as a low CMC, low toxicity and high emulsifying activity (Silva et al. 2014).

Various literatures describe bacteria of the genera *Pseudomonas* and *Bacillus* as great biosurfactant producers (Silva et al. 2014). However, most biosurfactants of a bacterial origin are inadequate for use in the food industry due to their possible pathogenic nature (Pattanath et al. 2008). *Candida bombicola* and *Candida lipolytica* are among the most commonly studied yeasts for the production of biosurfactants. A key advantage of using yeasts, such as *Yarrowia lipolytica, Saccharomyces cerevisiae* and *Kluyveromyces lactis*, resides in their "Generally regarded as safe (GRAS)" status. Organisms with GRAS status do not offer the risks of toxicity or pathogenicity, which allows their use in the food and pharmaceutical industries (Campos et al. 2013). Table 1 displays a list of microorganisms that produce biosurfactants.

Table 1. Classes of biosurfactants and respective producing microorganisms.

Biosurfactant class	Biosurfactant producing microorganisms
Glycolipids	*Acinetobacter calcoaceticus, Alcanivorax borkumensis, Arthrobacter paraffineus, Arthrobacter* sp., *Candida antarctica, Candida apicola, Candida batistae, Candida bogoriensis, Candida bombicola, Candida ishiwadae, Candida lipolytica, Lactobacillus fermentum, Nocardia* sp., *Pseudomonas aeruginosa, Pseudomonas* sp., *Rhodococcus erythropolis, Rhodotorula glutinous, Rhodotorula graminus, Serratia marcescens, Tsukamurella* sp. and *Ustilago maydis*
Polymeric surfactants	*Acinetobacter calcoaceticus, Bacillus stearothermophilus, Candida lipolytica, Candida utilis, Halomonas eurihalina, Mycobacterium thermoautotrophium* and *Sphingomonas paucimobilis*
Lipopeptides	*Acinetobacter* sp., *Bacillus licheniformis, Bacillus pumilus, Bacillus subtilis, Candida lipolytica, Gluconobacter cerinus, Pseudomonas fluorescens, Serratia marcescens, Streptomyces sioyaensis* and *Thiobacillus thiooxidans*
Fatty acids	*Arthrobacter paraffineus, Capnocytophaga* sp., *Corynebacterium insidibasseosum, Corynebacterium lepus, Nocardia erythropolis, Penicillium spiculisporum* and *Talaramyces trachyspermus*
Phospholipids	*Acinetobacter* sp., *Aspergillus* sp. and *Corynebacterium lepus*
Particulate biosurfactants	*Acinetobacter calcoaceticus, Cyanobacteria* and *Pseudomonas marginalis*

3. Metabolic Pathways of Biosurfactant Production

Hydrophilic substrates are primarily used by microorganisms for cell metabolism and the synthesis of the polar moiety of a biosurfactant, whereas hydrophobic substrates are used exclusively for the production of the hydrocarbon portion of the biosurfactant (Desai and Banat 1997, Tan 2000). Diverse metabolic pathways are involved in the synthesis of precursors for biosurfactant production and depend on the nature of the main carbon sources employed in the culture medium. For instance, when carbohydrates are the only carbon source for the production of a glycolipid, the carbon flow is regulated in such a way that both lipogenic pathways (lipid formation) and the formation of the hydrophilic moiety through the glycolytic pathway are suppressed by the microbial metabolism (Haritash and Kaushik 2009).

A hydrophilic substrate, such as glucose or glycerol, is degraded until the formation of Glycolytic pathway intermediates, such as Glucose-6-phosphate, which is one of the main precursors of carbohydrates found in the hydrophilic moiety of a biosurfactant. For the production of lipids, glucose was oxidized to pyruvate through glycolysis which was then converted to acetyl-CoA, which produces malonyl-CoA when united with oxaloacetate, followed by conversion into a fatty acid, which is one of the precursors for the synthesis of lipids (Hommel and Huse 1993). When a hydrocarbon is used as the carbon source, however, the microbial mechanism is mainly directed to the lipolytic pathway and Gluconeogenesis (the formation of glucose through different hexose precursors), thereby allowing its use for the production of fatty acids or sugars. The Gluconeogenesis pathway is activated for the production of sugars. This pathway consists of the oxidation of fatty acids through β-oxidation to acetyl-CoA (or propionyl-CoA in the case of odd numbered chain fatty acids). Beginning with the formation of acetyl-CoA, the reactions involved in the synthesis of polysaccharide precursors, such as Glucose-6-phosphate, are essentially the inverse of those involved in Glycolysis. However, reactions catalyzed by pyruvate kinase and phosphofructokinase-1 are irreversible. Thus, other enzymes exclusive to the process of Gluconeogensis are required to avoid such reactions (Tokumoto et al. 2009).

4. Biosynthesis of Biosurfactants

In their amphiphilic structure, the hydrophobic moiety is either a long-chain fatty acid, a hydroxy fatty acid, or α-alkyl β-hydroxy fatty acid, and the hydrophilic moiety may be a carbohydrate, carboxylic acid, phosphate, amino acid, cyclic peptide, or alcohol. Two primary metabolic pathways, namely, hydrocarbon and carbohydrate, are involved in the synthesis of their hydrophobic and hydrophilic moieties, respectively. The pathways for the synthesis of these two groups of precursors are diverse and utilize specific sets of enzymes. In many cases, the first enzymes for the synthesis of these precursors are regulatory enzymes; therefore, in spite of the diversity, there are some common synthesis and regulation features. The detailed biosynthetic pathways for the major hydrophobic and hydrophilic moieties have been extensively investigated and are well documented; however, a brief account by Hommel and Ratledge (1993) may be useful. According to Syldatk and Wagner (1987), the following possibilities exist for the synthesis of different moieties of biosurfactants and their linkages: (a) the hydrophilic and hydrophobic moieties are synthesized de novo by two independent pathways; (b) the hydrophilic moiety is synthesized de novo while the synthesis of the hydrophobic moiety is induced by the substrate; (c) the hydrophobic moiety is synthesized de novo, while the synthesis of the hydrophilic moiety is substrate dependent; and (d) the synthesis of both the hydrophobic and hydrophilic moieties is substrate dependent. Examples of all the above possibilities have been well documented by Syldatk and Wagner (1987) and Desai et al. (1996).

5. Substrates used for Biosurfactant Production

Over the past decade, a number of cheap waste materials have been explored as substrates for biosurfactant production, thus bringing forth an effective cost-cutting strategy coupled with the much-needed waste management (Banat et al. 2014, Satpute et al. 2017). Immense scope was found for a variety of renewable and cheap industrial wastes to be used for biosurfactant production. Prominent among these are food and agroindustry based residues. Use of industrial waste for valuable compound production assumes importance in recent times not only in economizing any commercial production process but also in establishing a sustainable effort for effective management of the unprecedented waste generated (Patil and Rao 2015). However, the argument is not limited to just the cost of raw materials but also the availability, stability and variability of each component are also critical factors to be considered. Moreover, the amount to be used, form (solid or liquid), particle size, texture, packaging, transportation, storage, stability and purity, all play a critical role in the final selection and formulation of any substrate for biosurfactant production.

Different relatively cheap and abundant substrates are currently available for use as carbon sources from various industrial sectors (Table 2). Many of these substrates have been reported as suitable substrates for growth and production of a wide range of microbial amphiphilic molecules. These substrates are described in detail as follows.

Table 2. Various cheaper renewable substrates available from different industrial sectors.

Source industry	Wastes as cheap renewable substrate
Agro-industrial waste and crop residues	Bran, Beet molasses, Sugarcane bagasse, Wheat straw, Cassava, Cassava flour, Waste water, Rice paddy straw, Soy hull, Corn and Sugarcane molasses
Animal fat	Waste
Coffee processing residues	Coffee processing residues, Coffee pulp, Coffee husks and Free groundnut spent
Crops	Cassava, Potato, Sweet potato, Soybean, Sweet sugar beet and Sorghum
Dairy industry	Curd whey, Cheese whey and Whey waste
Distillery industry	Industrial effluents
Food processing industry	Frying edible oils and fats, Olive oil, Potato peels, Rape seed oil, Sunflower and Vegetable oils
Fruit processing industry	Banana waste, Pomace of apple and grape, Carrot industrial waste and Pine apple waste
Oil processing mills	Coconut cake, Canola meal, Olive oil mill waste water, Palm oil mill, Peanut cake, Effluent, Soybean cake, Soapstock and Waste from lubricating oil

Source: Ibrahim Banat et al. (2014)

5.1 *Olive Oil Mill Effluent (OOME)*

Olive Oil Mill Effluent (OOME) is concentrated black liquor with a water-soluble portion of ripe olives and water that is used for the extraction of olive oil. OOME has polyphenols that represent a challenge in terms of disposal into the environment. However, it also contains nitrogen compounds (12 to 24 g/L), sugars (20 to 80 g/L), residual oil (0.3 to 5 g/L) and organic acids (5 to 15 g/L). Olive oil extraction involves an intensive consumption of water and produces large amounts of olive oil mill wastewater, thus causing deleterious environmental effects. OOME is a black liquor and consists of a high organic matter content (20–60 kg COD/m^3), depending on the olive oil extraction procedure (Marques 2001). OOME not only contains toxic substances such as polyphenols (Hamman et al. 1999, Marques 2001), but also valuable organic substances such as sugars, nitrogen compounds, organic acids and residual oils (Mercade et al. 1993). Mercade et al. (1993) successfully employed OOME for the strain *Pseudomonas* sp. to produce Rhamnolipids.

5.2 *Animal Fat*

Animal fat and tallow can be obtained in large quantities from meat processing industries and have been used as a cooking media for foods. However, these fats have recently lost most of their market share to vegetable oils owing to health concerns. Deshpande and Daniels (1995) used animal fat for the production of Sophorolipid biosurfactant with the yeast, *Candida bombicola*. When only fat was provided as a sole carbon source, the growth was poor. The mixture of 10% glucose and 10% fat gave the highest level of growth. Sophorolipid was produced at levels of 97 g/L and 12 g/L with and without pH control, respectively. Using animal fat and corn steep liquor, Santos et al. (2014) achieved maximum Glycolipids production by the yeast *Candida lipolytica*. The authors also report that the product has uses in bioremediation as well as oil mobilization and recovery.

5.3 *Vegetable Oils and Frying Oils*

Vegetable oils such as sunflower and soybean oil were used for the production of Rhamnolipid, Sophorolipid and Mannosylerythritol lipid biosurfactants by various microorganisms (Kim 2006). Apart from various vegetable oils, oil wastes from vegetable oil refineries those from the food industry were also reported as good substrates for biosurfactant production. Furthermore, various waste oils with their origins at the domestic level, in vegetable oil refineries or soap industries were found to be suitable for microbial growth and biosurfactant production (Haba 2000). Vegetable oil is a common medium used in large quantities for biosurfactant production. It can act as an effective and inexpensive raw material for biosurfactant production. Off different vegetable oils, mustard oil has been used by Vandana and Peter (2014) for biosurfactant production.

Frying oil and edible fats are considered great carbon sources for biosurfactant production. Vegetable oils constitute a lipid carbon source and are mainly comprised of saturated or unsaturated fatty acids with chains of 16 to 18 carbon atoms (Makkar et al. 2011). Different oils are adequate substrates for biosurfactants. Babassu oil (5% v/v) with a carbon source (1% glucose w/v) is a good medium for biosurfactant production. Sarubbo et al. (1999) found that two strains of *Candida lipolytica* produce biosurfactants toward the final exponential growth phase and onset of the stationary phase. Sunflower and olive oils have proven to be adequate energy and carbon sources for the production of biosurfactants. *Pseudomonas aeruginosa* strains produce a biosurfactant on residue from corn, soybean and canola oil plants (Raza et al. 2007). Canola oil residue and sodium nitrate have been reported adequate for microbial growth and production of up to 8.50 g/L of Rhamnoipids. The combination of glucose and canola oil has been used for the successful production of a biosurfactant by *Candida lipolytica* (Sarubbo et al. 2007). Used edible oils and fats are considered as a problematic waste, contributing to the environmental pollution. It is well known that microorganisms are able to grow on vegetable oils or fats and produce new products with potential industrial application such as lipase (Haba et al. 2000) and biodiesel (Alcantara et al. 2000, Cvengros and Cvengrosova 2004).

5.4 *Soap Stocks*

Soap stock is a gummy, amber-coloured by-product of oilseed processing. It is produced when hexane and other chemicals are used to extract and refine edible oil from seeds. Oil cakes or Soapstocks are produced from oilseed processing involving the refining of seed based edible oils with the use of chemicals (Bednarski et al. 2004). Soapstock has been used together with sunflower oil, olive oil or soybean oil as substrates to produce Rhamnolipids. Yields as high as 15.9 g/L have been reported using *Pseudomonas aeruginosa* in a Soapstock medium (Benincasa et al. 2002). Soapstock and oil refinery wastes have been used with *Candida antarctica* or *Candida apicola* for biosurfactant production, achieving a greater yield than that in the medium without oil refinery residue (Bednarski et al. 2004). Shabtai (1990) reported the production of two extracellular capsular Heteropolysaccharides, Emulsan and Biodispersan by *Acinetobacter calcoaceticus* using Soapstock as a carbon source. Emulsan forms and stabilizes the oil water emulsion (Kim et al. 2000), whereas Biodispersan disperses the large solid limestone granules, forming micrometer sized water suspensions (Rosenberg et al. 1988).

5.5 *Molasses and Soy Molasses*

Molasses is a co-product of the sugar industry generated during sugar manufacturing from either sugarcane or sugar beet and is a rich source of available carbon. It is defined as the runoff syrup from the final stage of crystallization, in which further crystallization of sugar is uneconomical. Average values for the constituents of cane molasses (75% dry matter) are 48–56% (total sugar), 9–12% (organic matter excluding sugar), 2.5% (protein), 1.5–5.0% (potassium), 0.4–0.8% (calcium), 0.06% (magnesium), 0.06–2.0% (phosphorus), 1.0–3.0 mg/kg (biotin), 15–55 mg/kg (pantothenic acid), 2,500–6,000 mg/kg (inositol) and 1.8 mg/kg (thiamine).

Soy molasses are generated during soybean processing and composed of carbohydrates, minerals, fats, lipids and others. Soy molasses contain a mixture of sugars and have been used for Rhamnolipids production using *Pseudomonas aeruginosa*, giving a Rhamnose concentration of 6.9 g/L and biosurfactant concentration of 11.7 g/L (Rodrigues et al. 2017). Molasses distillery wastewater was used as a substrate for Rhamnolipids production by *Pseudomonas aeruginosa* and the yield reached 2.6 g/L (Li et al. 2011). *Pseudomonas aeruginosa* strain was able to use distillery waste from the alcohol industry and curd whey waste from the milk industry as substrates for Rhamnolipids production and the yields reached 0.91 and 0.92 g/L, respectively (Dubey and Juwarkar 2001).

Sugarcane molasses is the final effluent of sugar refinement and comprises approximately 40% (w/w) sugars. Molasses is normally used as an ingredient in some food products and it has an impact on the immune system (Rahiman and Pool 2016). Adding molasses into the medium is able to increase Rhamnolipids production by *Pseudomonas fluorescens* and enhances phenol degradation (Suryanti et al. 2015). Other bacteria such as *Bacillus subtilis* were reported to utilize sugarcane molasses for biosurfactant production at 45°C (Makkar and Cameotra 1997). The yield of Rhamnolipids reached 0.24 g/L when *Pseudomonas aeruginosa* GS3 was grown in a medium that contained molasses and corn steep liquor (Patel and Desai 1997). *Bacillus licheniformis* and *Bacillus subtilis* grown on molasses produced biosurfactant at 3.3 and 3.78 g/L, respectively (Dos Santos Lopes et al. 2017). In addition, exploded sugarcane bagasse has been utilized to co-produce Rhamnolipids (9.1 g/L) and ethanol (8.4 g/L) using *Saccharomyces cerevisiae*, *Pseudomonas aeruginosa* and Crude Enzyme Complexes (CECs) (Dos Santos Lopes et al. 2017). The CECs were produced by *Aspergillus niger* in Solid state fermentation using different levels of exploded sugarcane bagasse, rice bran and corn cob as substrates (Dos Santos Lopes et al. 2017).

Ghurye and Vipulanandan (1994) used activated sludge from wastewater treatment as a source of microorganisms for biosurfactant production. A molasses concentration of 20 g/L was used as a carbon source. The production of biosurfactants appeared to be associated with growth since the Critical Micelle Dilution (CMD) and emulsification capacity increased with increasing biomass. Biosurfactants might consist of proteins or peptides moieties because pronase lowers the emulsification capacity of the cell-free broth.

Bacillus subtilis isolates MTCC 2423 and *Bacillus subtilis* MTCC 1427 were cultivated using molasses (2% total sugar) as a carbon source and incubated in thermophilic conditions (45°C). Maximal biosurfactant production as evidenced by surface tension lowering was achieved from both strains in the late stationary phase. However, strain MTCC 2423 produced greater biosurfactant content than strain MTCC 1427. As a result of biosurfactant accumulation, the surface tension of the medium was lowered to 29 and 31 dynes/cm by MTCC 2423 and MTCC 1427, respectively. Additionally, oil recovery from a sand pack column was 34% for MTCC 1427 and 38.46% for MTCC 2423, indicating the potential use of these biosurfactants in enhanced oil recovery (Makkar and Cameotra 1997).

Patel and Desai (1997) used molasses and corn steep liquor as the primary carbon and nitrogen sources to produce rhamnolipid biosurfactants from *Pseudomonas aeruginosa*. The biosurfactant production (quantified by measuring the interfacial tension and expressing Rhamnolipids in term of rhamnose) reached the maximum when 7% (v/v) of molasses and 0.5% (v/v) of corn steep liquor were used. Maximal surfactant production occurred after 96 hrs of incubation, when the cells reached the stationary growth phase. A rhamnose concentration of 0.25 g/L and a reduction of interfacial tension between surfactant and crude oil to up to 0.47 mN/m were achieved .

5.6 *Whey*

The dairy industry produces large quantities of whey, such as whey waste, cheese, curd and lactic whey, all of which can be used as substrates for the microbial production of metabolites (Banat et al. 2014). A high amount of lactose (approximately 75%) is found in lactic whey. Other components, such as proteins, vitamins and organic acids, are good sources for microbial growth and biosurfactant production (Thompson et al. 2000). Moreover, whey disposal represents a major pollution problem, especially in countries that depend on a dairy economy (Helmy et al. 2011). Thus, the disposal of this by-product represents the waste of a largely available substrate and an environmental hazard.

The effluent from the dairy industry, known as dairy wastewater, supports good microbial growth and is used as a cheap raw material for biosurfactant production. Dubey and Juwarkar (2004) cultivated *Pseudomonas aeruginosa* on whey waste; within 48 hrs of incubation the yield of biosurfactant obtained was 0.92 g/l. *Pseudomonas aeruginosa* strain produced a crystalline biosurfactant as the secondary metabolite and its maximal production occurred after the onset of nitrogen limiting conditions. The isolated biosurfactant possessed potent surface-active properties, as it effectively reduced the surface tension of water from 72 to 27 mN/m and formed a 100% stable emulsion of a variety of water insoluble compounds. *Lactobacillus pentosus* was capable of producing biosurfactants in a medium containing cheese whey at 1.4 g/L (Rodrigues et al. 2006, Banat et al. 2014). *Yarrowia lipolytica*, *Micrococcus luteus* and *Burkholderia cepacia* produced biosurfactants in whey wastewaters from milk factories (Yilmaz et al. 2009). Whey tofu-a waste during tofu production was shown to be a viable medium for Rhamnolipids production by *Pseudomonas fluorescens* (Suryanti et al. 2016).

5.7 *Corn Steep Liquor*

The agro-industry of corn-based products through wet processing results in both solid and liquid by-products, which, when disposed improperly, become a source of contamination and harm to the environment. Corn steep liquor is a by-product of washing water and soaking kernels as well as fractionating into starch and germen oil that contain 40% of solid matter. This by-product consists of 21% to 45% of proteins, 20% to 26% of lactic acid and approximately 8% of ash (containing Ca^{2+}, Mg^{2+}, K^{+}), approximately 3% of carbohydrates and a low fat content (0.9% to 1.2%) (Helmy et al. 2011). Nut oil refinery residue and corn steep liquor are low-cost nutrients for the production of Glycolipids by *Candida sphaerica*. The biosurfactant of this strain mobilises and removes up to 95% of motor oil on sand, making it useful for bioremediation (Luna et al. 2015). Silva et al. (2014) also report the production of a biosurfactant from *Pseudomonas cepacia* grown in a mineral medium supplemented with 2.0% of corn steep liquor and 2.0% of soybean waste frying oil.

5.8 *Starchy Substrates*

Abundant starch-based substrates also constitute renewable carbon sources. The potato processing industry produces significant amounts of starch-rich wastes that are adequate for biosurfactant production. Besides the approximate 80% water content, potato waste has carbohydrates (17%), proteins (2%) and fats (0.1%) as well as inorganic minerals, trace elements and vitamins (Helmy et al. 2011). Fox and Bala (2000) investigated a commercially prepared potato starch in a mineral salt medium for the production of a biosurfactant by *Bacillus subtilis*. Cassava wastewater, which is another carbohydrate rich waste product generated in large amounts, has been used for the production of Surfactin by *Bacillus subtilis* (Hatha et al. 2007). Other starchy wastes, such as rice water and cereal processing wastewater have the potential to permit microbial growth and biosurfactant production (Muthusamy et al. 2008).

Potato substrates were evaluated as a carbon source for surfactant production by *Bacillus subtilis*. Surface tensions dropped from 71.3 mN/m to 28.3 mN/m and 27.5 mN/m when potato and mineral salt media were used, respectively. A CMC of 0.10 g/L was obtained from a methylene chloride extract of the potato solid medium (Fox and Bala 2000). Furthermore, high solids and low solids potato process effluents were used as substrates for Surfactin production by *Bacillus subtilis* (Thompson et al. 2000). Surfactin production from LS potato effluent gave a yield (0.39 g/l) greater than that from the HS potato effluent (0.097 g/l). The Fourier Transform Infrared Spectroscopy (FTIR) of commercial Surfactin and the LS precipitate revealed that the biosurfactant produced was Surfactin.

Cassava flour wastewater mixed with nutrient broth served as a medium for Rhamnolipids production by *Pseudomonas fluorescens* (Suryanti et al. 2016). When solid residues were removed from cassava wastewater and autoclaved for use as a growth medium for various *Pseudomonas aeruginosa* strains, Rhamnolipids yields between 169.9 and 300.3 mg/L were achieved (Costa et al. 2009). One study showed an integrated process in which 10.5 g/L of biosurfactant was first produced by *Pseudozyma tsukubaensis*, and the resulting microbial cells were used to synthesize galactooligosaccharides from lactose (Fai et al. 2015).

5.9 Sugarcane Vinasse

Sugarcane vinasse, a residue from bioethanol production is a common waste during fermentation using a substrate from sugar crops. It also contains sugars and is able to serve as a substrate for Rhamnolipids production (Naspolini et al. 2017). *Pseudomonas aeruginosa* could produce 2.7 g/L of Rhamnolipids when Sugarcane vinasse was used as a substrate in submerged fermentation (Naspolini et al. 2017). *Pseudomonas luteola* and *Pseudomonas putida* grown on an autoclaved medium consisting of sugarcane beet molasses mixed with distilled water, gave maximum Rhamnolipids production at 72 hrs (Onbasli and Aslim 2009). Other strains such as *Bacillus subtilis* can also use Sugarcane vinasse as sources for biosurfactant and energy production (Azran et al. 2014). In addition to vinasse and molasses, sugar cane refining by-products such as sweet water have also been used as carbon sources for *Pseudomonas aeruginosa* with Rhamnolipids yields of 4.0–4.7 g/L (Li et al. 2014).

5.10 Lignocelluloses

Lignocellulose is present in agricultural products, which makes many related wastes or byproducts attractive low cost substrates for biosurfactant production. Lignocellulosic biomass has been used as an alternative cost-effective substrate for the microbial production of Rhamnolipids as it can be converted into fermentable sugars. It has been noted that pretreatment is required to convert lignocellulosic biomass to fermentable sugars because several enzymes are required for the cellulose degradation. Converting cellulose to fermentable sugars has been well studied. Pretreatment—a step to obtain cellulose is required, which is a critical step for the later enzymatic process as the lignin and other components may affect enzymatic activities or prevent enzymes from accessing cellulose (Li et al. 2014). Cell-degrading enzymes or strains will be mixed with cellulose to obtain fermentable sugars which can serve as carbon sources for microorganisms (Bras et al. 2011).

5.11 *Wheat Straw*

Pretreatment of wheat straw with sulphuric acid, phosphoric acid and ammonia is then followed by enzymatic hydrolysis with Cellulases from *Trichoderma reesei* to obtain sugars (Prabu et al. 2015). The resulting sugars were used to produce Rhamnolipids (9.38 g/L) by *Pseudomonas aeruginosa* (Prabu et al. 2015). Similar to wheat straw, lignocelluloses containing wastes such as barley pulp have been used for Rhamnolipids production using *Pseudomonas aeruginosa* and the yield of Rhamnolipids reached 2.4 g/L (Kaskatepe et al. 2017). Addition of glycerol to the media could increase Rhamnolipids yield to 9.3 g/L (Kaskatepe et al. 2017).

5.12 *Fruit Product Wastes*

In addition to wheat straw, yellow cashew fruit bagasse was crushed into powder and mixed with basal mineral medium for Rhamnolipids production using *Pseudomonas aeruginosa* (Iroha et al. 2015). Cashew apple juice was shown to serve as a medium for *Acinetobacter calcoaceticus* growth and biosurfactant production (Rocha et al. 2006). Potato peel mixed with urea was able to affect biosurfactant production positively (Ansari et al. 2014). *Pseudomonas aeruginosa* could produce 9.18 g/L of Rhamnolipids using orange peels as carbon sources (George and Jayachandran 2009). Fruit processing waste and sugar industry effluents were also found to be viable substrates for biosurfactant fermentation by *Kocuriaturfanesis* and *Pseudomonas aeruginosa* (Dubey et al. 2012). As part of whole waste recycling, vineyard pruning waste was collected and acid hydrolysis was performed to remove the hemicellulosic sugars (Vecino et al. 2017). The remaining Lignocellulosic fraction underwent delignification. The cellulosic fraction was then subjected to enzymatic hydrolysis using cellulase and β-glucosidase to obtain cellulosic sugars. The sugars were then used in fermentation as a low cost carbon source for biosurfactant production by *Lactobacillus paracasei* (Vecino et al. 2017). Distilled grape marc was also discovered to be a low cost feedstock of sugars for biosurfactant production by *Lactobacillus pentosus*, with values of relative emulsion volume close to 50% and stabilizing capacity values to maintain the emulsion at 99% (Jamal et al. 2014).

6. Advantages and Disadvantages of Cheaper Substrates in Biosurfactant Production

6.1 *Advantages of Cheaper Substrates in Biosurfactant Production*

- Commercial production cost can be reduced.
- Many cheaper or renewable substrates are available.
- Substrates are available in huge quantities.
- Enhance the yield of biosurfactant or bioemulsifier.
- Basic functional properties of the product do not change.

- Does not prove harmful to microorganisms.
- All components are eco-friendly and safe.

6.2 Disadvantages of Cheaper Substrates in Biosurfactant Production

- Substrates contain undesired compounds.
- Processing or treatment of the substrates is required to use them as carbon, nitrogen, or energy sources.
- The final product itself gets color or carries impurities from the substrates (e.g., Molasses).
- Special purification techniques need to be employed to obtain the pure product, thus increasing the production cost subsequently.
- Continuous supply of raw material with the same composition may vary.
- Raw substrates may be very specific for different organisms.
- A large quantity of raw substrates is essential, which may be difficult to get for a continuous supply for industrial processes.

7. Economics of Biosurfactant Production

To overcome the high cost constraints associated with biosurfactant production, two basic strategies are generally adopted worldwide to make it cost effective: (i) The use of inexpensive and waste substrates for the formulation of fermentation media which lower the initial raw material costs involved in the process; (ii) Development of efficient and successfully optimized bioprocesses, including optimization of the culture conditions and cost effective recovery processes for maximum biosurfactant production and recovery. As millions of tons of hazardous and non-hazardous wastes are generated each year throughout the world, a great need exists for their proper management and utilization. The residues from tropical agronomic crops such as cassava peels, soybeen (hull) (Lima et al. 2009), sugar beet (Onbasil and Aslim 2009), sweet potato (peels and stalks), potato (peels and stalks), sweet sorghum (Makkar and Cameotra 2002), rice and wheat bran and straw (Nadia Krieger et al. 2010), hull soy, corn and rice; bagasse of sugarcane and cassava, residues from the coffee processing industry such as coffee pulp, coffee husks, spent coffee grounds; residues of the fruit processing industries such as pomace and grape, wastes from pineapple and carrot processing, banana waste; wastes from oil processing mills such as coconut cake, soybean cake, peanut cake, canola meal and palm oil mill waste; saw dust, corn cobs, carob pods, tea waste, chicory roots among others have been reported as substrates for biosurfactant production (Pandey et al. 2000). Additional substrates used for biosurfactant production include water-miscible wastes, molasses, whey milk or distillery wastes (Rocha et al. 2007).

8. Global Biosurfactant Market

Surfactants have a huge demand worldwide. The global market for surfactants was estimated to be USD 30–64 Billion in 2016 and was predicted to grow to USD 39–86 Billion by 2021 (Kumar et al. 2016). With more stringent regulations on greener processes and catering to the huge demand, biosurfactants form a major share of the surfactants market. The global biosurfactant market in 2013 was 34,4068 tons and is expected to reach 46,1991 tonnes by 2020, growing at a CAGR of 4.3% from 2014 to 2020 (Gudina et al. 2016). Biosurfactant market revenue generation was over $1.8 Billion in 2016 and is expected to reach USD 2.6 Billion by 2023 (540 kilo tons by 2024) with the Rhamnolipid market set to witness a gain of over 8% (Geetha et al. 2018). Another market research projected the global biosurfactant market at over 5.52 billion by 2022, at a CAGR of 5.6% from 2017 to 2022 (Ma et al. 2016). Europe is emerging and is expected to continue to grow as the biggest market (around 53%) followed by the United States mainly due to stricter regulatory guidelines in the region. However, increasing awareness and infrastructure in Asian countries is making them a rising consumer of biosurfactants. From among biosurfactants, Sophorolipids were found to have the largest global market share with the detergent industry leading the product application sector. BASF Cognis (Germany) and Ecover (Belgium) have emerged as the two top surfactant manufacturers to venture into the biosurfactant market. Other producers in the market are MG Intobio, Urumqui Unite, Saraya, Sun Products Corporation, Akzo Nobel, Croda International PLC, Evonik Industries (Germany), Mitsubishi Chemical Corporation and Jeneil Biosurfactant (Kalyani et al. 2014). However, in spite of the huge market demand, production of a biosurfactant is not as competitive as its synthetic counterparts. Synthetic surfactants are priced around $2 kg^{-1}$ in the market (Santos et al. 2017). Hence, economizing the biosurfactant production process assumes significance in order to sustain the market for these compounds in the current environmentally fragile scenario and long term sustainable development.

9. Conclusion

In the era of global industrialisation, the exploration of natural resources has served as a source of experimentation for science and advanced technologies, giving rise to the manufacturing of products with a high aggregate value in the world market, such as biosurfactants. The use of cheaper, renewable substrates from various industries such as agricultural (sugars, molasses, plant oils, oil wastes, starchy substances, lactic whey), distillery wastes, animal fat, oil industries have been reported and reviewed thoroughly by several researchers. Over the past decade, a number of cheap waste materials have been explored as substrates for biosurfactant production, thus bringing forth an effective cost-cutting strategy coupled with the much-needed waste management. Immense scope was found for a variety of renewable and cheap industrial wastes to be used for biosurfactant production. Various cheaper substrates such as soybean oil not only act as nutrients for microbial growth but also act as an important source for isolation of potential Biosurfactants producing microorganisms.

Biosurfactants can be easily produced by using cheap agro industrial wastes and use of oil as substrates. Biosurfactant production can be realized from waste oils such as edible or motor oil as a carbon source. Bacterial consortia can be more efficient for the production of biosurfactants as compared to a single microbe. Some studies proposed the replacement of harsh chemical surfactants with the green ones. The agricultural wastes can also be used for biosurfactant production. Novel recombinant varieties of these microorganisms, which can grow on a wide range of cheap substrates and produce biosurfactants at high yields, can potentially bring the required breakthrough in the biosurfactant production process. Although a large number of biosurfactant producers have been reported in the literature, biosurfactant research, particularly related to production enhancement and economics, has been confined mostly to a few genera of microorganisms such as *Bacillus*, *Pseudomonas* and *Candida*. A judicial and effective combination of these strategies might, in the future, lead the way towards large-scale profitable production of biosurfactants. This will make biosurfactants highly sought after biomolecules for present and future applications as fine specialty chemicals, biological control agents, and new generation molecules for pharmaceutical, cosmetic and health care industries.

References

Alcantara, R., J. Amores, L. Canoira, E. Fidalgo, M.J. Franco et al. 2000. Catalytic production of biodiesel from soy bean oil used frying oil and tallow. Biomass Bioene. 18: 515–527.

Ansari, F.A., S. Hussain, B. Ahmed, J. Akhter, E. Shoeb et al. 2014. Use of potato peel as cheap carbon source for the bacterial production of biosurfactants. Int. J. Biol. Res. 2: 27–31.

Azran, M., N. Fazielawanie, J. Saidin, M. Wahid, K. Bhubalan et al. 2014. Bioconversion of cane sugar refinery by-products into Rhamnolipid by marine *Pseudomonas aeruginosa*. International Conference on Beneficial Microbes (ICOBM); Penang, Malaysia.

Banat, I.M., S.K. Satpute, S.S. Cameotra, R. Patil, N.V. Nyayanit et al. 2014. Cost effective technologies and renewable substrates for biosurfactants production. Front. Microbiol. 5: 1–18.

Bednarski, W., M. Adamczak and J. Tomasik. 2004. Application of oil refinery waste in the biosynthesis of Glycolipids by yeast. Biores. Techn. 95: 15–18.

Benincasa, M., J. Contiero, M.A. Manresa and J.O. Moraes. 2002. Rhamnolipid productions by *Pseudomonas aeruginosa* LBI growing on soap stock as the sole carbon source. J. Food En. 54: 283–288.

Bras, J.L., A. Cartmell, A.L. Carvalho, G. Verze, E.A. Bayer et al. 2011. Structural insights into a unique cellulase fold and mechanism of cellulose hydrolysis. Proc. Natl. Acad. Sci. USA. 108: 5237–5242.

Cameotra, S. and R. Makkar. 2004. Recent applications of biosurfactants as biological and immunological molecules. Curr. Opin. Microbiol. 7: 262–266.

Campos, J.M., T.L.M. Stamford, L.A. Sarubbo, J.M. Luna, R.D. Rufino et al. 2013. Microbial biosurfactants as additives for food industries. Biotechnol. Prog. 29: 1097–1108.

Costa, S.G., F. Lepine, S. Milot, E. Deziel, E.M. Nitschke et al. 2009. Cassava wastewater as a substrate for the simultaneous production of Rhamnolipids and polyhydroxyalkanoates by *Pseudomonas aeruginosa*. J. Ind. Microbiol. Biotechnol. 36: 1063–1072.

Cvengros, O. and Z. Cvengrosova. 2004. Used frying oils and fats and their utilization in the production of methyl ester of higher fatty acids. Biom. Bioen. 27: 173–181.

Danyelle Khadydja, F., Santos, Raquel D., Rufino, Juliana M., Luna, Valdemir A. et al. 2016. Biosurfactants: Multifunctional biomolecules of the 21st century. Int. J. Mol. Sci. 17(401): 1–13.

Das, P., S. Mukherjee and R. Sen. 2008. Antimicrobial potential of a lipopeptides biosurfactant derived from a marine *Bacillus circulans*. J. Appl. Microbio. 3(7): 225–236.

De Lima, C.J.B., E.J. Ribeiro, E.F.C. Sérvulo, M.M. Resende, V.L. Cardoso et al. 2009. Biosurfactant production by *Pseudomonas aeruginosa* grown in residual Soybean Oil. Appl. Biochem. Biotechnol. 152: 156–168.

Desai, J.D. and I.M. Banat. 1997. Microbial production of surfactants and their commercial potential. Microbiol. Mol. Bio. Rev. 61: 47–64.

Deshpande, M. and L. Daniels. 1995. Evaluation of sophorolipid biosurfactant production by *Candida bombicola* using animal fat. Biores. Tech. 54: 143–150.

Deziel, E., G. Paquette, R. Villemur, F. Lepine, J. Bisaillon et al. 1996. Biosurfactant production by soil *Pseudomonas* strain growing on polycyclic aromatic hydrocarbons. Appl. Environ. Microbiol. 62: 1908–1912.

Dos Santos Lopes, V., J. Fischer, T.M.A. Pinheiro, B.V. Cabral, V.L. Cardoso et al. 2017. Biosurfactant and ethanol co-production using *Pseudomonas aeruginosa* and *Saccharomyces cerevisiae* co-cultures and exploded sugarcane bagasse. Renew. Ener. 109: 305–310.

Dubey, K. and A. Juwarkar. 2001. Distillery and curd whey wastes as viable alternative sources for biosurfactant production. World J. Microbiol. Biotechnol. 17: 61–69.

Dubey, K. and A. Juwarkar. 2004. Determination of genetic basis for biosurfactant production in distillery and curd whey wastes utilizing *Pseudomonas aeruginosa* strain BS2. Ind. J. Biotech. 3: 74–81.

Dubey, K.V., P.N. Charde, S.U. Meshram, S.K. Yadav, S. Singh et al. 2012. Potential of new microbial isolates for biosurfactant production using combinations of distillery waste with other industrial wastes. Pet. Environ. Biotechnol. 12: 1–11.

Fai, A.E.C., A.P.R. Simiqueli, C.J. De Andrade, G. Ghiselli, G.M. Pastore et al. 2015. Optimized production of biosurfactant from *Pseudozyma tsukubaensis* using cassava wastewater and consecutive production of galactooligosaccharides: an integrated process. Biocatal. Agric. Biotechnol. 4: 535–542.

Fox, S.L. and G.A. Bala. 2000. Production of surfactant from *Bacillus subtilis* ATCC 21332 using potato substrates. Biores. Tech. 75: 325–240.

Geetha, S.J., I.M. Banat and S.J. Joshi. 2018. Biosurfactants: production and potential applications in Microbial Enhanced Oil Recovery (MEOR). Biocatal. Agri. Biotechnol. 14: 23–32.

George, S. and K. Jayachandran. 2009. Analysis of rhamnolipid biosurfactant produced trough submerged fermentation using orange fruit peelings as sole carbon source. Appl. Biochem. Biotech. 158(3): 694–705.

Ghurye, G.L. and C. Vipulanandan. 1994. A practical approach to biosurfactant production using non-aseptic fermentation of mixed cultures. Biotechnol. Bioeng. 44: 661–666.

Gudina, E.J., A.I. Rodrigues, V. de Freitas, Z. Azevedo, J.A. Teixeira et al. 2016. Valorization of agro-industrial wastes towards the production of Rhamnolipids. Biores. Tech. 212: 144–150.

Haba, E. 2000. Screening and production of Rhamnolipids by *Pseudomonas aeruginosa* 47T2 NCIB 40044 from waste frying oils. J. Appl. Microbiol. 88: 379–387.

Haba, E., O. Bresco, C. Ferrer, A. Marques, M. Busquets et al. 2000. Isolation of lipase screening bacteria by developing used frying oil as selective substrate. Enzym. Microb. Technol. 26: 40–44.

Hamman, O.B., T. de la Rubia and J. Martinez. 1999. Decolorization of olive oil mill wastewater by *Phanerochaeteflavido-alba*. Environ. Toxicol. Chem. 18: 2410–2415.

Haritash, A.K. and C.P. Kaushik. 2009. Biodegradation aspects of Polycyclic Aromatic Hydrocarbons (PAHs): A review. J. Hazard. Mater. 169: 1–15.

Hatha, A.A.M., G. Edward and K.S.M.P. Rahman. 2007. Microbial biosurfactants—Review. J. Mar. Atmos. Res. 3: 1–17.

Helmy, Q., E. Kardena, N. Funamizu and W. Wisjnuprapto. 2011. Strategies toward commercial scale of biosurfactant production as potential substitute for its chemically counterparts. Int. J. Biotechnol. 12: 66–86.

Henkel, M., M.M. Müller, J.H. Kügler, R.B. Lovaglio, J. Contiero et al. 2012. Rhamnolipids as biosurfactants from renewable resources: concepts for next generation rhamnolipid production. Process Biochem. 47: 1207–1219.

Hommel, R.K. and C. Ratledge. 1993. Biosynthetic mechanisms of low molecular weight surfactants and their precursor molecules. pp. 3–63. *In*: Kosaric, N. (ed.). Biosurfactants: Production, Properties, Applications. Marcel Dekker, Inc., New York, N.Y.

Hommel, R.K. and K. Huse. 1993. Regulation of sophorose lipid production by *Candida* (*torulopsis*) *apicola*. Biotechnol. Lett. 15: 853–858.

Ibrahim, M. Banat, Surekha K. Satpute, Swaranjit S. Cameotra, Rajendra Patil, Narendra V. Nyayanit et al. 2014. Cost effective technologies and renewable substrates for biosurfactants production. Front. Microbiol. 5(697): 1–18.

Iroha, O.K., O.U. Njoku, V.N. Ogugua and V.E. Okpashi. 2015. Characterization of biosurfactant produced from submerged fermentation of fruits bagasse of yellow cashew (*Anacardium occidentale*) using *Pseudomonas aeruginosa*. Afr. J. Environ. Sci. Tech. 9: 473–481.

Jamal, P., S. Mir, M.Z. Alam and W.M.F. Wan Nawawi. 2014. Isolation and selection of new biosurfactant producing bacteria from degraded palm kernel cake under liquid state fermentation. J. Oleo. Sci. 63: 795–804.

Kalyani, A., G. Naga Sireesha, A. Aditya, G. Girija Sankar, T. Prabhakar et al. 2014. Production optimization of Rhamnolipid biosurfactant by *Streptomyces coelicoflavus* (NBRC 15399T) using Plackett-Burman design. Eur. J. Biotechnol. Biosci. 1: 07–13.

Kaskatepe, B., S. Yildiz, M. Gumustas and S.A. Ozkan. 2017. Rhamnolipid production by *Pseudomonas putida* IBS036 and *Pseudomonas pachastrellae* LOS20 with using pulps. Curr. Pharm. Anal. 13: 138–144.

Kim, H. 2006. Extracellular production of a Glycolipid biosurfactant, mannosylerythritol lipid by *Candida* sp. SY16 using Fedbatch fermentation. Appl. Microbiol. Biotech. 70: 391–396.

Kim, P., D.K. Oh, J.K. Lee, S.Y. Kim, J.H. Kim et al. 2000. Biological modification of the fatty acid group in an Emulsan by supplementing fatty acids under conditions inhibiting fatty acid biosynthesis. J. Biosci. Bioeng. 90: 308–312.

Kim, S.H., E.J. Lim, S.O. Lee, J.D. Lee, T.H. Lee et al. 2000. Purification and characterization of biosurfactants from *Nocardia* sp. Biotech. Appl. Biochem. 31: 249–253.

Kumar, A.P., A. Janardhan, B. Viswanath, K. Monika, J.Y. Jung et al. 2016. Evaluation of orange peel for biosurfactant production by *Bacillus licheniformis* and their ability to degrade naphthalene and crude oil. Biotech. 6: 1–10.

Li, A., M. Xu, W. Sun and G. Sun. 2011. Rhamnolipid production by *Pseudomonas aeruginosa* GIM 32 using different substrates including molasses distillery wastewater. Appl. Biochem. Biotechnol. 163: 600–611.

Li, Q., W.T. Ng and J.C. Wu. 2014. Isolation, characterization and application of a cellulose degrading strain *Neurosporacrassa* S1 from oil palm empty fruit bunch. Microb. Cell Fact. 13: 157.

Lima, C.J.B., E.J. Ribeiro, E.F.C. Sérvulo, M.M. Resende, V.L. Cardoso et al. 2009. Biosurfactant production by *Pseudomonas aeruginosa* grown in residual soybean oil. Appl. Biochem. Biotech. 152: 156–168.

Luna, J.M., R.D. Rufino, A.M.T. Jara, P.P.F. Brasileiro, L.A. Sarubbo et al. 2015. Environmental applications of the biosurfactant produced by *Candida sphaerica* cultivated in low cost substrates. Coll. Surf. Physicochem. Eng. Asp. 480: 413–418.

Ma, K.Y., M.Y. Sun, W. Dong, C.Q. He, F.L. Chen et al. 2016. Effects of nutrition optimization strategy on rhamnolipid production in a *Pseudomonas aeruginosa* strain DN1 for bioremediation of crude oil. Biocat. Agri. Biotechnol. 6: 144–151.

Makkar, R.S. and S.S. Cameotra. 1997. Utilization of molasses for biosurfactant production by two *Bacillus* strains at thermophilic conditions. J. Am. Oil Chem. Soc. 74: 887–889.

Makkar, R.S. and S.S. Cameotra. 2002. An update on use of unconventional substrates for biosurfactants production and their new applications. Appl. Microbio. Biotech. 58: 428–434.

Makkar, R.S., S.S. Cameotra and I.M. Banat. 2011. Advances in utilization of renewable substrates for biosurfactant production. Appl. Microbiol. Biotechnol. 1: 1–19.

Marchant, R., S. Funston, C. Uzoigwe, P.K. Rahman, M. Banat et al. 2014. Production of biosurfactants from non-pathogenic bacteria. *In*: Biosurfactants; Taylor & Francis: New York, USA, pp. 73–81.

Marques, I.P. 2001. Anaerobic digestion treatment of olive mill wastewater for effluent reuse in irrigation. Desalination 137: 233–239.

Mercade, M.E., M.A. Manresa, M. Robert, M.J. Espuny, C. de Andres et al. 1993. Olive oil mill effluent (OOME)—New substrate for biosurfactant production. Biores. Tech. 43: 1–6.

Mukherjee, S., P. Das and R. Sen. 2006. Towards commercial production of microbial surfactants. Trends Biotech. 24: 509–514.

Muthusamy, K., S. Gopalakrishnan, T.K. Ravi and P. Sivachidambaram. 2008. Biosurfactants: Properties, commercial production and application. Curr. Sci. 94: 736–747.

Nadia Krieger, C. Doumit and A.M. David. 2010. Production of microbial biosurfactants by solid state cultivation. Adv. Exp. Med. Bio. 672: 203–210.

Naspolini, B.F., A.C. De Oliveira Machado, W.B. Cravo Junior, D.G.M. Freire, M.C. Cammarota et al. 2017. Bioconversion of sugarcane vinasse into high added value products and energy. Bio. Med. Res. Int. 2: 111–119.

Nitschke, M. and S. Sousa Silva. 2018. Recent food applications of microbial surfactants. Crit. Rev. Food Sci. Nutr. 58: 631–638.

Onbasli, L. and B. Aslim. 2009. Biosurfactant production in sugar beet molasses by some *Pseudomonas* spp. J. Environ. Biol. 30: 161–163.

Pandey, A., C.R. Soccol and D. Mitchell. 2000. New developments in solid state fermentation: I-bioprocesses and products. Process Biochem. 35: 1153–1169.

Patel, R. and A. Desai. 1997. Biosurfactant production by *Pseudomonas aeruginosa* GS3 from molasses. Lett. Appl. Microbiol. 25: 91–94.

Patil, Y. and P. Rao. 2015. Industrial waste management in the era of climate change—a smart sustainable model based on utilization of active and passive biomass. pp. 2079–2092. *In*: Walter Leal, F. (ed.). Handbook on Climate Change Adaptation. Berlin Heidelberg, Germany: Springer-Verlag.

Pattanath, K.M., K.S. Rahman and E. Gakpe. 2008. Production, characterization and applications of biosurfactants. Biotech. 7: 360–370.

Prabu, R., A. Kuila, R. Ravishankar, P.V.C. Rao, N.V. Choudary et al. 2015. Microbial rhamnolipid production in wheat straw hydrolysate supplemented with basic salts. RSC Adv. 5: 51642–51649.

Prasad, B., H.P. Kaur and S. Kaur. 2015. Potential biomedical and pharmaceutical applications of microbial surfactants. World J. Pharm. Sci. 4: 1557–1575.

Rahiman, F. and E.J. Pool. 2016. The effect of sugar cane molasses on the immune and male reproductive systems using *in vitro* and *in vivo* methods. Iran J. Basic Med. Sci. 19: 1125–1130.

Raza, Z.A., A. Rehman, M.S. Khan and Z.M. Khalid. 2007. Improved production of biosurfactant by a *Pseudomonas aeruginosa* mutant using vegetable oil refinery wastes. Biodegradation 18: 115–121.

Rocha, M.V., A.H. Oliveira, M.C. Souza and L.R. Goncalves. 2006. Natural cashew apple juice as fermentation medium for biosurfactant production by *Acinetobacter calcoaceticus*. World J. Microbiol. Biotechnol. 22: 1295–1299.

Rocha, M., M. Souza and S. Benedicto. 2007. Production of biosurfactant by grown on cashew apple juice. Biochem. Biotech. 53: 136–140.

Rodrigues, L., A. Moldes, J. Teixeira and R. Oliveira. 2006. Kinetic study of fermentative biosurfactant production by *Lactobacillus* strains. Biochem. Eng. J. 28: 109–116.

Rodrigues, M.S., F.S. Moreira, V.L. Cardoso and M.M. de Resende. 2017. Soy molasses as a fermentation substrate for the production of biosurfactant using *Pseudomonas aeruginosa* ATCC 10145. Environ. Sci. Pollut. Res. 24: 18699–18709.

Rosenberg, E., C. Rubinovitz, A. Gottlieb, S. Rosenhak, E.Z. Ron et al. 1988. Production of biodispersan by *Acinetobacter calcoaceticus* A2. Appl. Environ. Microbiol. 54: 317–322.

Santos, D.K.F., Y.B. Brandão, R.D. Rufino, J.M. Luna, A.A. Salgueiro et al. 2014. Optimization of cultural conditions for biosurfactant production from *Candida lipolytica*. Biocatal. Agric. Biotechnol. 3: 48–57.

Santos, D.K.F., H.M. Meira, R.D. Rufino, J.M. Luna, L.A. Sarubbo et al. 2017. Biosurfactant production from *Candida lipolytica* in bioreactor and evaluation of its toxicity for application as a bioremediation agent. Process Biochem. 54: 20–27.

Sarubbo, L.A., M.C. Marcal, M.L.C. Neves, A.L.F. Porto, G.M. Campos Takaki et al. 1999. The use of babassu oil as substrate to produce bioemulsifiers by *Candida lipolytica*. Can. J. Microbiol. 45: 1–4.

Sarubbo, L.A., C.B.B. Farias and G.M. Campos Takaki. 2007. Co-utilization of canola oil and glucose on the production of a surfactant by *Candida lipolytica*. Curr. Microbiol. 54: 68–73.

Satpute, S.K., G.A. Płaza and A.G. Banpurkar. 2017. Biosurfactants production from renewable natural resources: Example of innovative and smart technology in circular bioeconomy. Manag. Sys. Prod. Eng. 25: 46–54.

Sen, R. 1997. Response surface optimization of the critical media components for production of Surfactin. J. Chem. Tech. Biotech. 68: 263–270.

Sen, R. and T. Swaminathan. 1997. Application of response-surface methodology to evaluate the optimum environmental conditions for the enhanced production of Surfactin. Appl. Microbiol. Biotechn. 47: 358–363.

Sen, R. and T. Swaminathan. 2004. Response surface modeling and optimization to elucidate the effects of inoculum age and size on Surfactin production. Biochem. Engi. 21: 141–148.

Shabtai, Y. 1990. Production of exopolysaccharides by *Acinetobacter* strains in a controlled fed-batch fermentation process using Soapstock oil (SSO) as carbon source. Int. J. Biol. Macromol. 12: 145–152.

Shekhar, S., A. Sundaramanickam and T. Balasubramanian. 2015. Biosurfactant producing microbes and their potential applications: a review. Crit. Rev. Environ. Sci. Technol. 45: 1522–1554.

Silva, R.C.F.S., D.G. Almeida, J.M. Luna, R.D. Rufino, V.A. Santos et al. 2014. Applications of biosurfactants in the petroleum industry and the remediation of oil spills. Int. J. Mol. Sci. 15: 12523–12542.

Singh, P. and S.S. Cameotra. 2004. Potential applications of microbial surfactants in biomedical sciences. Trends Biotech. 22: 142–146.

Suryanti, V., S.D. Marliyana and A. Wulandari. 2015. Biosurfactant production by *Pseudomonas fluorescens* growing on molasses and its application in phenol degradation. *In*: International Conference of Chemical and Material Engineering (ICCME). American Institute of Physics: 1699.

Suryanti, V., D. Handayani, S. Marliyana and S. Suratmi. 2016. Physicochemical properties of biosurfactant produced by *Pseudomonas fluorescens* grown on Whey Tofu. In International conference on advanced materials for better future. IOP Conf. Series: Mat. Sci. Engi. 176: 012–023.

Syldatk, C. and F. Wagner. 1987. Production of biosurfactants. pp. 89–120. *In*: Kosaric, N., W.L. Cairns and N.C.C. Gray (ed.). Biosurfactants and Biotechnology. Marcel Dekker, Inc., New York.

Tan, H.M. 2000. Biosurfactants and their roles in bioremediation. Cheong Jit Kong, 1–12.

Thompson, D.N., S.L. Fox and G.A. Bala. 2000. Biosurfactants from potato process effluents. Appl. Biochem. Biotechnol. 84: 917–930.

Tokumoto, Y., N. Nomura, H. Uchiyama, T. Imura, T. Morita et al. 2009. Structural characterization and surface-active properties of a succinoyl trehalose lipid produce by *Rhodococcus* sp. SD-74. J. Oleo Sci. 58: 97–102.

Usman, M.M., A.N. Dadrasnia, K.T. Lim, A.F. Mahmud and S. Ismail. 2016. Application of biosurfactants in environmental biotechnology; remediation of oil and heavy metal. AIMS Bioeng. 3: 289–304.

Vandana, P. and J.K. Peter. 2014. Production, partial purification and characterization of biosurfactant from *Pseudomonas fluorescens*. Int. J. Adv. Tech. Engi. Sci. 2(7): 258–264.

Vecino, X., L. Rodriguez Lopez, E. Gudina, J. Cruz, A. Moldes and L. Rodrigues. 2017. Vineyard pruning waste as an alternative carbon source to produce novel biosurfactants by *Lactobacillus paracasei*. J. Ind. Eng. Chem. 55: 40–49.

Yilmaz, F., A. Ergene, E. Yalcin and S. Tan. 2009. Production and characterization of biosurfactants produced by microorganisms isolated from milk factory wastewaters. Environ. Technol. 30: 1397–1404.

2

Microbial Biosurfactants Methods of Investigation, Characterization, Current Market Value and Applications

P Saranraj,[1,*] *R Z Sayyed,*[2] *P Sivasakthivelan,*[3] *Mustafa Salah Hasan,*[4] *Abdel Rahman Mohammad Al-Tawaha*[5] and *K Amala*[1]

1. Introduction

Biosurfactants are amphiphilic compounds produced on living surfaces, mostly microbial cell surfaces, or excreted extracellularly containing hydrophobic and hydrophilic moieties that reduce surface tension and interfacial tensions between individual molecules at the surface and interface, respectively. Since biosurfactants and bioemulsifiers both exhibit emulsification properties, bioemulsifiers are often categorized with biosurfactants, although emulsifiers may not lower surface tension. A biosurfactant may have one of the following structures: mycolic acid, glycolipids, polysaccharide–lipid complex, lipoprotein or lipopeptide, phospholipid, or the microbial cell surface itself (Vandana and Peter 2014).

Biosurfactants are most economically sought after 21st century biotechnological compounds. However, inefficient bioprocessing has mitigated the economical

[1] Department of Microbiology, Sacred Heart College (Autonomous), Tirupattur, Tamil Nadu, India.
[2] Department of Microbiology, PSGVP Mandal's Arts, Science and Commerce College, Shahada – 425 409, Maharashtra, India.
[3] Department of Agricultural Microbiology, Faculty of Agriculture, Annamalai University, Annamalai Nagar, Tamil Nadu, India.
[4] University of Fallujah, College of Veterinary Medicine, Department of Internal and Preventive Medicine, Fallujah, Iraq.
[5] Department of Biological Sciences, Al - Hussein Bin Talal University, Maan, Jordan.
* Corresponding author: microsaranraj@gmail.com

commercial production of these compounds. Although much work is being done on the use of low-cost substrates for their production, a paucity of literature exists on the upcoming bioprocess optimization strategies and their successes and potential for economical biosurfactant production (Zhang et al. 2012). Biosurfactants have many advantages as compared to their chemically synthesized counterparts. Unlike synthetic surfactants, microbially produced compounds are easily biodegradable and thus particularly suited for environmental applications such as bioremediation and dispersion of oil spills. The chemical diversity of naturally produced amphiphiles offers a wider selection of surface-active agents with properties closely related to specific applications. In addition, microbial processes from raw materials, which are available in large quantities, can produce biosurfactants. Biosurfactants can also be produced from industrial wastes an area of particular interest for bulk production and can be efficiently used in handling industrial emulsions, control of oil spills, biodegradation and detoxification of industrial effluents and in bioremediation of contaminated soil (Noudeh et al. 2010).

The surfactants of bacterial, fungal, and yeast origin are referred to as microbial biosurfactants. Synthetic surfactants are widely used in industrial applications because of their availability in commercial quantities, unlike microbial surfactants. The use of synthetic surfactants in various industries is highly associated with huge environmental impact and undesired ecological disturbances. These drawbacks can be resolved through the use of biosurfactants in place of their synthetic congeners in addition to their favorable competitiveness and greener value. With properties, such as eco-friendliness, specificity, low toxicity, stability in varying environmental conditions, and chemical diversity, microbial biosurfactants stand the chance of replacing synthetic surfactants in industrial applications (such as the petroleum industry, bioremediation, agriculture, pharmaceuticals, food industry, laundry, cosmetics, and energy-saving technology) in the near future. The most widely used biosurfactants are rhamnolipids (from *Pseudomonas*), sophorolipids (mainly from *Torulopsis*), mannosylerythritol lipid (mainly from *Candida*), surfactin (from *Bacillus*), and emulsan (from *Acinetobacter*). Their use in different biotechnological applications will reduce environmental pollution that is currently caused by synthetic surfactants, thereby engendering sustainability. Currently, only a few small industries are producing microbial biosurfactants for commercial use in the global sector. Large industries should take a solid step to incorporate microbial biosurfactants in their commercial products to enhance their use in the global market. Although biosurfactants are thought to be eco-friendly, few research findings indicate that under certain circumstances they can be reversed to being toxic to the environment. Nevertheless, careful and controlled use of these interesting surface-active molecules will surely help in the enhanced cleanup of toxic environments and provide us with a clean environment.

2. Methods for Investigation of the Production of Biosurfactants

2.1 *Drop Collapse Assay*

The Drop collapse test is an easy and fast procedure to asses a microbe for biosurfactant production. This test does not require any special equipment, but only a small volume of the microbial sample. Drops of oil are placed on the slide and then 10 µl of the microbial sample is added by piercing the drop using a micropipette without disturbing the dome shape of the oil. If the drop collapses within 1 min it is considered to be positive for the Drop collapse test (Das and Chandran 2010, Vandana and Peter 2014). The drop collapse assay is rapid and easy to carry out, requires no specialized equipment but just a small volume of sample. In addition, it can be performed in microplates. This assay has been applied several times for screening purposes. However, it displays a relatively low sensitivity since a significant concentration of surface active compounds must be present in order to cause a collapse of the aqueous drops on the oil or glass surfaces.

The Drop collapse assay was originally developed by Jain et al. (1991). The drop collapse assay is quite simple and requires no specialized equipment and just a small volume of sample. This assay relies on the destabilization of liquid caused by the presence of surfactants. Therefore, drops off a cell suspension or a culture supernatant is placed on an oil coated, solid surface. If the liquid does not contain surfactants, the polar water molecules are repelled from the hydrophobic surface and the drops remain stable. If the liquid contains surfactants, the drops spread or even collapse because the force or interfacial tension between the liquid drop and the hydrophobic surface is reduced. The stability of drops is dependent on surfactant concentration and correlates with surface and interfacial tensions.

Drop collapse method is one of the qualitative methods used to determine the presence of biosurfactant. Tabatabaee et al. (2005) conducted experiments to confirm the reliability of the method using polystyrene microwell plate; oil-coated wells collapse was observed when the culture broth contained biosurfactant and there was no change in the shape of the droplets in the absence of biosurfactant.

2.2 *Surface or Interface Activity*

Surface tension measurement by a Du Nouy Ring-TypeTensiometer (Krüss, K10T) is one of the simplest techniques used. The surface tension measurement is carried out at room temperature after dipping the platinum ring in the solution for a while in order to attain equilibrium. A higher biosurfactant concentration in the test sample provides a lower surface tension until the CMC is reached (Crosman et al. 2002). This is widely used because of its accuracy, ease of use and ability to provide a fairly rapid measurement of surface and interfacial tensions.

2.3 *Thin Layer Chromatography*

Thin layer chromatography (TLC) is also used in preliminary characterization of the biosurfactant where the cell free extract containing biosurfactant is separated on a silica gel plate using chloroform: methanol: water (70:10:0.5, v/v/v); this is then followed by using colour developing reagents. Lipopeptide biosurfactant showed red spots in the presence of ninhydrin reagent, while glycolipid biosurfactant is detected as yellow spots when Anthrone is used as the colour reagent (Yin et al. 2009).

2.4 *Emulsification Index (E24)*

The emulsifying capacity is evaluated by an emulsification index (E24). The Standard method for determining E24 of the culture is by adding 2 ml of kerosene and 2 ml of the cell-free broth in a test tube, followed by vortexing for 2 min. The mixture is allowed to stand for 24 hrs. The E24 index can be defined as the percentage of the height of emulsified layer (cm) divided by the total height of the liquid column (cm) (Tabatabaee et al. 2005). The Emulsification index (E24) can be calculated by using the following equation (Wei et al. 2005, Sarubbo et al. 2006). The emulsification capacity of biosurfactants was actually developed by Cooper and Goldenberg (1987).

$$\textit{Emulsification index (E24)} = \textit{Height of emulsion formed} \times \textit{100/Total height of solution}$$

Another methodology has been reported in which the Emulsification activity was measured by adding 5 ml of mineral oil to 5 ml of supernatant in a graduated tube and vortexed vigorously. The test tube containing the mixture was then detained for 24 hours and the Emulsification index (E24%) was determined using above formula (Noudeh et al. 2010).

2.5 *Parafilm M Test*

The bacterial supernatant is mixed with 1% xylene cyanol and drops of the mixture are added onto the surface of Parafilm M, which is hydrophobic in nature. The shape of the drop is checked after 1 min. This is followed by evaluating the diameters of the drops. The spread out drops signify the presence of surfactants (Morita et al. 2007).

2.6 *Blood Agar Hemolysis Assay*

Blood agar hemolysis assay is used for preliminary screening of microorganisms for the ability to produce Biosurfactants. Blood agar plate containing 5% Sheep blood is used for the test of hemolytic activity. Positive strains will cause lysis of the blood cells and exhibit a colourless, transparent ring around the colonies. Therefore, those microorganisms which show positive blood hemolysis are considered as biosurfactant producers. The basic principle approach followed is to screen biosurfactants for causing lysis of erythrocytes. The assay also predicts the surface activity of biosurfactant producing microorganisms (Vandana and Peter 2014). As per recommendation the Blood agar method is a preliminary screening method which should be supported by other techniques based on surface activity measurements (Mulligan et al. 1984).

2.7 *Oil Displacement Activity*

The oil spreading assay was developed by Morikawa et al. (2000). For this assay, 10 μl of crude oil is added to the surface of 40 ml of distilled water in a petridish to form a thin oil layer. Then, 10 μl of culture or culture supernatant are gently placed on the centre of the oil layer. If biosurfactant is present in the supernatant, the oil is displaced and a clearing zone is formed. The diameter of this clearing zone on the oil surface correlates to surfactant activity, also called Oil displacement activity.

2.8 *Bacterial Adherence to Hydrocarbon (BATH)*

Bacterial Adhesion to Hydrocarbon Assays was developed and proposed by Rosenberg et al. (1988). It is a simple method for determining the surface hydrophobicity characteristic of bacterial cells.

2.9 *CTAB Agar Plate Method*

The CTAB (Cethyl Trimethyl Ammonium Bromide) agar plate method was developed by Siegmund and Wagner (1991). It is a semi-quantitative assay for the detection of extracellular Glycolipids or other anionic surfactants. The CTAB agar assay is a comfortable screening method, specific for anionic biosurfactants. The disadvantage of this method is that CTAB is harmful as it inhibits the growth of some microbes. CTAB contains the cationic surfactant. In the process microbes growing on the plate secrete anionic surfactants which form a dark blue, insoluble ion that pairs with CTAB and Methylene blue. Thus, productive colonies are surrounded by dark blue halos.

2.10 *Microplate Assay*

The surface activity of individual strains can be determined qualitatively with the Microplate assay developed and patented by Vaux and Cottingham (2001). This assay is based on the change in optical distortion that is caused by surfactants in an aqueous solution. Pure water in a hydrophobic well has a flat surface. The presence of surfactants causes some wetting at the edge of the well and the fluid surface becomes concave and takes the shape of a diverging lens. For this assay, a 100 µl sample of the supernatant of each strain is taken and put into a microwell of a 96-mircowell plate. The plate is viewed using a baking sheet of paper with a grid. The optical distortion of the grid provides a qualitative assay for the presence of surfactants. The Microplate assay is easy, rapid and sensitive and allows an instantaneous detection of surface-active compounds (Chen et al. 2007). Just a small volume (100 µl) of sample is needed. Furthermore, the method is suitable for automated high throughput screening. Chen et al. (2007) demonstrated the efficiency of the Microplate method for high throughput screening purposes.

2.11 *Penetration Assay*

Maczek et al. (2007) developed another assay suitable for high throughput screening, called the Penetration assay. This assay relies on the contacting of two insoluble phases which leads to a colour change. For this assay, the cavities of a 96 well microplate are filled with 150 µl of a hydrophobic paste consisting of oil and silica gel. The paste is covered with 10 µl of oil. Then, the supernatant of the culture is colored by adding 10 µl of a red staining solution to 90 µl of the supernatant. The colored supernatant is placed on the surface of the paste. If biosurfactant is present, the hydrophilic liquid breaks through the oil film barrier into the paste. The silica enters the hydrophilic phase and the upper phase changes from clear red to cloudy white within 15 minutes. The described effect relies on the phenomenon that silica gel enters the hydrophilic phase from the hydrophobic paste much faster if biosurfactants are present. Biosurfactant free supernatant turns cloudy but stays red. The penetration assay is a simple, qualitative technique for screening large amounts of potential isolates. It can be applied in high throughput screening. The assay was described as recently as 2007 and to our knowledge there has been no further report of its application by now.

2.12 *Solubilization of Crystalline Anthracene*

Willumsen and Karlson (1997) developed an assay based on the solubilization of crystalline anthracene. This screening method is based on the solubilization of a highly hydrophobic, crystalline compound, anthracene, by the biosurfactants. Therefore, crystalline anthracene is added to the culture supernatant and incubated on a shaker at 25°C for 24 hrs. The concentration of the dissolved hydrophobic anthracene is measured photometrically at 354 nm and correlates to the production of biosurfactant. This is a simple and rapid screening method, but the anthracene might be toxic to some microbes. To our knowledge there have been no further reports on its application.

3. Characterization of Biosurfactants

The characterization of the biosurfactants is usually done based on the purified product obtained after extraction and purification. Characterization helps in analyzing the characteristics of the biosurfactant produced. This can be useful for figuring out their application. For example, as reported by Lin et al. (2011), by characterizing the novel biosurfactant produced in their study, they could suggest its application in skin care products. A few of the characterization techniques used for analyzing biosurfactant properties are listed in Table 1.

4. Current Market Value and Future Trends of Biosurfactants

Over the past decade, the growth of the global market of biosurfactants market has been enormous. The major reason for this growth is the demand for such products because of their eco-friendly nature, even though the cost of production is still higher as compared to the synthetic surfactants. Consumers are becoming more aware of the hazards caused by the synthetic agents and this awareness is paving the way for an increased demand of microbial surfactants. It has been reported in Transparency Market Research, among all the geographical regions, the leading continent in terms of production and consumption of biosurfactants is Europe followed by North America. It is also reported that detergents and personal care products will be contributing to more than 56.8% of the global biosurfactants market in 2018. In terms of quantity, it is expected that the volume of global biosurfactants market will be 476,512.2 tons by 2018 and 21% of this volume will be consumed by developing Asian countries. Even the companies who were selling surfactant based products have also turned to the use of microbial surfactants. Some of them are BASF-Cognis and Ecover. However, BASF-Cognis was ahead with over 20% of the market in 2011 (Albany 2012).

Table 1. Techniques used for characterization of biosurfactants.

Biosurfactant type	**Techniques**	**Microorganisms**	**References**
Rhamnolipids	IR analysis	*Pseudomonas aeruginosa*	Rikalovic et al. (2012)
	HPLC analysis	*Pseudomonas* sp.	Abdel Mawgoud et al. (2009)
	HPLC and FAB-MS analysis	*Pseudomonas aeruginosa*	George and Jayachandran (2013)
Sophorolipids	FTIR analysis	*Candida bombicola*	Shah and Prabhune (2007)
	FTIR and GC-MS analysis	*Rhodotorula muciliginosa* and *Candida rugosa*	Das et al. (2011)
	FTIR, NMR and LC-MS analysis	*Candida bombicola*	Daverey and Pakshirajan (2009)
	HPLC, ESI-MS, 1H NMR and 13C NMR analysis	*Candida bombicola*	Ashish et al. (2011)
Trehalolipids or Trehalose Lipids	NMR, MALDI-TOF and GC-MS analysis	*Rhodococcus* sp.	Tokumoto et al. (2009)
	1H NMR Spectroscopy and Mass Spectrometry	*Rhodococcus opacus*	Niescher et al. (2006)
Lipopeptides	TLC and GC-MS analysis	*Bacillus aureus*	Seghal et al. (2010)
	FT-IR analysis	*Candida tropicalis*	Ashish et al. (2011)
	HPLC and ESI-MS analysis	*Bacillus subtilis*	Chen et al. (2008)
	HPLC and ESI-Q-TOF-MS analysis	*Bacillus amyloliquefaciens*	Zhang et al. (2012)
Phospholipids	HPLC, TLC, API-MS and LC-MS analysis	*Sphingobacterium* sp.	Burgos Diaz et al. (2011)
Flavolipids	NMR, FAB-MS and ESI-MS	*Flavobacterium* sp.	Bodour et al. (2004)

Basically, the market for biosurfactants is of two types: Highly expensive (value added) and less expensive (commodity) biosurfactants. Highly expensive biosurfactants generally, are applicable for medical use. Many medical applications of biosurfactants have already been enlisted in the applications section in this paper. Most of the biosurfactants produced currently, find their application in value added products like pharmaceuticals or personal care products which are produced in small volumes. It is because the cost of these finished products is usually high as compared to the cost of carbon sources used for feeding the microorganisms. However, use of biosurfactants as commodity surfactants is still questionable. It is because the production cost is sufficiently higher than the selling price of general commodities (Garcia Becerra et al. 2010). This problem can be alleviated by researching and higher usage of cost free waste products as major carbon sources for the production of microbial surfactants.

5. Advantages and Disadvantages of Biosurfactants

5.1 *Advantages of Biosurfactants*

Researches have shown that biosurfactants exhibit many advantages over chemically synthesized surfactants. The following are some of the advantages of biosurfactants (Mulligan and Wang 2006).

- **Eco-friendly and biodegradable:** Biosurfactants are easily degraded by bacteria and other microscopic organisms; hence they do not pose much threat to the environment. Biodegradability is a very important issue in relation to environmental pollution. Biosurfactants do not create much problem to the environment as they are easily biodegraded there after reducing pollution. They are suited for environmental applications such as bioremediation and dispersion of oil spills (Mulligan 2005).
- **Low toxicity:** For instance Glycolipids from *Rhodococcus* species were 50% less toxic than Tween 80 in naphthalene solubilization tests. Chemically derived surfactants possess higher toxicity than biosurfactants. It was also reported that biosurfactants showed higher EC50 values (effective concentration to decrease 50% of test population) than synthetic surfactants (Desai and Banat 1997).
- **Biocompatibility and digestibility:** Biosurfactants are biocompatible in nature (Rosenberg and Ron 1999). Living organisms are able to perform their physiological activity as per their potential. When these biological surfactants interact with living organisms they do not change the bioactivity of the organisms. Due to this property, they have their application in pharmaceuticals, cosmetics and as functional food additives. This ensures their application in cosmetics, pharmaceuticals and as functional food additives.
- **Availability of raw materials:** Biosurfactants can be produced from cheap raw materials that are available in large quantities. Hydrocarbons, carbohydrates and lipids, can be used separately or in combination with each other as a carbon source (Kosaric 2001).
- **Acceptable production economics:** Depending on their applications, biosurfactants can also be produced from industrial wastes and by-products and this is of particular interest for bulk production.
- **Use in environmental control:** Biosurfactants can be efficiently used in handling industrial emulsions, control of oil spills, biodegradation and detoxification of industrial effluents and bioremediation of contaminated soil.
- **Specificity:** Biosurfactant producing microbes are very specific with their substrates. Biosurfactants being complex organic molecules with specific functional groups are often specific in their action. This would be of particular interest in detoxification of specific pollutants, deemulsification of industrial emulsions, specific cosmetic, pharmaceutical and food applications.
- **Physical factors:** Environmental factors such as pH, temperature and ionic strength cannot affect most of the biosurfactants. *Bacillus licheniformis* strain produces Lichenysin which is not affected by a pH range of 4.5–9.0, temperature

ranges of up to 50°C, NaCl and Calcium concentrations of 50 g/L and 25 g/L respectively (Krishnaswamy Muthusamy et al. 2008).

5.2 *Disadvantages of Biosurfactants*

Despite the numerous advantages that biosurfactants have been known to exhibit, they are known to have the following associated demerits too (Kosaric 2001).

- Large scale production of biosurfactants may be expensive. However this problem could be overcome by coupling the process of utilization of waste substrates, combating their polluting effects at the same time which balances the overall costs. A sterilized medium is essential for biosurfactant production therefore their large scale production is quite difficult and expensive.
- There is difficulty in obtaining pure substances (biosurfactants), of particular importance in pharmaceutical, food and cosmetic applications. This is because downstream processing of diluted broths involved may require multiple consecutive steps.
- Over producing strains of bacteria are rare and those found generally display a very low productivity. In addition, complex media need to be applied to the sample.
- The regulation of biosurfactant synthesis is hardly understood, seemingly it represents "secondary metabolite" regulation. Thus considering a batch culture, secondary metabolite production begins when the culture is stressed due to the depletion of a nutrient. This phenomenon is closely correlated with the transition phase-slow growth rate of culture and with the morphological changes that this phase implies. Among others O_2 limitation has been described as an essential parameter to govern biosurfactant production.
- An improvement in the production yield is hampered by the strong foam formation. Consequently, diluted media have to be applied and only immobilized systems provide an increased productivity.
- During the process of biosurfactant production raw substrates are result in lower yields as compared to the processed substrate materials. Raw substrates takes more time for the production of biosurfactants.
- During product recovery and purification, the acquisition of products with a high degree of purity is difficult because of several consecutive purification steps required for the metabolic broth.
- In the fermentation process, to increase the productivity, formation of foam has complications. To overcome the problem, a diluted medium is required.
- Lack of knowledge on Biosurfactants producing microorganisms is one of the major disadvantages. Known microbial species are not capable of producing large amounts of surfactant yields and they also require a complex culture medium. In this context, exploration of potential microorganisms for biosurfactant production is today's need. The regulation of biosurfactant synthesis is not fully understood till now as to whether, these biomolecules may be produced as secondary metabolites or in association with microbial growth.

6. Application of Biosurfactants

6.1 Biosurfactants in Oil Industry

6.1.1 Microbial Enhanced Oil Recovery (MEOR)

Biosurfactants are of much interest in petroleum-related industries because of their role in Enhanced Oil Recovery (EOR). If the primary methods like pumping are used for recovering oil from reservoirs, the recovery is only 30%. However, the addition of biosurfactants lowers the surface as well as interfacial tensions of the oil which helps in facilitating oil flow and making the recovery operations easier (Kosaric 2001). This method is also referred to as Microbial Enhanced Oil Recovery (MEOR).

An area of considerable potential for biosurfactant application is the field of Microbial Enhanced Oil Recovery (MEOR). Enhanced oil recovery methods were devised to recover oil remaining in reservoirs after primary and secondary recovery procedures. It is an important tertiary recovery technology, which utilizes microorganisms and their metabolites for residual oil recovery (Banat 1995). In MEOR, microorganisms in reservoirs are stimulated to produce polymers and surfactants, which aid MEOR by lowering interfacial tension at the oil-rock interface. This reduces the capillary forces preventing oil from moving through rock pores. There are several reports that describe various methods used in laboratory studies of MEOR.

Biosurfactants can also aid oil emulsification and assist in the detachment of oil films from rocks (Banat 1995). *In situ* removal of oil is due to multiple effects of the microorganisms on both the environment and oil. These effects include gas and acid production, reduction in oil viscosity, plugging by biomass accumulation, reduction in interfacial tension by biosurfactants and degradation of large organic molecules. These are all factors responsible for decreasing the oil viscosity and making its recovery easier (Jain et al. 1992).

6.1.2 Oil Spills Cleanup and Oil Bioremediation

Petroleum is one of the principal sources of energy on a global scale. This source of energy is usually transferred via pipelines or transported on ships through oceans and seas to different parts of the world. Petroleum leakage, during transfer into oceans happens very often. It is therefore one of the major environmental pollutants. Also the large amount of oil sludge and oily waste materials generated by the oil refineries pose a threat to our ecosystem. The expensive disposal methods are the main cause of this threat. The latest example of this type of environmental disaster is Rena oil spill disaster. The fuel on board the Rena consisted of 1,700 tonnes of heavy fuel oil and 200 tonnes of diesel fuel (Taylor 2011). This oil spill caused a huge loss of flora and fauna. In many such incidents, the recovery is done by chemical technology. However, that also imposes threats to the flora and fauna to some extent. Natural surfactants can find utility in oil remediation and oil spill clean-up. Biosurfactants can be used for oil spills cleanup or enhanced oil recovery because they can reduce the oil water interfacial tension leading to emulsification. The stability of emulsions formed, is because of the ability of biosurfactants to lower interfacial tension between interfaces and oil (Banat 2000).

Bioemulsifiers and biosurfactants are a novel group of molecules. They are the most powerful and versatile by products of modern microbial technology. It can be exploited effectively in fields such as enzymes and biocatalysts for petroleum upgrading, bio fouling degradation and biocorrosion of hydrocarbons within oil reservoirs (Perfumo et al. 2010). Biosurfactants also plays an important role in petroleum extraction, upgrading and refining and petrochemical manufacturing.

6.2 *Biosurfactants in Binding of Heavy Metals*

Rhamnolipid biosurfactant has been shown to be capable of removing heavy metals like Cadmium, Lead and Zinc in soil. The mechanism by which Rhamnolipid reduces metal toxicity may involve a combination of Rhamnolipid complexation of Cadmium and Rhamnolipid interaction with the cell surface to alter Cadmium uptake. High molecular weight polysaccharide emulsifiers interact with metals binding them as has been shown for the binding of Uranium by Emulsan.

6.3 *Biosurfactants in Sludge Tank Clean-up*

The process of cleaning up pipelines, is very expensive and tedious. Conventional pumps fail to remove the sludge and waste oil deposits which result in the accumulation of these by products in the storage tanks. Manual labour is expensive and time consuming and also very hazardous. Use of biosurfactants is one of the economically viable alternatives. Not only does it help in sludge removal but it also beneficial due to its ability to recover oil from these wastes (Banat et al. 2000). Also recently oil tank bottom sludge was treated with a novel biosurfactant produced by an Actinobacteria *Gordonia* sp. And it could efficiently clean up the tanks. The dispersion activity by *Gordonia* sp. was reported to be better than that of its chemical counterparts or surfactants having a plant origin (Matsui et al. 2012).

6.4 *Biosurfactants as Emulsifiers*

The ability of the Biosurfactant to emulsify hydrocarbon—water mixture has been very well documented. This property has been demonstrated to increase hydrocarbon degradation significantly and is thus potentially useful for oil spill management.

6.5 *Biosurfactants as Commercial Laundry Detergents*

Surfactants, which are an important component in commercial laundry detergents exert toxicity into fresh water living organisms due to their chemical synthesization. Public awareness about environmental pollution by chemical surfactants is increasing day by day. It has stimulated the search for natural substitutes of chemical surfactants in laundry detergents which are eco-friendly in nature (Das and Mukherjee 2005). They are capable of good emulsion formation with vegetable oils and also demonstrate excellent stability and compatibility with commercial laundry detergents favoring their inclusion in laundry detergent formulations (Das and Mukherjee 2007). All surfactants, are an essential component current business clothing cleansers, which are synthetically integrated and pose a danger to new water living beings. Developing open mindfulness about the ecological dangers and those related with compound surfactants has invigorated the scan for eco-friendly, characteristic substitutes of surfactant concoctions in clothing cleansers. Biosurfactants, for example, Cyclic Lipopeptides (CLP) are steady over a a wide pH extent (7.0–12.0) and warming them to high temperatures does not bring about any loss in their surface-dynamic property. They show great emulsion development capacity with vegetable oils and brilliant similarity and strength with business clothing cleansers for them to be considered for use in clothing cleansers.

6.6 Biosurfactants for Phytoremediation

Heavy metals are the main problem among the inorganic pollutants. At higher concentration they are toxic as they form free radicals and cause oxidative stress. Additionally, they disturb the ordinary action of some fundamental proteins and shades by supplanting them. In any case, by utilizing both metal safe and biosurfactant creating microscopic organisms, the capacity of the plant can be expanded for phytoremediation. For instance, biosurfactant producing *Bacillus* sp. strain can build the effectiveness of the plant growth development of sundangrass, tomato and maize and furthermore take-up of cadmium in the contaminated soil. From this examination it is clear that the species taken for this reason posess the root colonization action. So an organism helping the phytoremediation process is produced for the remediation of overwhelming metals (Kugler et al. 2015).

6.7 Biosurfactants in Mining

Biosurfactants may be used for the dispersion of inorganic minerals in mining and manufacturing processes. Rosenberg et al. (1988) described the production of *Acinetobacter calcoaceticus*, an anionic polysaccharide called biodispersan, which prevented flocculation and dispersed 10% limestone in a water mixture. Biodispersan served two functions: dispersant and surfactant and catalyzed the fracturing of limestone into smaller particles. To elucidate the mechanism of this action, Rosenberg and Ron (1998) suggested that the pH should be alkaline (9 ± 9.5) during the grinding process, and that biodispersan is an anionic polymer at that pH. The polymer enters microdefects in the limestone and lowers the energy required for cleaving the microfractures. Kao Chemical Corporation (Japan) used *Pseudomonas*, *Corynebacterium*, *Nocardia*, *Arthrobacter*, *Bacillus* and *Alcaligenes* to produce biosurfactants for the stabilization of coal slurries to aid the transportation of coal. Similarly, Polman et al. (1994) tested biosurfactants for solubilization of coal and achieved partial solubilization of North Dakota Beulah Zap lignite coal using a crude preparation of biosurfactants from *Candida bombicola* (Breckenridge and Polman 1994).

6.8 Biosurfactants in Biofilm Formation

Biofilms are aggregates of microorganisms in which microbial cells adhere to a surface and to each other. They include bacteria and a mixture of microbial extracellular material produced at the surface and any other material trapped within the resulting matrix. In the food industry bacterial biofilms present on surfaces mostly represent sources of contamination, which are often associated with disease transmission and food spoilage. Therefore, minimizing microbial adherence to surfaces in contact with food is an essential step in providing quality safe products to consumers. Biosurfactants' involvement in microbial adhesion and detachment from surfaces has been investigated. Meylheuc et al. (2006) reported on the preconditioning of stainless steel surfaces with a *Pseudomonas fluorescens* anionic biosurfactant as a method to reduce the number of *Listeria* cells adhering to the surfaces that favored disinfectants' bactericidal activities. The bioconditioning of surfaces through the use of microbial surfactants has been suggested as a new strategy to reduce adhesion.

Nitschke et al. (2009) also reported on the effect of Rhamnolipids and Surfactin biosurfactants use on the adhesion of several food pathogens to polypropylene and stainless steel surfaces. The use of Surfactin in particular caused a reduction in the number of adhering cells on preconditioned stainless steel polypropylene and polystyrene surfaces and delayed bacterial adhesion of both growing and non-growing cells. Rhamnolipids and other plant biosurfactants have also recently been reported to have some role in the inhibition of complex biofilms and as adjuvants to enhance some antibiotic microbial inhibitors (Brzozowski et al. 2011). These properties suggest that biosurfactants could be considered as suitable compounds for developing strategies to delayor prevent microbiological colonization of industrial plant surfaces used for food stuff preparations. Other possible future uses may be for biofilms prevention or reduction for oral cavities which are the main surfaces exposed to food stuffs within the human body. Whether such components may eventually become a part of hygiene products or simply involved in food stuffs such as chewing gums remains to be seen.

6.9 CO_2 Migration through Biosurfactants

The Greenhouse effect is a natural process that is responsible for heating the earth's surface. The gases present in the atmosphere, e.g., CO_2, methane, etc., have the ability to absorb infra-red radiations emitted from the surface of the earth. Certain studies have reported the utility of microbial surfactants for reduction of CO_2 emissions into the atmosphere. The biosurfactants may not help in total elimination but may play a significant role in reducing the amount of this Green House Gas (GHG) present in the atmosphere (Gakpe et al. 2008). Patel (2003) concluded, based on his study, that the increased production and use of biological surfactants should be part of an overall GHG (Green house gas) emission reduction strategy consisting of a whole range of measures addressing both energy demand and supply.

7. Conclusion

Biosurfactants have led to considerable interest for present and future application due to a non-toxic and eco-friendly nature. Successful commercialization of every biotechnological product depends largely on its bioprocess economics. At present, the prices of microbial surfactants are not competitive with those of the chemical surfactants due to their high production costs and low yields. Hence, they have not been commercialized extensively. To produce commercially viable biosurfactants, process optimization at the biological and engineering level needs to be improved.

Improvement in the production technology of biosurfactants has already enabled a 10–20 fold increase in productivity, although further significant improvements are required. However, the use of cheaper substrates and optimal growth and production conditions coupled with novel and efficient multi-step downstream processing methods and the use of recombinant and mutant hyper producing microbial strains can make biosurfactant production economically feasible.

Microbial derived surfactants have special advantages over their chemically manufactured counterparts because of their lower toxicity, biodegradable nature and effectiveness at extreme temperatures and pH values. Despite having so many advantages biosurfactants use in industry is still limited due to the high cost involvement in their production processes. It has been established that biosurfactants are eco-friendly products suitable for the use in medical, environmental, cosmetics and agricultural fields. It is also noticed that the biosurfactants are a very good source for biodegradation of oils, controlling environmental pollution, oil fields, cosmetic industry, oil industry, removal of oil pollution, applications in agriculture and agrochemical industries. As these biosurfactants have antimicrobial properties they can also be used as biocontrol agents.

References

Abdel Mawgoud, A.M., M.M. Aboulwafa and N.A.H. Hassouna. 2009. Characterization of rhamnolipid produced by *Pseudomonas aeruginosa* Isolate Bs20. Appl. Biochem. Biotech. 157: 329–345.

Albany, L. 2012. Biosurfactants industry is expected to Reach USD 2,210.5 Million Globally in 2018: Transparency Market Research. Retrieved from Transparency Market- Research.

Ashish, G., N. Gupta and M. Debnath. 2011. Characterization of biosurfactant production by mutant strain of *Candida tropicalis*. 2nd International Conference on Environmental Science and Technology (pp. IPCBEE Vol. 6), IACSIT Press, Singapore.

Banat, I.M. 1995. Biosurfactants production and possible uses in microbial enhanced oil recovery and oil pollution remediation—a review. Bioresour. Technol. 51: 1–12.

Banat, I.M. 2000. Potentials for use of biosurfactants in oil spills cleanup and oil bioremediation. Water Studies, 177–185.

Bodour, A.A., C. Guerrero Barajas, B.V. Jiorle, M.E. Malcomson, A.K. Paull et al. 2004. Structure and characterization of flavolipids, a novel class of biosurfactants produced by *Flavobacterium* sp. strain MTN11. Appl. Environ. Micro. 70(1): 114–120.

Breckenridge, C.R. and J.K. Polman. 1994. Solubilization of coal by biosurfactants derived from *Candida bombicola*. Geomicrobiol. J. 12: 285–288.

Brzozowski, B., W. Bednarski and P. Golek. 2011. The adhesive capability of two *Lactobacillus* strains and physicochemical properties of their synthesized biosurfactants. Food Technol. Biotechnol. 49: 177–186.

Burgos Diaz, C., R. Pons, M.J. Espuny, F.J. Aranda, J.A. Teruel et al. 2011. Isolation and partial characterization of a biosurfactant mixture produced by *Sphingobacterium* sp. isolated from soil. J. Collo. Interf. Sci. 361(1): 195–204.

Chen, C., S. Baker and R. Darton. 2007. The application of a high throughput analysis method for the screening of potential biosurfactants from natural sources. J. Microbiol. Met. 70: 503–510.

Chen, H., L. Wang, C. Su, G.H. Gong, P. Wang, Z.L. Yu et al. 2008. Isolation and characterization of lipopeptide antibiotics produced by *Bacillus subtilis*. Lett. Appl. Micro. 47(3): 180–186.

Cooper, D.G. and B.G. Goldenberg. 1987. Surface active agents from two *Bacillus* species. Appl. Environ. Micro. 53: 224–229.

Crosman, J.T., R.J. Pinchuk and D.G. Cooper. 2002. Enhanced biosurfactant production by *Corynebacterium alkanolyticum* ATCC 21511 using self-cycling fermentation. J. App. Micro. Biotech. 79(5): 467–472.

Das, K. and A.K. Mukherjee. 2005. Characterization of biochemical properties and biological activities of biosurfactants produced by *Pseudomonas aeruginosa* mucoid and non-mucoid strains. Appl. Microbiol. Biotechnol. 69: 192–199.

Das, K. and A.K. Mukherjee. 2007. Crude petroleum-oil biodegradation efficiency of *Bacillus subtilis* and *Pseudomonas aeruginosa* strains isolated from petroleum oil contaminated soil from north-east India. Biores. Tech. 98: 1339–1345.

Das, N. and P. Chandran. 2010. Microbial degradation of petroleum hydrocarbon contaminants: An overview. Biotech. Res. Int. 11: 1–13.

Das, K., P. Chandran and S. Nilanjana. 2011. Characterization of Sophorolipid biosurfactant produced by yeast species grown on diesel oil. Int. J. Sci. Natu. 2(1): 63–71.

Daverey, A. and K. Pakshirajan. 2009. Production, characterization, and properties of Sophorolipids from the yeast *Candida bombicola* using a low-cost fermentative medium. Appl. Biochem. Biotech. 158(3): 663–674.

Desai, J.D. and I.M. Banat. 1997. Microbial production of surfactants and their commercial potential. Microbiol. Mol. Biol. Rev. 61: 47–64.

Gakpe, P., K.S. Rahman and K. Edward. 2008. Production, characterization and applications of biosurfactants. Rev. Biotech. 7: 360–370.

Garcia Becerra, F.Y., D.G. Allen and E.J. Acosta. 2010. Surfactants from waste biomass. pp. 167–189. *In*: Mikael Kjellin, I.J. (ed.). Surfactants from Renewable Sources, John Wiley & Sons Ltd.

George, S. and K. Jayachandran. 2013. Production and characterization of rhamnolipid biosurfactant from waste frying coconut oil using a novel *Pseudomonas aeruginosa*. J. Appl. Microbio. 114(2): 373–383.

Jain, D.K., D.L. Collins Thompson, H. Lee, J.T. Trevors et al. 1991. A drop-collapsing test for screening biosurfactant producing microorganisms. J. Microbio. Methods. 13(4): 271–279.

Jain, D.K., H. Lee and J.T. Trevors. 1992. Effect of addition of *Pseudomonas aeruginosa* UG2 inocula or biosurfactants on biodegradation of selected hydrocarbons in soil. J. Ind. Microbiol. 10: 87–93.

Kosaric, N. 2001. Biosurfactants and their applications for Soil Bioremediation. Food Tech. Biotech. 39: 295–304.

Krishnaswamy Muthusamy, Subbuchettiar Gopalakrishnan, Thiengungal Kochupappy Ravi and Panchaksharam Sivachidambaram. 2008. Biosurfactants: Properties, commercial production and application. Cur. Sci. 94(25): 739–747.

Kugler, J.H., M. Le Roes Hill, C. Syldatk and R. Hausmann. 2015. Surfactants tailored by the class *Actinobacteria*. Front. Microbiol. 6: 212.

Lin, T.C., C.Y. Chen, T.C. Wang and Y.S. Chen. 2011. Characterization of biosurfactant from a diesel oil degradation bacterium and application potential in beauty care products. 2011 3rd International Conference on Chemical, Biological and Environmental Engineering (pp. 114–118).

Maczek, J., S. Junne and P. Gotz. 2007. Examining biosurfactant producing bacteria—an example for an automated search for natural compounds. Appl. Note Cy. Bio. 11: 201–207.

Matsui, T., T. Namihira, T. Mitsuta and H. Saeki. 2012. Removal of oil tank bottom sludge by novel biosurfactant, JE1058BS. J. Japan Petrol. Inst. 55(2): 138–141.

Meylheuc, T., M. Renault and M.N. Bellon Fontaine. 2006. Adsorption of a biosurfactant on surfaces to enhance the disinfection of surfaces contaminated with *Listeria monocytogenes*. Int. J. Food Microbiol. 109: 71–78.

Morikawa, M., Y. Hirata and T. Imanaka. 2000. A study on the structure-function relationship of lipopeptide biosurfactants. Biochimica. Biophysica. Acta. 1488(3): 211–218.

Morita, T., M. Konishi, T. Fukuoka, T. Imura, D. Kitamoto et al. 2007. Physiological differences in the formation of theglycolipid biosurfacants, mannosylerythritol lipids, between *Pseudozyma antartica* and *Pseudozyma aphidis*. Appl. Microbio. Biotech. 74: 307–315.

Mulligan, C.N. 2005. Environmental applications for biosurfactants. Environ. Pollut. 133(2): 183–198.

Mulligan, C., D. Cooper and R. Neufeld. 1984. Selection of microbes producing biosurfactants in media without hydrocarbons. J. Ferm. Tech. 62(4): 311–314.

Niescher, S., V. Wray, S. Lang, S.R. Kaschabek, M. Schlömann et al. 2006. Identification and structural characterization of novel trehalose dinocardiomycolates from n-alkane grown *Rhodococcusopacus* 1CP. Appl. Microbio. Cell Physiol. 70: 605–611.

Nitschke, M., L.V. Araujo, S.G. Costa, R.C. Pires, A.E. Zeraik et al. 2009. Surfactin reduces the adhesion of food-borne pathogenic bacteria to solid surfaces. Lett. Appl. Microbiol. 49: 241–247.

Noudeh, G.D., A.D. Noodeh, M.H. Moshaf, E. Behravan, M.A. Afzadi et al. 2010. Investigation of cellular hydrophobicity and surface activity effects of biosynthesed biosurfactant from broth media of PTCC 1561. Afr. J. Microbio. Res. 4(17): 1814–1822.

Patel, M. 2003. Surfactants based on renewable raw materials. J. Ind. Ecol. 7: 47–62.

Perfumo, A., I. Rancich and I.M. Banat. 2010. Possibilities and challenges for biosurfactants use in petroleum industry. pp. 135–145. *In*: Sen, R. (ed.). Biosurfactants Advances in Experimental Medicine and Biology, New York, NY: Springer.

Polman, J.K., K.S. Miller, D.L. Stoner and C.R. Brackenridg. 1994. Solubilization of bituminous and lignite coals by chemically and biologically synthesized surfactants. J. Chem. Tech. Biotechnol. 61: 11–17.

Rikalovic, M.G., G. Gojgic-cvijovic, M.M. Vrvic and I. Karadzic. 2012. Production and characterization of Rhamnolipids from *Pseudomonas aeruginosa*. J. Serbian Chem. Soc. 77(1): 27–42.

Rosenberg, E., A. Rubinovitz, A. Gottlieb, S. Rosenhak, E. Ron et al. 1988. Production of biodispersan by *Acientobacter calcoaceticus* A2. Appl. Environ. Microbiol. 54: 317–322.

Rosenberg, E. and E.Z. Ron. 1999. High and low molecular mass microbial surfactants. Appl. Microbio. Biotech. 52(2): 154–162.

Sarubbo, L. 2006. Production and stability studies of the Bioemulsifier obtained from a strain of *Candida glabrata* UCP 1002. J. Biotech. 9(4): 400–406.

Seghal, K.G., T.A. Thomas, J. Selvin, B. Sabarathnam, A.P. Lipton et al. 2010. Optimization and characterization of a new lipopeptide biosurfactant produced by marine *Brevibacteriumaureum* MSA13 in solid state culture. Biores. Tech. 101: 2389–2396.

Shah, S. and A. Prabhune. 2007. Purification by silica gel chromatography using dialysis tubing and characterization of sophorolipids produced from *Candida bombicola* grown on glucose and arachidonic acid. Biotech. Lett. 29(2): 267–272.

Siegmund, I. and F. Wagner. 1991. New method for detecting Rhamnolipids excreted by *Pseudomonas* spp. during growth on mineral agar. Biotech. Tech. 5: 265–268.

Tabatabaee, S., M.M. Assadi, A.A. Noohi and V. Sajadian. 2005. Isolation of biosurfactant producing bacteria from oil reservoirs. Iranian J. Environ. Health Sci. Engi. 2(1): 6–12.

Taylor, A. 2011. Oil Spill Disaster on New Zealand. The Atlantic. New Zealand.

Tokumoto, Y., N. Nomura, H. Uchiyama, T. Imura, T. Morita et al. 2009. Structural characterization and surface-active properties of a succinoyl trehalose lipid produced by *Rhodococcus* sp. SD-74. J. Oleo Sci. 58(2): 97–102.

Vandana, P. and J.K. Peter. 2014. Production, partial purification and characterization of biosurfactant from *Pseudomonas fluorescens*. Int. J. Adv. Tech. Engi. Sci. 2(7): 258–264.

Vaux, D. and M. Cottingham. 2001. Method and apparatus for measuring surface configuration. Patent number WO 2007/039729 A1.

Wei, Y.H., C. Chou and J.S. Chang. 2005. Rhamnolipid production by indigenous *Pseudomonas aeruginosa* J4 originating from petrochemical waste water. Biochem. Engi. J. 24: 146–154.

Willumsen, P. and U. Karlson. 1997. Screening of bacteria, isolated from PAH contaminated soils, for production of biosurfactants and bioemulsifiers. Biodeg. 7(5): 415–423.

Yin, H., J. Qiang, Y. Jia, J. Ye, H. Peng et al. 2009. Characteristics of biosurfactant produced by *Pseudomonas aeruginosa* S6 isolated from oil containing wastewater. Process Biochem. 44: 302–308.

Zhang, S.M., Y.X. Wang, L.Q. Meng, J. Li, X.Y. Zhao et al. 2012. Isolation and characterization of antifungal lipopeptides produced by endophytic *Bacillus amyloliquefaciens* TF28. Afri. J. Microbio. 6(8): 1747–1755.

3

Molecular Approaches Towards the Synthesis of Biosurfactants

Vidya Kothari[1,*] and *Arpana Jobanputra*[2]

1. Introduction

1.1 *Important Features Relevant to Bio-surfactant Production in Microorganisms*

The secret behind microbes synthesizing bio-surfactants is still unknown. The rationale behind it is the presence of various hydrophobic substrates. This reduces the surface and interfacial tension between two immiscible fluid phases. Few microbes produce bio surfactants on water soluble substrates (Banat 1995, Mulligan 2005). Different organisms having various biosynthetic pathways and specific enzymes are involved (Banat et al. 2000). Synthesis takes place by *de novo* pathways or assemblies from substrates (Benincasa et al. 2004).

1.2 *Quorum Sensing in Bio-surfactant Production*

Bio-surfactants are produced by bio-film producing bacteria. The bacterial communities are bio-film producers that are regulated by multiple signal processing (Peele et al. 2016). Bio-surfactants are exopolymers that can tolerate extreme PH and temperature conditions. These characterizes favours the bacteria to sustain under adverse conditions. A bacterium produces many kinds of molecules that allow bacteria to communicate population size. These molecules are generally regarded as auto inducer peptides that serve as signal carriers. Bio-surfactant producing quorum sensing regulatory bacteria use acyl carrier proteins and pass them onto the

[1] Department of Microbiology, Shardabai Pawar Mahila Arts, Commerce & Science College. Shardanagar, Malegaon Bk., Tal. Baramati, Dist. Pune 413115.

[2] Department of Microbiology, PSGVPM's S. I. Patil Arts, G. B. Patel Science & S. T. S. K V. S. Commerce College, Shahada, 425409, Maharashtra, India.
Email: arpana_j12@rediffmail.com

* Corresponding author: vidya6782@gmail.com

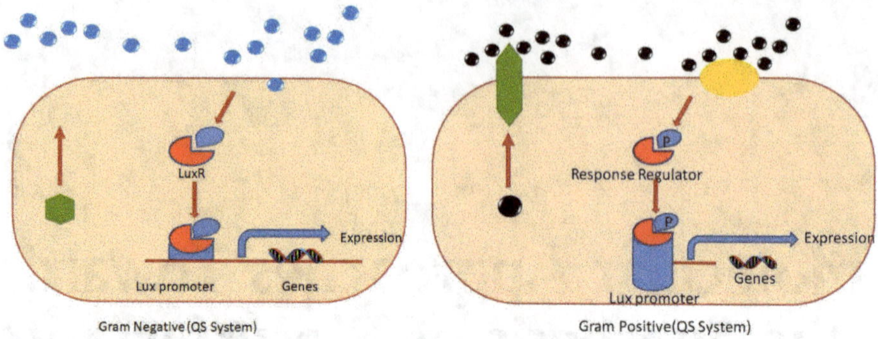

Figure 1. Showing Quorum sensing in bio-surfactant and biofilm formation (Sankar 2018).

homoserine moiety. Quorum-sensing signalling may be extracellular or intracellular and is often mediated by N-acyl-homoserine lactone (AHL) concentration. Quorum sensing reveals the fact that bacteria have the capacity to access a number of other components, they can activate once the threshold number is reached. Acyl homoserine lactones are present mainly in Gram-negative bacteria and they control their own synthesis. Oligopeptide molecules are present mainly in Gram-Positive bacteria; their synthesis is dependent on ribosomes. It has been hypothesized that the production of auto inducing peptides and bio-film formation are interlinked (Peele et al. 2016).

1.3 *Pseudomonas Species*

Pseudomonas species synthesize different bio-surfactant like, glycolipid rhamnolipids, cyclic lipopeptides-putisolvins, and lipopolysaccharides. Glycolipids are acidic crystalline glycolipid 1-rhamnose and *l*-β-hydroxydecanoic acid from *P. aeruginosa* (Jarvis et al. 1949). Two types of cyclic lipopeptides (putisolvins I and II) are produced by *P. putida*. These compounds are similar to polymers with higher rhamnose-hydroxyacid ratio (Bergstrom et al. 1947), which possesses surfactant activity and also plays a significant role in biofilm formation and degradation. *P. aeruginosa* synthesizes mono as well as di-rhamnolipids including different rhamnolipid derivatives which include 3-(3-hydroxyalkanoyloxy-) alcanoic acid (HAA), mono-rhamnolipid (l-rhamnosyl-3-hydroxydecanoyl-3-hydroxydecanoate) (Rendell et al. 1990, Deziel et al. 2003) and di-rhamnolipid (l-rhamnosyl-l-rhamnosyl-3-hydroxydecanoyl-3-hydroxydecanoate) (Rahim et al. 2001). The involvement of the quorum-sensing System (QSS) for rhamnolipid biosynthesis in *Pseudomonas* spp. Figure 2 diagrammatically represent that different components are involved in rhamnolipid biosynthesis. Two QSS regulating rhamnolipid syntheses are present in two different regions of the chromosome (Stover et al. 2000). Formation of mono and di-rhamnolipids is mediated through two different transferases like, rhamnosyltransferase I and II. Rhamnolipid synthesis is coupled with nitrogen limitations to the cell (Mulligan et al. 1989). To enhance biosurfactant biosynthesis phosphate limiting conditions are required (Bazire et al. 2005). Rhamnosyl transferase I, contain four genes like, *rhl*A, *rhl*B, *rhl*R, *rhl*I. In heterologous hosts plasmids encode four genes sufficient to produce rhamnolipids

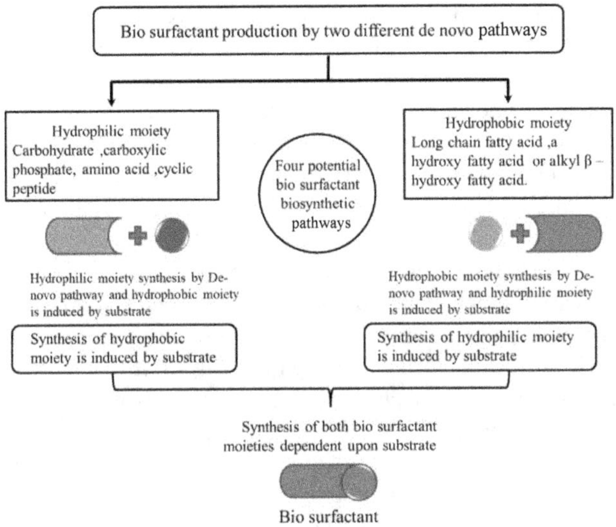

Figure 2. Potential bio surfactant biosynthetic pathways in microorganisms: biosurfactant molecule. Probable biosurfactant biosynthetic pathways operating in different microorganisms. Based on Syldatk and Wagner (1987) four assumptions.

(Ochsner et al. 1995). Genes *rhl*A, *rhl*B are placed upstream while *rhl*R, *rhl*I are located downstream of the structural genes (Fig. 3).

The *rhl*A and *rhl*B genes code for active rhamnosyltransferase I and are transcribed together as a bicistronic RNA (Sullivan 1998, Ochsner et al. 1994, Pesci et al. 1997). Structural proteins are encoded by *rhl*B and present in the periplasm.

Inner membrane proteins required for synthesis, transport, or solubilizations of rhamnosyltransferase are encoded on *rhl*A (Ochsner et al. 1995). In the first QSS, genes *rhl*A, *rhl*B are positively regulated by rhlR. The transcriptional activator and autoinducer are encoded by *rhl*R and *rhl*I respectively. Two signal molecules like, N-butanoyl-Lhomoserine (PAI-2) and hexanoyl-l-homoserine lactone are produced by *rhl*I. The transcriptional activator produced by *rhl*R binds to autoinducer PAI-2 and this active complex causes transcriptional activation of *rhl*A and *rhl*B that encode rhamnosyltransferase I. The second QSS contains two genes namely *las*R and *las*I (Bodour 2002, Sullivan 1998). In this system the autoinducer is encoded by *las*I namely N-(3-oxododecanoyl)-l-homoserine-Lactone (PAI-1) RhlR regulatory protein requires autoinducers N-butyryl-HSL and N-(3-oxohexanoyl)-HSL for its activity (Lazdunski et al. 2004). Induction of second QSS occurs by cyclic AMP levels as indicated by the presence of the *las*R promoter region of both lux-box and a binding consensus sequence of a cyclic AMP receptor protein. The transcription of *rhl*R system is positively regulated by *las* system (Bodour et al. 2002). The *rhl* system is posttranslationally controlled by the *las* system by hindrance of PAI-2 by PAI-1 from binding to RhlR. This situation is created till enough PAI-2 and/or PAI-1 are produced to create a blockage effect (Bodour et al. 2002). Figure 2 illustrates the regulation of rhamnolipid synthesis in *Pseudomonas* spp. It is proved that rhlR expression is strongly influenced by environmental factors and is partially LasR-independent under certain culture conditions. Different regulatory proteins like Vfr sigma factor σ54 and RhlR themselves regulate expression of rhlR (Latifi et al. 1996, Medina et al. 2003). The *rhl*I negative mutant is unable to produce rhamnolipids on its own. It initiates biosurfactant production by mutant by the addition of synthetic N-acylhomoserine lactone (signal molecule). Bio-surfactant genes are expressed in unsaturated porous media containing hexadecane and play a role in the biodegradation process. The *gfp* reporter gene was integrated with either the promoter region of *pra*, which encodes for the emulsifying PA protein and to the promoter of the transcriptional activator *rhlR*. It was found that GFP was produced in culture, which indicated that the *rhlR* and *pra* genes are both transcribed in unsaturated porous media. The *gfp* expression was localized at the hexadecane-water interface. Pamp and Tolker-Nielsen (2007) demonstrated that the BS produced by *P. aeruginosa* has an additional role in structural biofilm development. The mutant deficient in *rhlA* cannot synthesize BS. The main role of *rhlA* in BS is biosynthesis and biofilm development. The protein AlgR2 responsible for the regulation of nucleoside diphosphate kinase also down regulates rhamnolipid production in *P. aeruginosa* (Pamp et al. 2007, Schlictman et al. 1995) in 2005, Lequette and Greenberg worked on identifying the role of QSS responsible for rhamnolipid biosynthesis on biofilm architecture. A*rhlA*-*gfp* fusion into a neutral site in the *P. aeruginosa* genome significantly lightened the activity of *rhlAB* promoter in rhamnolipid-producing biofilms. Campos-Garci′ et al. (1998) discovered a new *gene rhl*G which is a homologue of the *fabG* gene encoding NADPH-dependent β-ketoacyl acyl carrier protein (ACP) reductase. This is required for the synthesis of FA. This gene *rhl*C is obligatory for the synthesis of b-hydroxy acid moiety of rhamnolipids and partly contributes to production of poly-β-hydroxyalkanoate (PHA). This study proved that different pathways are involved in synthesis of FA moiety of rhamnolipids than and those for general FA synthetic pathways. Till the year 2001, it was obvious that, rhamnosyltransferase 1 (RhlAB) catalyses the synthesis of mono-rhamnolipids from dTDP-l-rhamnose and β-hydroxydecanoyl-β-hydroxydecanoate, whereas di-rhamnolipids are produced from mono-rhamnolipids and dTDP-l-rhamnose. For the first time, Rahim et al., proved in 2001 that rhamnosyltransferase catalyas TDP-1, a gene critical to the synthesis of mono rhamnolipid from dTDP-1(l-rhamnose-l-rhamnose-β-hydroxydecanoyl-β-hydroxydecanoate) production in *P. aeruginosa* contains 325 amino acids (35.9 kDa), RhlC. The *rhlC* gene is located with an upstream gene (PA 1131) in an operon with an unknown function. A σ54-type promoter for the PA1131-*rhlC* operon was identified and a single transcriptional start site was mapped. The biological role of RhlC was confirmed by insertional mutagenesis studies and allelic replacement. Inhibition of QSS was demonstrated by work with mutants. On the basis of northern blot analysis, *P. aeruginosa* strain PR1-E4: a *lasR* deletion mutant revealed that over production of the *P. aeruginosa* DksA homologue down regulated transcription of the autoinducer synthase gene *rhlI* thereby inhibiting QSS (Campos-Garci′ et al. 1998). Generally hydrocarbons utilizing microbes produce BS *P. putida* PCL1445, which possess surfactant activity and also plays a significant role in biofilm formation and degradation. Mutants from Tn5*luxAB* library of strain PCL1627 defective in BS production contained transposon inserted in a *dnaK* homologue located downstream of *grpE* and upstream of *dnaJ* indicating positive regulation of these genes in biosurfactant synthesis. Two-component signaling system GacA/

GacS was involved in BS synthesis (Branny et al. 2001). Generally hydrocarbons utilizing microbes producing BS. *P. aeruginosa* degrade hexadecane only if it can produce rhamnolipids (Dubern et al. 2005, Noordman and Janssen 2002, Raza et al. 2007). Mutated *Pseudomonas* spp. produces low rhamnolipid BS (Raza et al. 2006, Mulligan and Gibbs 1989). Whereas, rhamnolipid defective mutants grow very poorly on hydrocarbons (Beal and Betts 2000, Shrive et al. 1995). Figure 3 shows that hydrophobic substrates effect biosurfactant production. The uptake of hydrocarbon can be improved by addition of BS in the growth medium.

This concept was proved by Koch et al. (1991). He constructed a transposon TN5-GM induced mutant of *P. aeruginosa* PG201 which could not grow on minimal medium with hexadecane. With rhamnolipid supplementation, the same culture grew well. Al-Tahhan et al. (2000). The emulsifier makes the cell surface more hydrophobic through release of lipopolysaccharide (LPS). All these studies clearly suggest a role of BS in the survival of microbes on hydrophobic substrates. Natural or chemical mutations are employed to improve quality and yield of bio surfactants from microorganisms (Tahzibi et al. 2004). Iqbal et al. (1995) demonstrated hyper-production of biosurfactant, by, high biodegradation of crude oil by an EBN-8 a gamma ray induced mutant of *P. aeruginosa*.

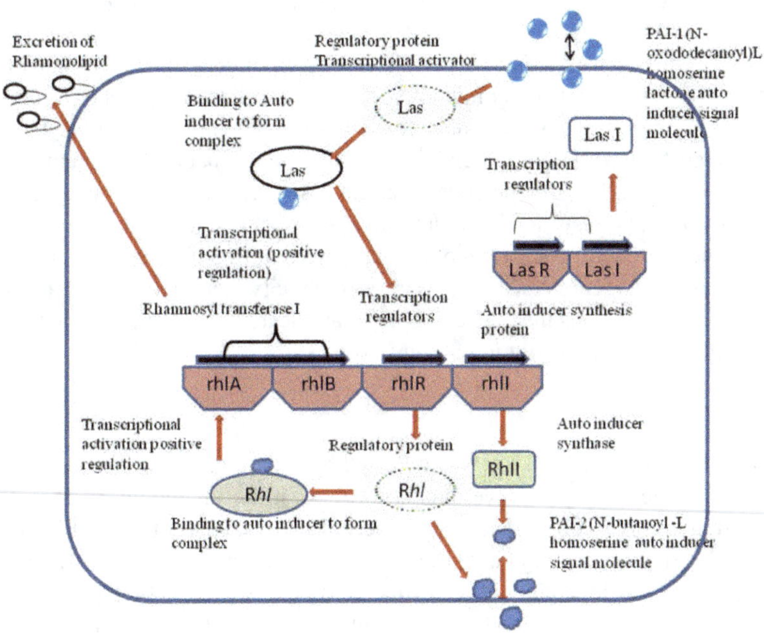

Figure 3. Rhamnolipid synthesis in *Pseudomonas* spp. by two quorum sensing system: Pictorial representation of two quorum sensing system (QSS) present at different regions of *Pseudomonas* spp. chromosome. Thick black bold arrows: Genes on chromosome of *Pseudomonas*; Black arrows: Protein synthesis from gene; Dotted oval indicates inactive regulatory protein; Continuous oval: Active complex of regulatory protein and autoinducer 16, 17, 18. The same mutant produced 4.1 and 6.3 g/L of rhamnolipids when grown on hexadecane and paraffin oil respectively. Another gamma ray induced *P. putida* 300-B mutant gave a high yield of rhamnolipid (4.1 g l–1) on soybean waste frying oil as a carbon source and glucose as a growth initiator over the wild type strain (Raza et al. 2006). Koch et al. (1988) constructed a lactose utilizing strain of *P. aeruginosa* by insertion of *E. coli* lac Y genes.

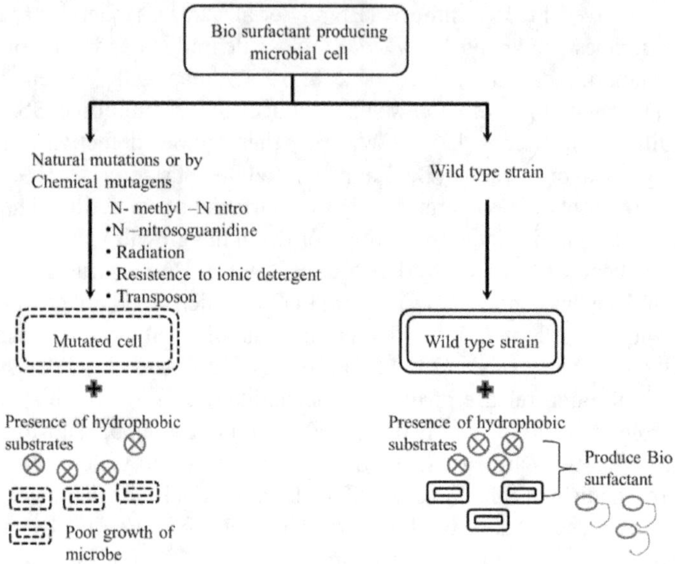

Figure 4. Effect of biosurfactant production on growth in presence of hydrophobic substrates.

Two systems, lacZY and lux4B, were incorporated into a chromosome of *P. aeruginosa* UG2. This recombinant strain could utilize lactose and produced BS efficiently. Similar studies were also carried out by Flemming et al. (1994). Their work proved to be efficient in sensitive detection and quantitative enumeration of *P. aeruginosa* UG2Lr (spontaneous rifampin-resistant derivative) using supportive data from antibiotic resistance, bioluminescence and PCR analyses.

Ochsner et al., constructed recombinant strains of *P. putida* and *P. Fluorescence* by knocking down genes responsible for pathogenicity thereby produce harmless BS producing stains. This is the best example of application of molecular knowledge in producing biotechnologically improved strains.

1.4 *Bacillus Species*

Surfactin produced by *Bacillus* was reported by Arima et al. (1968), i.e., surfactin. Surfactin is the most effective bio surfactant, it has reducing surface tension (72–27 dynes/cm) (Cooper and Goldenberg 1987, Nakano et al. 1988). It has a low CMC (critical micelle concentration) value and finds potential applications in biotechnology and medicine. Most of the researchers proved that more than 70% of research on biosurfactants has accounted for *Bacillus* spp. alone. Surfactin production, structure, enzymes involved in biosynthesis, organization and genetics of production has been reviewed in great measure. Due to the great potential of surfactin and its diverse applications it became necessary to study the underlying genetic mechanisms.Kluge et al. (1988) laid the foundation for molecular studies by proposing a non-ribosomal mechanism of surfactin synthetase. Table 1 contains a brief summary of genetic mechanisms involved in surfactin synthesis.

Surfactin contains β-hydroxyl FA, usually β-hydroxytetradecanoic acid, synthesized by a 27 kb *srfA* operon. All these act under regulation of QSS. First QSS

Table 1. Genetic mechanism involved in surfactin synthesis form *Bacillus* sp.

Operon/Genes/Operator/ Promtor/Protein	Role	References
	Quorum sensing system I	
srfAA†‡	Amino acid activating domain for Glu, Leu, DLeu Expression of *com*S gene	D'Souza et al. 1993, Galli et al. 1994, Nakano et al. 1988
SrfAC†	Encodes a thioesterase of a Type I motif responsible for peptide termination	De ferra et al. 1997
sfp	Surfactin production	Nakano et al. 1992
Sfp†	Activation of surfactin synthetase by post translational modification	Nakano et al. 1992
	Quorum sensing system I	
ComQ	Modification of *com*X to form signal peptide ComX	Solomon et al. 1996
ComP (Membrane bound protein)	Gets autophosphorylated upon stimulation and transfers its phosphate group to ComA	Solomon et al. 1996
Phosphorylated ComA ComS (located within and out of frame srfA gene)	Binds comA-box and initiate transcription of surfactin peptide synthetase, *srf*AA-AD operon and *com*S Development of competence	Magnuson et al. 1994
ComX (Signal peptide)	Controls expression of *srf*A and interaction with • Membrane bound histidine kinase ComP • Response regulator ComA	Lazazzera et al. 1997
*Spo*OK (Oligopeptide permease) RapC	Transfer of Competence stimulating factor (CSF)through the cell membrane; Phosphotransferase activity	Lazazzera et al. 1997
ComR (Polynucleotide phosphorylase)	Enhances *srf*A expression posttranscriptionally	Luttinger et al. 1996
SinR (Transcriptional regulator)	Negatively controls *srf*A possibility by regulating *com*R	Liu et al. 1996

involves nonribosomal peptide synthetases with four open reading frames (ORFs) in the *srf*A operon (Kluge et al. 1988, D'Souza et al. 1994). Operon *srf*A catalyses three multifunctional enzymes for surfactin synthesis (49). These modular building blocks are called surfactin synthetases encoded by *srfA*, *srfB* and *srfC*. The *srf*A locus plays an important role in surfactin production; *srf*A locus by cloning the DNA flanking srfA::Tn917 insertions followed by chromosome walking (Nakano and coworkers) (Cosmina et al. 1993). An operon (> 25 kb) and the gene *srfA* codes for template enzymes while another gene *Sfp* located downstream of the *srfA* operon encodes for 4′- phosphopantotheinyl transferase. This gene product alters enzymes to their functional forms for their transcription (Nakano et al. 1991, Hsieh et al. 2004). Tn9171ac mutations confirmed that surfactin production required both the intact 5′ as well as 3′ ends of *srfA*. The 5′ region was responsible for sporulation and competence for DNA uptake along with surfactin production and contains 20,535 bp. This region contains *srfA* promoter and two ORFs *srfAA* and *srfAB* encoding surfactin synthetase I and II. The *srfAA* contains three amino acid activating domains for Glu, Valand Leu, while *srfAB* peptide synthesizing domain contains domains for activating Val, Asp and d-Leu. Gene *srfC* contains activating regions for Leu (Fuma et al. 1993, Nakano et al. 1988) and encodes thioesterase Type I motif responsible for termination of peptide.

A third locus within *srfA* operon, in the *srfB gene* is required for surfactin production. *srfB* is also essential for expression of *srfA-lacZ* and is identical to a competence gene *comA*. Surfactin production is under *ComA*(*SrfB*)-dependent regulation operating at the transcriptional level *srfA* and is positively regulated by product of *srfB* (Galli et al. 1994, Nakano et al. 1991). Subsequently, SrfD stimulates the initiation process (Nakano and Zuber 1989) but, release of surfactin is still not detected. There is a hypothesis that passive diffusion releases surfactin across the cytoplasm membrane (Steller et al. 2004). Once the cell reaches log phase, ComX gets accumulated in the medium and interacts with membrane bound histidine kinase ComP and the response regulator ComA (Stein 2005). Further, after phosphorylation, by ComP, ComA binds to promoter *srf*A and transcription begins. Competence stimulating factor (CSF), a signal peptide influences *srf*A expression (Fabret et al. 1995). It is transported across the membrane and interacts with at least two different intracellular receptors depending upon its concentration. Mutation in ComA inhibits development of competence indicating that, the *com*A gene is responsible for expression of *srf* and other com genes (Hamon et al. 2001). In addition to all these proteins, ComR and SinR also influence *srf*A expression (Liu et al. 1996). ComA is regulated positively as well as negatively by ComP under the control of the ComX pheromone (Cosby et al. 1998). The authors also suggested that *srf* expression requires SpoOK and another, as yet unidentified, extracellular factor under variable pH conditions. The gene *spoOK* codes for an oligopeptide permease that functions in cell-density-dependent control of sporulation and competence (Liu et al. 1996, Perego et al. 1991, Rudner et al. 1991). Thus molecular mechanism ensures appropriate surfactin synthesis.

The sfp locus plays an important regulatory role at the transcriptional level. The *sfp* locus from a producer strain *B. subtilis* ATCC 21332 was transferred to a standard *B. subtilis* (Liu et al. 1996) and further subjected to transposon mutagenesis. Over all studies proved that, *B. subtilis* with a *sfp* genotype contains some genes required for surfactin synthesis; *sfp* locus responsible for surfactin production alters the transcriptional regulation of *srf* (Borchert et al. 1994). A *gsp* gene with sequence homology to *sfp* gene from Gramicidin operon of *B. brevis* complemented, a defect in the *sfp* gene in trans and was able to initiate surfactin synthesis in a non producer strain *B. subtilis* JH642 with *sfp* phenotypes (Lee et al. 2005). Additionally, the *Sfp* gene is also responsible for hydrocarbon degradation, since it was successfully integrated in chromosome of *B. subtilis* to enhance bioavailability of hydrophobic liquids (Morikawa et al. 1992). Sequencing of *sfp* gene revealed 100% sequence homology to amino acid sequences reported earlier by Nakano et al. A research team of Morikawa et al., worked on cloning and nucleotide sequencing of regulator gene in *B. pumilis*. Studies indicated that out of three large ORFs (ORF1, 2, 3), ORF3 was essential for surfactin synthesis. Additionally, production of antimicrobial substances or other secondary metabolites is associated with resistance to the producing organism. Tsuge et al. (2001) proposed a function of gene *yerP* as a determinant of self resistance to surfactin in *B. subtilis* (Yakimov et al. 1995). YerP was homologous to the resistance, nodulation and cell division (RND) family of proteins, which confers resistance to a wide range of noxious compounds secreted by the organism. Mutagenesis with mini-Tn10 transposon indicated that it had inserted itself in the *yerP* gene in surfactin susceptible mutant. The molecular mechanism for BS synthesis in *B. licheniformis* is similar to that in surfactin synthesis Yakimov et al. (1997), Yakimov et al. (2000). A recombinant strain of *B. licheniformis* KGL11 was constructed by inserting the surfactin synthetase enzyme. This mutant produced 12 times the Biosurfactant of parent strain (Lin et al. 1998, Mulligan et al. 1989). With a better understanding of the molecular phenomena, many attempts were aimed to enhance BS/BE production. Mulligan et al. (1989) were successful in obtaining a threefold higher BS production over wild types employing recombinant *B. subtilis* with modified peptide synthetase. A plasmid pC112 with lpa-14, a gene was used to construct a recombinant strain of *B. subtilis* MI113. A high yield of surfactin was achieved by fermentation technology (Ohno et al. 1995). Another recombinant strain of *Bacillus subtilis* MI113 (pC115), was constructed from *B. subtilis* RB14C. This recombinant strain had a gene responsible for surfactin, iturin production and produced new surfactin variants along with usual surfactin when cultured in solid-state fermentation employing soybean curd residue (okara) as substrate (Carrera et al. 1993, Nakayama et al. 1997). Along with a large number of research papers published, enormous patents on BS production appear to date. Carrera et al. (1993) filed U.S. patents (5,264,363; 5,227,294) on *B. subtilis* ATCC 55033 mutant strain which produced 4-6 times better BS over wild types. Another US patent (7,011,969) on *B. subtilis* SD901 strain mutated with N-methyl-N'-nitro-N-nitrosoguanidine resulted in 4–25 times more surfactin production (Yoneda et al. 2006). Such studies are opening arrays for improved BS production technologies. Various mutant/recombinant strains of *Bacillus* spp. have been constructed for better quality and optimum quantity of surfactin production.

1.5 *Serratia* Species

Serratia also has capability to synthesise biosurfactants. *Serratia*, a Gram-negative organism produces extracellular surface active (Matsuyama et al. 1987) and surface translocating agents. *S. marcescens* synthesizes a cyclic lipopeptide BS 'Serrawettin' which contains a 3-hydroxy-C10 FA side chain. The mobility (swarming/sliding motility) and cell density of a population is monitored; depending on this study, regulatory systems control gene expression. This will help the microbial community in interacting with its surrounding (Williams et al. 2000, Matsuyama et al. 1995). The SpnIR QSS is responsible for regulation of flagellum-independent population surface migration and synthesis of BS (prodigiosin) in *S. marcescens* SS-1 (Horng et al. 2002). Wei et al. (2006) study revealed that *spnIR* quorum-sensing genes were located on a Tn3 family transposon, Tn*TIR*. They also proved that SpnR negatively regulated transposition frequency of Tn*TIR*. They reported evidence for the first time of the involvement of a *luxIR*-type QSS in the regulation of transposition frequency. BS production is controlled by an auto-induction system which helps in the swarming of cells (Matsuyama et al. 1995). *S. marcescens* ATCC 274 produces temperature dependant serrawettin W1[cyclo-(d-3- hydroxydecanoyl –l-seryl)2]. Presence of *swr*W gene encoding serrawettin W1 amino lipid synthetase was detected in *S. marcescens* 274 by transposon mutagenesis. The swrW had all four domains of nonribosomal peptide synthetase (NRP), responsible for condensation, adenylation, thiolation and thioesterisation. The swrW NRP is unimodular and specifies only lysine (Li 2005). Authors proposed a pathway for serrawettin synthesis. Parallell production of serrawettin and pigment production in *S. marcescens* 274 is coded by an ORF namely pswP. Synthesis of serrawettin is supposed to be through a non ribosomal peptide synthetases (NRPSs) system which is a product of the *pswP* gene. A single mutation in the gene can cause parallel disruption of both, pigment as well as BS production in *S. marcescens* (Sunaga et al. 2004). Screening of serrawettin W1 overproducing mini Tn5 insertional mutants suggested a down regulating mechanism for BS production. The transposon was inserted between the *hexS* gene. *hexS* is a suppressive gene controlling production, so insertion and deactivation enhances production of exolipids. Thus, target specific repression of *hexS* gene production in transcription is elucidated (Tanikawa et al. 2006). Such abortion of repression can be useful for large scale and economical production of surface active agents. Production of BS and the surface migration in *S. marcescens* SS-1 is controlled by N-acylhomoserine lactones (AHLs) of QSS located on a mobile transposon (Horng et al. 2002, Wei et al. 2006). Production of BS is under negative control. *S. marcescens* SS-1 produces four AHLs via spnI. The production is regulated by SpnR in *spnI/spnR* QSS. The SpnR is a homologue of the transcriptional regulator LuxR. Furthermore, deletion of this *spnR* gene to produce an isogenic mutant strain *S. marcescens* SMΔR was found to enhance BS activity. Upstream of *spnI* is a gene *spnT* encoding a 464 amino acid protein (Horng et al. 2002). The *spnT* is cotranscribed with *spnI* and also functions as a negative regulator of BS production and sliding motility. Thus mobility and horizontal transfer of these genes was proved by Wei et al., similar correlation of genes (*swr*/QS) and enzyme involvement in BS production and swarming motility exists in *S. liquefaciens* (Lindum et al. 1998, Rosenberg et al. 1989, Riedel et al. 2003). This interdependence is obligatory for *S. liquefaciens* MG1 to develop a swarming colony. The gene *swr*I encodes a similar putative AHL synthase for synthesis of extracellular signal molecules *N*-butanoyl-l-homoserine lactone (BHL) and *N*-hexanoyl-l-homoserine lactone. Expression of *swrA*, encoding serrawettin synthetase, is a homoserine lactone (HSL) and is dependent on QSS (Ullrich et al. 1991, Chopade et al. 1986). The flagellar master operon (*flhDC*) and AHL are involved in flagellar mobility and cell density regulation. A mutant strain of *S. Liquefaciens* was developed by transposon mutagenesis to construct a nonswarming mutant deficient in serrawettin W2 production. Sequence analysis indicated homology with gene *swrA* that encodes a putative peptide synthetase. Expression of *swrA* is controlled by QSS. Transposon mutagenesis involving the promoter less *lux*AB reporter confirmed action of *swrA* gene via QSS in the production of the lipopeptide BS. The gene *swrA* encodes a putative peptide synthetase (Shete et al. 2006). Microbes are able to change their cell surface hydrophobicity during different growth phases, morphogenesis and differentiation (Patil and Chopade 2001, Rosenberg and Kjelleberg 1986). Cell surface hydrophobicity is affected by cell bound and extracellular factors, viz., serraphobin (capacity to bind with hexadecane) and serratamolide (act as wetting agent). Serratamolide negative mutants revealed that serratamolide increases cell surface hydrophobicity (Foght et al. 1989).

1.6 *Acinetobacter* Species

Acinetobacter sp. are ubiquitous in nature, being isolated from various sources like soil, mud, marine water, fresh water and meat products (Belsky et al. 1979, Gorkovenko et al. 1997) and reported for production of BE (Rosenberg et al. 1979, Zuckerberg et al. 1979, Whitfield et al. 1999, Whitfield et al. 2003, Patil and Chopade 2001, Sar and Rosenberg 1983). *Acinetobacter* species are promising bacteria producing high molecular weight BSs. The best known marine BE, now exploited commercially as 'Emulsan' appeared in 1972. This emulsifier is produced by *A. calcoaceticus* RAG-1, isolated from the Mediterranean Sea. Emulsan produced by RAG-1 has a heteropolysaccharide backbone with a repeating trisaccharide of *N*-acetyl-d-galactosamine, *N*-acetylgalactosamineuronic acid and an unidentified *N*-acetyl amino sugar. Fatty acids (FA) are covalently linked to the polysaccharide through *o*-ester linkages (Nakar et al. 2001, Nakar et al. 2003). Different species of *Acinetobacter* have the ability to produce protein polysaccharide complexes. About 16% of the patents on BSs have been reported from *Acinetobacter* spp. alone (Nesper et al. 2003) which indicates the tremendous market potential of exopolysaccharides (EPS).

1.7 Emulsan

It is a complex polysaccharide (9.9 × 105) produced by *A. calcoaceticus* RAG-1 and stabilizes oil-water emulsions efficiently (Gorkovenko et al. 1997, Gorkovenko et al. 1995, Rosenberg et al. 1979). In spite of structural complexity, researchers have been identifying genes implicated in emulsan synthesis and emulsification phenomena. Polymer biosynthesis is accomplished by a single gene cluster of 27 kbp with 20 open reading frames (ORFs) called as *wee* regulon which contains *weeA* to *week* genes that accomplish polymer biosynthesis (Zhang et al. 1997, Johri et al. 2002, Whitfield and Roberts 1999). Putative proteins encoded by the *wee* cluster have been tabulated by Nakar and Gutnick in detail. These genes lead to the formation of a polysaccharide containing amino sugars, with *O*-acyl- and *N*-acyl-bound side chain of FA. Further addition of intermediates takes place as follows: WeeA converts UDP-*N*-acetyl-d-glucosamine into UDP-*N*-acetylmannosamine. Consequently, WeeB oxidizes the UDP-*N*-acetylmannosamine into UDP-*N*-acetylmannosaminuronic acid. This regulon possesses *wzb* and *wzc* genes which are responsible for biosynthesis of emulsan. Gene products Wzc and Wzb were over expressed, purified and a bulk of polysaccharides was produced successfully (Kaplan et al. 2004, Zosim et al. 1986). The *Wee*E or *Wee*F are possibly involved in formation of UDP-*N*-acetyl-l-galactosaminuronic acid. The gene *Wee*J further catalyses the formation of diamino 2, 4-diamino-6-deoxy-d-glucosamine, a component of the repeating unit, from UDP-4-keto-6-deoxy-d-glucosamine. The sequence of *Wee*K is similar to dTDP-glucose 4, 6-dehydratase and therefore could possibly be responsible for conversion of UDP-d-glucosamine into UDP-4-keto-6-deoxy-d-glucosamine. The over all process is summarized in detail by Nesper et al. The monomers gather on a lipid carrier on the cytoplasmic face of the inner membrane. They are transferred by Wzx protein to the periplasmic face of the membrane. Wzy polymerase further catalyzes the polymerization process. Finally, lipid intermediates lead to the formation of a protein-polysaccharide complex which is transported across the periplasm to the outer membrane. This assembly gets accumulated on the cell surface and is further excreted as a polymer complex in the exterior (Franco et al. 1998). Due to the complex nature of exopolymers, genetic studies remained at a nascent level for a long period. However, with the advent of recent technologies and innovations, bioengineering of BE producing microorganisms has become possible. The complex polysaccharide backbone of emulsan was altered by modifying the culture conditions for *A. venetianus* RAG-1 (Daniels et al. 1999, Dams-Kozlowska et al. 2007, Leahy et al. 1993). The emulsan structure was modified by transposon mutagenesis of FA moiety. Analysis of various factors like, yield, FA content, molecular weight and emulsification behavior demonstrated that the parent strain yielded high emulsan as compared to the mutant strain. The factors are dependent on the type of FA supplemented during the production process. However, cloning and sequencing of mutants with enhanced emulsifying activity indicated that they were involved in the biosynthesis of emulsan. The presence and composition of long chain FAs on the polysaccharide backbone influenced emulsification behaviour. Such studies are highly significant and open newer avenues for applications of amphiphiles in diverse fields (Reddy et al. 1989). Based on similar kinds of studies, an interesting U.S. patent (20040265340) on “Emulsan adjuvant immunization formulations” was filed by Kaplan et al., The emulsan analog and mutants of *A. calcoaceticus* RAG-1 were produced in the presence of different FA sources.

1.8 Apoemulsan

It is an extracellular, polymeric lipoheteropolysaccharide produced by *A. venetianus* RAG-1. Purified deproteinized emulsan (apoemulsan, 103 kDa) which consists of d-galactosamine, l-galactosamineuronic acid (pKa, 3.05) and a diamino, 2-desoxy *n*-acetylglucosamine 44. It retained emulsifying activity towards certain hydrocarbon substrates but was unable to emulsify relatively nonpolar, hydrophobic, aliphatic materials (Shabtai et al. 1986, Zosim et al. 1986). It is now known that polymers are synthesized from the Wzy pathway. However, there also appears a differing report which claims that the process is based on the presence of polysaccharide-copolymerase (PCP) (Franco et al. 1998, Daniels et al. 1999). However, recently Dams-Kozlowska and Kaplan proved that synthesis of this polymer was dependant on the Wzy pathway where, PCP protein controlled the length of the polymer. This was proved by inducing defined point mutations in the proline-glycine-rich region of apoemulsan PCP protein (Wzc). Five of the eight mutants produced higher weight BEs than the wild type while four had modified biological properties. This study demonstrated the functional effect of Wzc modification on the molecular weight of a polymer and the genetic system controlling apoemulsan polymerization. It has been suggested that emulsifying activity and release of the polymer is mediated via esterase gene *est* (34.5 kD). A study carried out by Leahy et al. (1993), proved that lipase is responsible for enhanced emulsification properties. Lipase negative mutants exhibited less emulsification activity. The gene *est* has been cloned and over expressed in *E. coli* BL21 (DE3) behind the phage T7 promoter with His tag system (Toren et al. 2002). Further Alon and Gutnick (1993) also proved that *est* gene encodes the protein that is located on the outer membrane the same gene was sequenced and expressed in *E. coli.* High amounts of esterase were found to be associated when the cell was grown in the presence of nitrogen. Variants resistant to cetyl trimethyl ammonium bromide (CTAB) showed enhanced emulsan production (Bekerman et al. 2005). Site directed mutagenesis revealed that esterase-defective mutants could not release emulsan. Defective mutant proteins were capable of enhancing apoemulsan-mediated emulsifying activity. Bach et al. (2003), carried out studies on emulsans from *A. venetianus* RAG-1. It was seen that apoemulsan and esterase are essential for the formation of stable oil-water emulsions (Toren et al. 2002, Rosenberg et al. 1988, Gutnick 1985).

1.9 *Alasan*

The polymer produced by *A. radioresistens* KA53 is designated as 'Alasan' and finds significant application in bioremediation (Rosenberg et al. 1989). Alasan is an alanine containing complex hetero-polysaccharide and protein polymer that stabilizes oil in water emulsions in n-alkanes with chain length 10 or higher and alkyl aromatics, liquid paraffins, soyabean, coconut oil and crude oils (Barkay et al. 1999). The proteins of alasan have been identified as AlnA, AlnB and AlnC. One of the alasan proteins (AlnA) of 45 kDa exhibiting the highest emulsification activity was purified and denoted high sequence homology to an OmpA-like protein from *Acinetobacter* spp. Four hydrophobic regions in AlnA forming specific structures on the surface of the hydrocarbon are responsible for surface activity (Rosenberg et al. 1988, Elkeles et al. 1994). The AlnB protein exhibited strong homology to perioxiredoxins (family of thiol—specific antioxidant enzymes). It was proposed that all three proteins may be released as a complex with AlnA entering the oil phase and Alnb forming a compact shell around the hydrocarbon, thereby forming stable emulsions (Kaplan et al. 1982). *A. calcoaceticus* RA grown on crude oil sludge possesses three plasmids, one of which pSR4, a 20 kb fragment was found to be essential for growth and emulsification of crude oil in liquid culture (Ilan et al. 1999).

1.10 *Biodispersan*

It is an extracellular, anionic polysaccharide produced by *A. calcoaceticus* A2 which acts as a dispersing agent for water-insoluble solids (Inge et al. 2007, Hommel et al. 1993, Rusansky et al. 1987). It is nondialyzable, with an average molecular weight of 51,400 and contains four reducing sugars, namely, glucosamine, 6-methylaminohexose, galactosamine uronic acid and an unidentified amino sugar (Solaiman et al. 2004). Rich protein was also secreted along with the extracellular polysaccharide. Protein defective mutants produced equal/enhanced biodispersions as compared to the parent strain (Ashby et al. 2006).

1.11 Exopolysaccharide (EPS)

A. calcoaceticus BD4, BD413 produces EPS with rhamnose and glucose. 80138 EPS production is mediated by proteins like Ptk (protein tyrosine kinases) and was also found in *A. johnsonii*. These proteins encode for virulence factors and may serve as a target for the development of new antibiotics (Zerkowski et al. 2007).

2. Genetics of Glycolipid Synthesis in Fungi and Yeast

2.1 Candida

Candida species synthesize Sophorolipids (SLs) which are one of the most common glycolipids (Van Bogaert et al. 2007, Esders et al. 1972). The composition of sophorolipids is sophorose disaccharide glycosidically linked to a hydroxy FA. Several researchers identified, characterized and cloned the genes involved in the biosynthesis of sophorolipids (Bucholtz et al. 1976, Inge et al. 2007, Konishi et al. 2007). Mono-oxygenase enzyme, cytochrome P450 dependant on NADPH (nicotinamide adeninedinucleotide phosphate) is essential for FA conversion. The *CPR* (cytochrome P450 reductase) gene of *Candida bombicola* was isolated using degenerate PCR and genomic walking. The CPR gene is composed of 687 amino acids. Heterologous expression in *Escherichia coli* proved the functionality of the gene. The recombinant protein had NADPH-dependent cytochrome *c* reducing activity (Haïnoen et al. 2003). The genes of cytochrome P450 are diverse and also within the genome of a single organism. The phenomenon responsible for induction and expression of these genes was unknown (Hsieh et al. 2004). Specific glycosyltransferase I lead to the coupling of glycosidic linkage of glucose and FA. Glycosyltranferase II carries out subsequent glycosidic coupling. Both glycosyltransferases have been partially purified (Whiteley et al. 1999, Morita et al. 2007). Like other microorganisms *C. bombicola* produces glycolipid when grown on alkanes. Cytochrome P450 monooxygenase obtains reducing equivalents from NADPH cytochrome P450 reductase (CPR). The *CPR* gene of *C. bombicola* was isolated, sequenced and expressed in *E. coli*. The recombinant protein shows NADPH-dependent 'cytochrome c' reducing activity (Morita et al. 2007, Inoh et al. 2004).

2.2 Mycobacterium, Corynebacteria, Rhodococcus

Trehalose lipid (TL) contains carbohydrates and long-chain aliphatic acids/hydroxy aliphatic acids and are the most effective BSs produced by *Mycobacteria, Corynebacteria* and *Rhodococcus* species. 27 Finerty198 studied genes responsible for glycolipid biosynthesis in *Rhodococcus* sp. H13-A.A Genomic library were generated using *E. coli-Rhodococcus* shuttle vector pMVS301. Tn917 transpositional mutagenesis in *Rhodococcus*, was employed for isolation and analysis of sporulation and developmental genes in strains of *Bacillus*.

2.3 Pseudozyma, Ustilago maydis

Mannosylerythritol lipids (MELs) are produced by genus *Pseudozyma*. A yeast strain *P. antarctica* produces MEL. Genetic study was conducted on prospective genes involved in MEL production (Ripp et al. 2000). Under nitrogen limitation, *Ustilago maydis*, a dimorphic basidiomycete produces two different classes of glycolipids, ustilagic acids and ustilipids. Ustilagic acids contain cellobiose linked O-glycosidically to 15, 16 dihydroxyhexadecanoic acid, while ustilipids are derived from β-d-mannopyranosyl-d-erythritol and belong to the class of mannosylerythritol lipids (Minast et al. 1993). The first report of molecular characterization of glycolipid production using mutants came very recentlyin 2005 by Hewald et al. (2005). They identified two genes *emt1* and *cyp1* responsible for the production of extracellular glycolipids by the fungus. Gene *cyp1* codes for cytochrome P450 monooxygenase and is involved in synthesing 15, 16 dihydroxyhexadecanoic acid. *U. maydis Emt1* codes for a protein which resembles eukaryotic prokaryotic glycosyltransferases and transfers GDP-mannose to form mannosyl-d-erythritol. DNA micro-array analysis revealed that *emt1* is part of a gene cluster which comprises five open reading frames. Three proteins namely Mac1, Mac2 and Mat1, contain short sequence motifs characteristic for acyl- and acetyltransferases. Mac1 and Mac2 are essential for MEL production and are involved in acylation of MEL. Enzyme Mat1 acts as an acetyl coenzyme which is dependent on acetyltransferase. Mat1 displays relaxed region selectivity and is able to acetylate MEL at both, the C-4 and C-6 hydroxyl groups. The fifth protein is an export protein of the major facilitator family. This is the first report on the presence of a gene cluster for production of extracellular glycol lipids in a fungus. With these studies, the authors introduce the possibility of the transfer of genes between species or recent progenitors, for secondary metabolite production in fungal species.

3. Uses of Biosurfactant Molecular Genetics in Biotechnological Applications

The inherent genetic mechanism controls phenotypic expression for any particular organism. Understanding of this molecular mechanism will play a vital role in altering efficient microbes for potential, economic products. There has been an increasing progress in biotechnology in recent years, which has generated excessive opportunities. Initially biotechnological tools were aimed at hyperproducing mutant/recombinant strains. Mutant of *P. aeruginosa* PTCC 1637 produced 10 times more BS compared to that of the wild type. Those of *B. subtilis* MI113 and *B. licheniformis* KGL11 enhanced production by 8 and 12 times respectively. Remarkably *B. subtilis* SD901 mutant produced 4–25 times higher yield (Panilaitis et al. 2002). Biotechnological applications have been recently extended to the initial screening methodology of BS producers. The best example is represented from the work by Hsieh et al., The *sfp* locus was used for PCR based detection of BS producing *B. Amyloliquefaciens* and *B. circulans*. Such types of methods would authenticate the conventional screening methods enlisted in the brief review of Bodour and Miller-Maier (2000) Onsimilar lines, *P. rugulosa* NBRC 10877 was identified as MEL producer on the basis of rDNA sequence. A direct search for genes involved (Fig. 5) would be faster and less laborious. Newer inventions like those of Whiteley et al. (1999) could be used to identify modulators and genes of QSS signals in bacteria. Novel indicator strains and vectors have been engineered. Techniques like

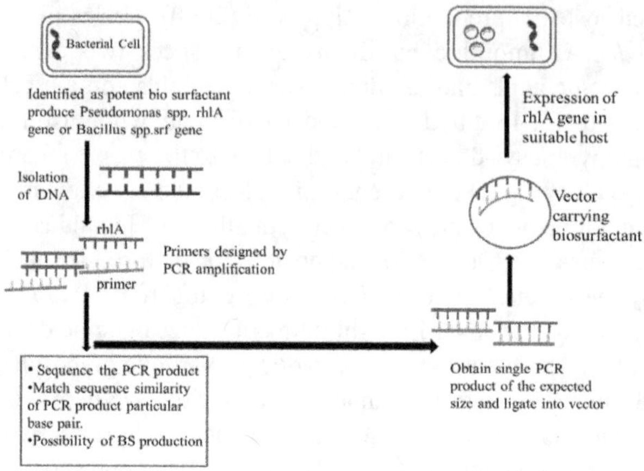

Figure 5. Molecular approach for screening of biosurfactant producers.

electroporation are useful in transformation studies and have been used successfully in *Pseudozyma.* The cationic liposome bearing MEL (produced by *C. antartica*) has been demonstrated to dramatically increase gene transfection efficiency into mammalian cells.

Similar studies have been reported by Inoh et al. (2004). Thus, molecular tools help to regulate and modify biosynthetic pathways to improve BS production technologies. Such significant findings can be used to upgrade laboratory scale studies towards field application. Advantages of techniques in identification, isolation and manipulation of structural genes involved in bio-surfactant biosynthesis have made it easier to improve existing bio-surfactant production technologies. The first genetically engineered bioluminescent strain *P. fluorescens* HK44, with a plasmid containing pUTK21 (naphthalene degradation), transposon and introduced *lux* gene fused within a promoter for naphthalene catabolic genes was released for bioremediation processes. The strain HK44 was capable of generating bioluminescence in response to soil hydrocarbon bioavailability. Authors suggested that *lux*-based bioreporter microorganisms can prove a practical alternative in determination of biodegradation *in situ,* with the process being well-monitored and controlled (Ripp et al. 2000). It is possible to use naturally occurring molecular tools for investigation purpose. Three cryptic plasmids from both *A. calcoaceticus* BD413, BD4 were isolated, characterized, sequenced and used in the construction of *E. coli* shuttle plasmids. Studies were done to clone and express the alcohol dehydrogenase regulon from *A. lwoffii* RAG-1. Gene expression and transformation in emulsan production and cell surface esterase activity in *A. lwoffii* RAG-1 were also analysed (Minast et al. 1993). The gene (*alnA*) was cloned, sequenced and over expressed in *E. coli.* The recombinant emulsifier protein (AlnA) exhibited 70% emulsifying activity as compared to that of native protein and 2.4 times more than that of the alasan complex. Thus, for the first time Toren et al., in the year 2002, successfully produced a recombinant surface-active protein using a defined gene. The existing molecular knowledge has opened gateways in drug discovery and manipulations. Protein products from microbes can be used for formulation of newer antibiotics and/or life saving drugs. Dams-Kozlowska and Kaplan, introduced a promising and new approach for bioengineering emulsan analogs which has novel application in the field of medicine as biological adjuvants for vaccine and drug delivery (Panilaitis et al. 2002, Castro et al. 2005). A research team of Symmank et al., genetically tailored peptide synthetase, which produced surfactin with reduced haemolytic activity. Rhamnolipid was synthesized in a heterologous host of *P. putida* by cloning *rhlAB* with *rhlRI* from the pathogenic producer strain *P. aeruginosa* (Cha 2008). These discoveries are highly commendable and certainly provide a promising approach towards conversion of pathogenic to a virulent strains. Thus, maximum exploitation of molecular mechanisms will not only add to our existing understanding of BS production but will also help bridge the gap between research and actual application.

4. Conclusion

With respect to the structural complexity, molecular mechanisms involved in polymer synthesis have been revealed. Among the low molecular weight bio-surfactants, the genetic mechanisms in *Pseudomonas* and *Bacillus* have been clearly illuminated. The bio-surfactant production in both microbes is under the influence of QSS. Different genes are involved and an interplay of these genes ensures efficient BS synthesis. Choosing desired substrates, an optimization of physicochemical parameters is not enough. Understanding the genetic mechanisms will help in accelerating research towards achieving economical production. Continued research is adding to the ever-expanding knowledge in this field and will certainly prove to be a boon for the surfactant industry. Although the utility of genetically modified organisms seems to be improbable due to environmental constraints, an understanding of the genetic mechanisms and molecular biology will improve the production of biosurfactants. This will help us in better understanding the production phenomena for further manipulation of conditions resulting in optimal and faster production of these surface active agents. More united efforts are needed for optimal utilization of generated information. A strong foundation of molecular mechanisms will help in an application-oriented outlook for the surfactant industry.

References

Alon, R.N. and D.L. Gutnick. 1993. Esterase from the oil degrading *Acinetobacter lwoffii* RAG-1: Sequence analysis and over expression in *Escherichia coli*. FEMS Microbiol. Lett. 12: 275–280.

Al-Tahhan, R.A., T.R. Sandrin and A.A. Bodour. 2000. Rhamnolipid-induced removal of lipopolysaccharide from *Pseudomonas aeruginosa*: Effect on cell surface properties and interaction with hydrophobic substrates. Appl. Environ. Microbiol. 66(8): 3262–3268.

Arima, K., A. Kakinuma and G. Tamura. 1968. Surfactin, a crystalline lipopeptide surfactant produced by *Bacillus subtilis*: Isolation, characterization and its inhibition of fibrin clot formation. Biochem. Biophys. Res. Commun. 31: 488–494.

Ashby, R.D., D. Solaiman and T.A. Foglia. 2006. The use of fatty acid-esters to enhance free acid sophorolipid synthesis. Biotechnol. Lett. 28: 253–260.

Bach, H., Y. Berdichevsky and D. Gutnick. 2003. An exocellular protein from the oil-degrading microbe Acinetobacter venetianus RAG-1 enhances the emulsifying activity of the polymeric bioemulsifier emulsan. Appl. Environ. Microbiol. 69(5): 2608–2615.

Banat, I.M. 1995. Biosurfactants characterization and use in pollution removal: State of the art. A review. Acta Biotechnol. 15: 25167.

Banat, I.M., R. Makkar and S. Cameotra. 2000. Potential commercial applications of microbial surfactants. Appl. Microbiol. Biotechnol. 53: 495–508.

Barkay, T., S. Navon-Venezia, E.Z. Ron and E. Rosenberg. 1999. Enhancement of solubilization and biodegradation of polyaromatic hydrocarbons by the bioemulsifier alasan. Appl. Environ. Microbiol. 65: 2697–2702.

Bazire, A., A. Dheilly, F. Diab, D. Morin, M. Jebbar, D. Haras and A. Dufour. 2005. Osmotic stress and phosphate limitation alter production of cell-to-cell signal molecules and rhamnolipid biosurfactant by *Pseudomonas aeruginosa*. FEMS Microbiol. Lett. 253(1): 125–131.

Beal, R. and W.B. Betts. 2000. Role of rhamnolipid biosurfactant in the uptake and mineralization of hexadecane in *Pseudomonas aeruginosa*. J. Bacteriol. 89: 158–168.

Bekerman, R., G. Segal, E.Z. Ron and E. Rosenberg. 2005. The AlnB protein of the bioemulsan alasan is a peroxiredoxin. Appl. Microbiol. Biotechnol. 66: 536–541.

Belsky, I., D.L. Gutnick and E. Rosenberg. 1979. Emulsifier of arthrobacter RAG-1: Determination of emulsifier-bound fatty acids. FEBS Lett. 101: 175–178.

Benincasa, M., A. Abalos, A. Oliveira and A. Manresa. 2004. Chemical structure, surface properties and biological activities of the biosurfactant produced by *Pseudomonas aeruginosa* LB1 from soapstock. Antonie Van Leeuwenhoek 85: 1–8.

Bergstrom, S., H. Theorell and H. Davide. 1947. On a metabolic product of Pseudomonas pyocyanea, pyolipic acid, active against Mycobacterium tuberculosis. Arkiv Kemi 23A(13): 1–15.

Bodour, A.A. and R. Miller-Maier. 2000. Biosurfactants: Types, screening methods and applications. pp. 750–770. *In*: Bitton, G. (ed.). Encyclopedia of Environmental Microbiology. 1st ed. Hoboken: John Wiley and Sons, Inc.

Bodour, A.A. and R.M. Maier. 2002. Biosurfactant: Types, screening methods and applications. pp. 750–770. *In*: Bitton, G. (ed.). Encyclopedia of Environmental Microbiology. New York: John Wiley & Sons.

Borchert, S., T. Stachelhaus and M.A. Arahiel. 1994. Induction of surfactin production in Bacillus subtilis by gsp, a gene located upstream of the gramicidins operon in Bacillus brevis. J. Bacteriol. 176(8): 2458–2462.

Branny, P., J.P. Pearson, E.C. Pesci, T. Kohler, B.H. Iglewski and C. VanDelden. 2001. Inhibition of quorum sensing by a Pseudomonas aeruginosa dksA homologue. J. Bacteriol. 183(5): 1531–1539.

Bucholtz, M.L. and R.J. Light. 1976. Acetylation of 13-sophorosyloxydocosanoic acid by an acetyltransferase purified from Candida bogoriensis. J. Biol. Chem. 251(2): 424–430.

Campos-Garci′, A.J., A.D. Caro, R. Najera, R.M. Miller-Maier, R. Al-Tahhan and G. Soberon. 1998. The Pseudomonas aeruginosa rhlG gene encodes an NADPH dependent β-Ketoacyl reductase which is specifically involved in rhamnolipid synthesis. J. Bacteriol 180(17): 4442–4451.

Carrera, P., P. Cosmina and G. Grandi. 1993. Eniricerche SPA., Milan, Italy. Mutant of Bacillus subtilis, United States Patent. Application No. 5264363.

Carrera, P., P. Cosmina and G. Grandi. 1993. Eniricerche SPA., Milan, Italy. Method of producing surfactin with the use of mutant of Bacillus subtilis. United States Patent. Application No. 5227294.

Castro, G.R., R.R. Kamdar, B. Panilaitis and D.L. Kaplan. 2005. Triggered release of proteins from emulsan-alginate beads. J. Control Release 109: 149–157.

Cha, M., N. Lee, M. Kim and S. Lee. 2008. Heterologous production of Pseudomonas aeruginosa EMS1 biosurfactant in Pseudomonas putida. Bioresour. Technol. 99(7): 2192–2199.

Chopade, B.A. 1986. Genetics of Antibiotic Resistance in Acinetobacter calcoaceticus. Ph.D. Thesis Submitted to University of Nottingham, England, Great Britain.

Cooper, D.G. and B.G. Goldenberg. 1987. Surface-active agents from two Bacillus species. Appl. Environ. Microbiol. 53: 224–229.

Cosby, W.M., D. Vollenbroich, O.H. Lee and P. Zuber. 1998. Altered srf expression in Bacillus subtilis resulting from changes in culture pH is dependent on the Spo0K oligopeptide permease and the ComQX system of extracellular control. J. Bacteriol. 180(6): 1438–1445.

Cosmina, P., F. Rodriguez, F. Ferra, G. Grandi, M. Perego, G. Venema and D. Sinderen. 1993. Sequence and analysis of the genetic locus responsible for surfactin synthesis in Bacillus subtilis. Mol. Microbiol. 8: 821–831.

D'Souza, C., M. Nakano, N. Corbell and P. Zuber. 1993. Amino acid site mutations in amino—acid-activating domains of surfactin synthetase; Effects on surfactin production and competence development in Bacillus subtilis. J. Bacteriol. 173(11): 3502–3510.

D'Souza, C., M.M. Nakano and P. Zuber. 1994. Identification of comS, a gene of the srfA operon that regulates the establishment of genetic competence in Bacillus subtilis. Proc. Natl. Acad. Sci. USA 91(20): 9397–9401.

Dams-Kozlowska, H. and D.L. Kaplan. 2007. Protein engineering of wzc to generate new emulsan analogs. Appl. Environ. Microbiol. 73(12): 4020–4028.

Daniels, C. and R. Morona. 1999. Analysis of Shigella flexneriwzz (Rol) function by mutagenesis and cross-linking: wzz is able to oligomerize. Mol. Microbiol. 34: 181–194.

De Ferra, F., F. Rodriguez, O. Tortora, C. Tosi and G. Grandi. 1997. Engineering of peptide synthetases key role of the thioesterase-like domain for efficient production of recombinant peptides. J. Biol. Chem. 272(40): 25304–25309.

Deziel, E., F. Lepine, S. Milot and R. Villemur. 2003. rhlA is required for the production of a novel biosurfactant promoting swarming motility in Pseudomonas aeruginosa: 3-(3-hydroxy-alkanoyloxy) alkanoic acids (HAAs), the precursors of rhamnolipids. Microbiol. 149: 2005–2013.

Dubern, J.F., E.L. Lagendijk, B.J.J. Lugtenberg and G.V. Bloemberg. 2005. The heat shock genes dnaK, dnaJ and grpE are involved in regulation of putisolvin biosynthesis in Pseudomonas putida PCL1445. J. Bacteriol. 187(17): 5967–5976.

Elkeles, A., E. Rosenberg and E.Z. Ron. 1994. Production and secretion of the polysaccharide biodispersan of Acinetobacter calcoaceticus A2 in protein secretion mutants. Appl. Environ. Microbiol. 60(12): 4642–4645.

Esders, T.W. and R.J. Light. 1972. Characterization and *in vivo* production of three glycolipids from Candida bogoriensis: 13-glucopyranosylglucopyranosyloxydocosanoic acid and its mono- and diacetylated derivatives. J. Lipid Res. 13: 663–671.

Fabret, C., Y. Quentin, A. Guiseppi, J. Busuttil, J. Haiech and F. Denizot. 1995. Analysis of errors in finished DNA sequences: The surfactin operon of Bacillus subtilis as an example. Microbiol. 141: 345–350.

Flemming, C.A., K.T. Leung, H. Lee, J.T. Trevors and C.W. Greer. 1994. Survival of lux-lac-marked biosurfactant-producing Pseudomonas aeruginosa UG2L in soil monitored by nonselective plating and PCR. Appl. Environ. Microbiol. 60(5): 1606–1613.

Foght, J.M., D.L. Gutnick and D.W.S. Westlake. 1989. Effect of emulsan on biodegradation of crude oil by pure and mixed bacterial cultures. Appl. Environ. Microbiol. 55: 36–42.

Franco, A.V., D. Liu and P.R. Reeves. 1998. The Wzz (cld) protein in *Escherichia coli*: Amino acid sequence variation determines O-antigen chain length specificity. J. Bacteriol. 180: 670–2675.

Fuma, S., Y. Fujishima, N. Corbell, C. Dsouza and M. Nakano. 1993. Nucleotide sequence of 5′ portion of srfA that contains the region required for competence establishment in Bacillus subtilis. Nucleic Acids Research 21(1): 93–97.

Galli, G., F. Rodriguez, P. Cosmina, C. Pratesic, R. Nogarotto, F. Ferra and G. Grandi. 1994. Characterization of the surfactin synthetase multi-enzyme complex. Biochem. Biophys. Acta 1205: 19–28.

Gorkovenko, A., J. Zhang, R.A. Gross, A.L. Allen and D.L. Kaplan. 1995. Biosynthesis of emulsan analogs: Direct incorporation of exogenous fatty acids. Proc. Am. Chem. Soc. Div. Polym. Sci. Eng. 72: 92–93.

Gorkovenko, A., J. Zhang, R.A. Gross, A.L. Allen and D.L. Kaplan. 1997. Bioengineering of emulsifier structure: Emulsan analogs. Can. J. Microbiol. 43: 384–390.

Hamoen, L.W., G. Venema and O.P. kuipers. 2003. Controlling competence in Bacillus subtilis; shared use of regulators. Microbiol. 149: 9–17.

Hamon, M.A. and B.A. Lazazzera. 2001. The sporulation transcription factor ApoOA is required for biofilm development in Bacillus subtilis. Mol. Microbiol. 42: 1199–1209.

Hewald, S., K. Josephs and M. Bolker. 2005. Genetic analysis of biosurfactant production in Ustilago maydis. Appl. Environ. Microbiol. 71(6): 3033–3040.

Hommel, R.K. and K. Huse. 1993. Regulation of sophorose lipid production by Candida apicola. Biotechnol. Lett. 33: 853–858.

Horng, Y.T., S.C. Deng, M. Daykin, P. Soo, J. Wei, K. Luh, S. Ho, S. Swift, H. Lai and P. Willams. 2002. The LuxR family protein SpnR functions as a negative regulator of N-acyl homoserine lactone-dependent quorum sensing in Serratia marcescens. Mol. Microbiol. 45: 1655–1671.

Hsieh, F.C., M.C. Li, T.C. Lin and S.S. Kao. 2004. Rapid detection and characterization of surfactin-producing Bacillus subtilis and closely related species based on PCR. Curr. Microbiol. 49: 186–191.

Ilan, O., Y. Bloch, G. Frankel, H. Ullrich, K. Geider and I. Rosenshine. 1999. Protein tyrosine kinases in bacterial pathogens are associated with virulence and production of exopolysaccharide. The EMBO J. 18: 3241–3248.

Inge, N.A., V. Bogaert, D. Develter, W. Soetaert and E. Vandamme. 2007. Cloning and characterization of the NADPH cytochrome P450 reductase gene (CPR) from Candida bombicola. FEMS Yeast Res. 7(6): 922–928.

Inoh, Y., D. Kitamoto, N. Hirashima and M. Nakanishi. 2004. Biosurfactant MEL dramatically increases gene transfection via membrane fusion. J. Control Release 94(2-3): 423–431.

Iqbal, S., Z.M. Khalid and K.A. Malik. 1995. Enhanced biodegradation and emulsification of crude oil and hyperproduction of biosurfactant by a gamma ray-induced mutant of Pseudomonas aeruginosa. Lett. Appl. Microbiol. 21(3): 176–179.

Jarvis, F.G. and M.J. Johnson. 1949. A glycolipid produced by Pseudomonas aeruginosa. J. Am. Chem. Soc. 71: 4124–4126.

Johri, A.K., W. Blank and D.L. Kaplan. 2002. Bioengineered emulsans from Acinetobacter calcoaceticus RAG-1 transposon mutants. Appl. Microbiol. Biotechnol. 59: 217–223.

Kaplan, D.L., J. Fuhrman and R.A. Gross. 2004. Emulsan adjuvant immunization formulations and use. United States Patent. Application No. 20040265340.

Kaplan, N. and E. Rosenberg. 1982. Exopolysaccharide distribution of and bioemulsifier production by Acinetobacter calcoaceticus BD4 and BD413. Appl. Environ. Microbiol. 44: 1335–1341.

Kluge, B., J. Vater, J. Salnikow and K. Eckart. 1988. Studies on the biosynthesis of surfactin, a lipopeptide antibiotic from Bacillus subtilis ATCC 21332. FEBS Lett. 231: 107–110.

Koch, A.K., J. Reiser, O. Kappeli and A. Fiechter. 1988. Genetic construction of lactose-utilizing strains of Pseudomonas aeruginosa and their application in biosurfactant production. Biotechnol. 6: 1335–1339.

Koch, A.K., O. Kappeli, A. Fiechter and J. Reiser. 1991. Hydrocarbon assimilation and biosurfactant production in Pseudomonas aeruginosa mutants. J. Bacteriol. 173(13): 4212–4219.

Konishi, M., T. Morita, T. Fukuoka, T. Imura, K. Kakugawa and D. Kitamoto. 2007. Production of different types of mannosylerythritol lipids as biosurfactant by the newly isolated yeast strains belonging to the genus Pseudozyma. Appl. Microbiol. Biotechnol. 75: 521–531.

Latifi, A., M. Foglino, K. Tanaka, P. Williams and A. Lazdunski. 1996. Hierachical quorum sensing cascade in Pseudomonas aeruginosa links the transcriptional activators LasR and RhlR (VsmR) to expression of the stationary-phase sigma factor RpoS. Mol. Microbiol. 21: 1137–1146.

Lazazzera, B.A., J.M. Solomon and A.D. Grossman. 1997. An exported peptide functions intracellularly to contribute to cell density signaling in B. subtilis. Cell. 89: 917–925.

Lazdunski, A.M., I. Ventre and J.N. Sturgis. 2004. Regulatory circuits and communication in Gram-negative bacteria. Nat. Rev. Microbiol. 2: 581–592.

Leahy, J.G., J.M. Jones-Meehan, E.L. Pullas and R. Colwell. 1993. Transposon mutagenesis in Acinetobacter calcoaceticus RAG-1. J. Bacteriol. 175: 1838–1840.

Lee, Y.K., S.B. Kim, C.S. Park, J.G. Kim, H.M. Oh, B.D. Yoon and H.S. Kim. 2005. Chromosomal integration of sfp gene in Bacillus subtilis to enhance bioavailability of hydrophobic liquids. Appl. Microbiol. Biotechnol. 67(6): 789–794.

Lequette, Y. and E.P. Greenberg. 2005. Timing and localization of rhamnolipid synthesis gene expression in Pseudomonas aeruginosa biofilms. J. Bacteriol. 187(1): 37–44.

Li, H., T. Tanikawa, Y. Sato, Y. Nakogawa and T. Matsuyama. 2005. Serratia marcescens gene required for surfactant serrawettin W1 production encodes putative aminolipid synthetase belonging to nonribosomal peptide synthetase family. Microbiol. Immunol. 49(4): 303–310.

Lin, S.C., K.G. Lin, C.C. Lo and Y.M. Lin. 1998. Enhanced biosurfactant production by a Bacillus licheniformis mutant. Enzyme Microb. Technol. 23: 267–273.

Lindum, P.W., U. Anthoni, C. Christophersen, L. Eberl, S. Molin and M. Givskov. 1998. N-acyl-l-homoserine lactone autoinducers control production of an extracellular lipopeptide biosurfactant required for swarming motility of Serratia liquefaciens MG1. J. Bacteriol. 180(23): 6384–6388.

Liu, L., M.M. Nakano, O.H. Lee and P. Zuber. 1996. Plasmid-amplified comS enhances genetic competence and suppresses sinR in Bacillus subtilis. J. Bacteriol. 178: 5144–5152.

Luttinger, A., J. Hahn and D. Dubnau. 1996. Polynucleotide phosphorylase is necessary for competence development in Bacillus subtilis. Mol. Microbiol. 19: 343–356.

Magnuson, R., J. Solomon and A.D. Grossman. 1994. Biochemical and genetic characterization of a competence pheromone from *B. subtilis.* Cell 77(2): 207–216.

Matsuyama, T., M. Sogawa and I. Yanot. 1987. Direct colony thin-layer chromatography and rapid characterization of Serratia marcescens mutants defective in production of wetting agents. Appl. Environ. Microbiol. 53(5): 1186–1188.

Matsuyama, T., A. Bhasin and R.M. Harshey. 1995. Mutational analysis of flagellum-independent surface spreading of Serratia marcescens 274 on a low-agar medium. J. Bacteriol. 177: 987–991.

Medina, G., K. Juarez, R. Diaz and G. Soberon-Chavez. 2003. Transcriptional regulation of Pseudomonas aeruginosa rhlR, encoding a quorum-sensing regulatory protein. Microbiol. 149: 3073–3081.

Minast, W. and D.L. Gutnick. 1993. Isolation, characterization and sequence analysis of cryptic plasmids from Acinetobacter calcoaceticus and their use in the construction of *Escherichia coli* shuttle plasmids. Appl. Environ. Microbiol. 59(9): 2807–2816.

Morikawa, M., M. Ito and T. Imanaka. 1992. Isolation of a new surfactin producer Bacillus pumilus A-1 and cloning and nucleotide sequence of the regulator gene, psf-1. J. Ferm. Bioeng. 74(5): 255–261.

Morita, T., H. Habe and T. Fukuoka. 2007. Gene expression profiling and genetic engineering of a basidiomycetous yeast, Pseudozyma antarctica, which produces multifunctional and environmentally-friendly surfactants (biosurfactant). The XXIIIrd International Conference on Yeast Genetics and Molecular Biology Melbourne. Australia, 1–6.

Morita, T., K. Konishi, T. Fukuoka, T. Imura and D. Kitamoto. 2007. Microbial conversion of glycerol in to glycolipid biosurfactant, mannosylerythritol lipids by basidiomycete yeast Pseudozyma antarctica, JCM 1037. J. Biosci. Bioeng. 104(1): 78–81.

Mulligan, C.N. and B.F. Gibbs. 1989. Correlation of nitrogen metabolism with biosurfactant production by *Pseudomonas aeruginosa.* Appl. Environ. Microbiol. 55: 3016–3019.

Mulligan, C.N., G. Mahmourides and B.F. Gibbs. 1989. Biosurfactant production by chloramphenicol-tolerant strain of Pseudomonas aeruginosa. J. Biotechnol. 12: 37–44.

Mulligan, C.N., T.Y.K. Chow and B.F. Gibbs. 1989. Surfactin production by a Bacillus subtilis mutant. Appl. Microbiol. Biotechnol. 31: 486–489.

Mulligan, C.N. 2005. Environmental application for biosurfactants. Environ. Pollut. 133: 183–198.

Nakano, M.M., M.A. Marahiel and P. Zuber. 1988. Identification of a genetic locus required for biosynthesis of the lipopeptide antibiotic surfactin in Bacillus subtilis. J. Bacteriol. 170(12): 5662–5668.

Nakano, M.M. and P. Zuber. 1989. Cloning and characterization of srfB, a regulatory gene involved in surfactin production and competence in Bacillus subtilis. J. Bacteriol. 171(10): 5347–5353.

Nakano, M.M., R. Magnuson, A. Myers, J. Curry, A. Grossman and P. Zuber. 1991. srfA is an operon required for surfactin production, competence development and efficient sporulation in Bacillus subtilis. J. Bacteriol. 173(5): 1770–1778.

Nakano, M.M., L.A. Xia and P. Zuber. 1991. Transcription initiation region of the srfA operon, which is controlled by the comP-comA signal transduction system in Bacillus subtilis. J. Bacteriol. 173(17): 5487–5493.

Nakano, M.M., N. Corbell, J. Besson and P. Zuber. 1992. Isolation and characterization of sfp: a gene that functions in the production of the lipopeptide biosurfactant, surfactin, in Bacillus subtilis. Mol. Gen. Genet. 232(2): 313–321.

Nakar, D. and D.L. Gutnick. 2001. Analysis of the wee gene cluster responsible for the biosynthesis of the polymeric bioemulsifier from the oil-degrading strain Acinetobacter lwoffii RAG-1. Microbiol. 47: 1937–1946.

Nakar, D. and D.L. Gutnick. 2003. Involvement of a protein tyrosine kinase in production of the polymeric bioemulsifier emulsan from the oil-degrading strain Acinetobacter lwoffii RAG-1. J. Bacteriol. 185(3): 1001–1009.

Nakayama, S., S. Takahashi, M. Hirai and M. Shoda. 1997. Isolation of new variants of surfactin by a recombinant Bacillus subtilis. Appl. Microbiol. Biotechnol. 48: 80–82.

Nesper, J., C.M. Hill, A. Paiment, G. Harauz, K. Beis, J. Naismith and C. Whitefield. 2003. Translocation of group 1 capsular polysaccharide in Escherichia coli serotype K30. Structural and functional analysis of the outer membrane lipoprotein Wza. J. Biol. Chem. 278: 49763–49772.

Noordman, W.H. and D.B. Janssen. 2002. Rhamnolipid stimulates uptake of hydrophobic compounds by Pseudomonas aeruginosa. Appl. Environ. Microbiol. 68: 4502–4508.

Ochsner, U.A., A. Fiechter and J. Reiser. 1994. Isolation, characterization and expression in *Escherichia coli* of the *Pseudomonas aeruginosa* rhlA genes encoding rhamnosylatransferas involved in rhamnolipid biosurfactant sysnthesis genes. J. Biol. Chem. 269: 19787–19795.

Ochsner, U.A., A.K. Koch, A. Fiechter and J. Reiser. 1994. Isolation and characterization of a regulatory gene affecting rhamnolipid biosurfactant synthesis in Pseudomonas aeruginosa. J. Bacteriol. 176(7): 2044–2054.

Ochsner, U.A., J. Reiser, A. Fiechter and B. Witholt. 1995. Production of Pseudomonas aeruginosa rhamnolipid biosurfactant in heterologous hosts. Appl. Environ. Microbiol. 61: 3503–3506.

Ochsner, U.A. and J. Reiser. 1995. Autoinducer-mediated regulation of rhamnolipid biosurfactant synthesis in Pseudomonas aeruginosa. Proc. Natl. Acad. Sci. USA 92: 6424–6428.

Ohno, A., T. Ano and M. Shoda. Production of a lipopeptide antibiotic, surfactin, by recombinant Bacillus subtilis in solid state fermentation. Biotechnol. Bioeng 47: 209–214.

Pamp, S.J. and T. Tolker-Nielsen. 2007. Multiple roles of biosurfactant in structural biofilm development by Pseudomonas aeruginosa. J. Bacteriol. 189: 2531–2539.

Panilaitis, B., A. Johri, W. Blank, D. Kaplan and J. Fuhrman. 2002. Adjuvant activity of emulsan, a secreted lipopolysaccharide from Acinetobacter calcoaceticus. Clin. Diagn. Lab. Immunol. 9: 1240–1247.

Patil, J.R. and B.A. Chopade. 2001. Studies on bioemulsifier production by Acinetobacter strains isolated from healthy human skin. J. Appl. Microbiol. 91(2): 290–298.

Peele, K. Abraham, Ravi Teja Ch and Vidya P. Kodali. 2016. Emulsifying activity of a biosurfactant produced by a marine bacterium. 3 Biotech. 6(2): 177.

Perego, M., C.F. Higgins, S.R. Pearce, M.P. Gallagher and J.A. Hoch. 1991. The oligopeptide transport system of Bacillus subtilis plays a role in the initiation of sporulation. Mol. Microbiol. 5: 173–185.

Pesci, E.C., J.P. Pearson, P.C. Seed and B.H. Iglewski. 1997. Regulation of las and rhl quorum sensing in Pseudomonas aeruginosa. J. Bacteriol. 179: 3127–3132.

Rahim, R., U.A. Ochsner, C. Olvera, M. Graninger, P. Messner, J. Lam and G. Soberon-Chavez. 2001. Cloning and functional characterization of the Pseudomonas aeruginosa rhlC gene that encodes rhamnosyltransferase 2, an enzyme responsible for di-rhamnolipid biosynthesis. Mol. Microbiol. 40(3): 708–718.

Raza, Z.A., M.S. Khan, Z.M. Khalid and A. Rehman. 2006. Production of Biosurfactant using different hydrocarbons by *Pseudomonas aeruginosa* EBN-8 mutant. Z. Naturforsch. 61c: 87–94.

Raza, Z.A., M.S. Khan and Z.M. Khalid. 2007. Evaluation of distant carbon sources in biosurfactant production by a gamma ray-induced Pseudomonas putida mutant. Process Biochem. 42(4): 686–692.

Reddy, P.G., R. Allon and M. Mevarech. 1989. Cloning and expression in Escherichia coli of an esterase-coding gene from the oil degrading bacterium *Acinetobacter calcoaceticus* RAG-1. Gene 76: 145–152.

Rendell, N.B., G.W. Taylor, M. Somerville, H. Todd, R. Wilson and P.J. Cole. 1990. Characterization of *Pseudomonas* rhamnolipids. Biochem. Biophys. Acta 16: 189–193.

Riedel, K., D. Talker-Huiber, M. Givskov, H. Schwab and L. Eberl. 2003. Identification and characterization of a GDS esterase gene located proximal to the swr quorum-sensing system of Serratia liquefaciens MG1. Appl. Environ. Microbiol. 69(7): 3901–3910.

Ripp, S., D.E. Nivens, C. Werner, J. Jarrell, J. Easter, C. Cox, R. Burlage and G. Sayler. 2000. Controlled field release of a bioluminescent genetically engineered microorganism for bioremediation process monitoring and control. Environ. Sci. Technol. 34: 846–853.

Rosenberg, E., A. Zuckerberg, C. Rubinovitz and D. Gutnick. 1979. Emulsifier of Arthrobacter RAG-1: Isolation and emulsifying properties. Appl. Environ. Microbiol. 37: 402–408.

Rosenberg, M. and S. Kjelleberg. 1986. Hydrophobic interactions in bacterial adhesion. Adv. Microb. Ecol. 9: 353–393.

Rosenberg, E., C. Rubinovitz, A. Gottlieb, S. Rosenhak and E. Ron. 1988. Production of biodispersan by Acinetobacter calcoaceticus A2. Appl. Environ. Microbiol. 54: 317–322.

Rosenberg, E., C. Rubinovitz, R. Legmann and E. Ron. 1988. Purification and chemical properties of Acinetobacter calcoaceticus A2 biodispersan. Appl. Environ. Microbiol. 54: 323–326.

Rosenberg, E., Z. Schwartz, A. Tenenbaum and E. Ron. 1989. Microbial polymer that changes the surface properties of limestone; effect of biodispersan in grinding limestone and making paper. J. Dispersion Sci. Technol. 10: 241–250.

Rudner, D.Z., J.R. Ledeaux, K. Ireton and A. Grossman. 1991. The spo0K locus of Bacillus subtilis is homologous to the oligopeptide permease locus and is required for sporulation and competence. J. Bacteriol. 173: 1388–1398.

Rusansky, S., R. Avigad, S. Michaeli and S. Zaki. 1987. Effects of mixed nitrogen sources on biodegradation of phenol by immobilized Acinetobacter sp. strain W-17. Appl. Environ. Microbiol. 53: 1918–1923.

Sankar Ganesh, P. and V. Ravishankar Rai. 2018. Alternative Strategies to Regulate Quorum Sensing and Biofilm Formation of Pathogenic Pseudomonas by Quorum Sensing Inhibitors of Diverse Origins Springer Link, pp. 33–61.

Sar, N. and E. Rosenberg. 1983. Emulsifier production by Acinetobacter calcoaceticus strains. Curr. Microbiol. 9: 309–314.

Schlictman, D., M. Kubo, S. Shankar and A. Chakrabarty. 1995. Regulation of nucleoside diphosphate kinase and secretable virulence factors in Pseudomonas aeruginosa: Roles of algR2 and algH. J. Bacteriol. 177(9): 2469–2474.

Shabtai, Y. and D.L. Gutnick. 1985. Exocellular esterase and emulsan release from the cell surface of Acinetobacter calcoaceticus. J. Bacteriol. 161: 1176–1181.

Shabtai, Y. and D.L. Gutnick. 1986. Enhanced emulsan production in mutants of Acinetobacter calcoaceticus RAG-1 selected for resistance to cetyltrimethylammonium bromide. Appl. Environ. Microbiol. 52: 146–151.

Shete, A.M., G.W. Wadhawa, I.M. Banat and B.A. Chopade. 2006. Mapping of patents on bioemulsifier and biosurfactant: A review. J. Sci. Indust. Res. 65: 91–115.

Shrive, G.S., S. Inguva and S. Gunnam. 1995. Rhamnolipid biosurfactant enhancement of hexadecane biodegradation by Pseudomonas aeruginosa. Mol. Marine Biol. Biotechnol. 4: 331–337.

Solaiman, D., R.D. Ashby and T.A. Foglia. 2004. Characterization and manipulation of genes in the biosynthesis of sophorolipids and poly (hydroxyalkanoates). *In*: Proceedings of the United States-Japan Cooperative program in natural resources, protein resources panel Annual Meeting, 215–219.

Solomon, J.M. and A.D. Grossman. 1996. Who's competent and when: regulation of natural genetic competence in bacteria. Trends Genet. 12: 150–155.

Solomon, J.M., B.A. Lazazzera and A.D. Grossman. 1996. Purification and characterization of an extracellular peptide factor that affects two different developmental pathways in Bacillus subtilis. Genes Dev 10: 2014–2024.

Stein, T. 2005. Bacillus subtilis antibiotics: structures, syntheses and specific functions. Mol. Microbiol. 56(4): 845–857.

Steller, S., A. Sokoll, C. Wilde. 2004. Initiation of surfactin biosynthesis and role of Srf-Dthioesterase protein. Biochem. 43: 11331–11343.

Stover, C.K., X.Q. Pham, A.L. Erwin et al. 2000. Complete genome sequence of *Pseudomonas aeruginosa* PAO1, an opportunistic pathogen. Nature 31: 959–964.

Sullivan, E. 1998. Molecular genetics of biosurfactant production. Curr. Opinion Biotechnol. 9: 263–269.

Sunaga, S., H. Li, Y. Sato et al. 2004. Identification and characterization of the pswP gene required for the parallel production of prodigiosin and serrawettin W1 in Serratia marcescens. Microbiol. Immunol. 48(10): 723–728.

Syldatk, C. and F. Wagner. 1987. Production of biosurfactants. pp. 89–120. *In*: Kosaric, N., W.L. Cairns and N.C.C. Gray (eds.). Bio-Surfactants and Biotechnology. New York: Marcel Dekker, Inc.

Tahzibi, A., F. Kamal and M.M. Assadi. 2004. Improved production of rhamnolipids by a Pseudomonas aeruginosa mutant. Iran Biomed. J. 8(1): 25–31.

Tanikawa, T., Y. Nakagawa and T. Matsuyama. 2006. Transcriptional downregulator HexS controlling prodigiosin and serrawettin W1 biosynthesis in Serratia marcescens. Microbiol. Immunol. 50(8): 587–596.

Toren, A., E. Orr, Y. Paitan et al. 2002. The active component of the bioemulsifier alasan from Acinetobacter radioresistens KA53 is an OmpA-Like protein. J. Bacteriol. 184(1): 165–170.

Toren, A., G. Segal, E. Ron et al. 2002. Structure–function studies of the recombinant protein bioemulsifier AlnA. Environ. Microbiol. 4(5): 257–261.

Tsuge, K., Y. Ohata and M. Shoda. 2001. Gene yerP, involved in surfactin self-resistance in Bacillus subtilis. Antimicrob. Agents Chemother. 45(12): 3566–3573.

Ullrich, C., B. Kluge, Z. Palacz et al. 1991. Cell-free biosynthesis of surfactin, a cyclic lipopeptide produced by Bacillus subtilis. Biochem. 30: 6503–6508.

Van Bogaert, I.N.N., D. Develter, W. Soetaert et al. 2007. Cloning and characterization of the NADPH cytochrome P450 reductase gene (CPR) from *Candida bombicola*. FEMS Yeast Res. 7(6): 922–928.

Wei, J., P.C. Soo, Y.T. Horng et al. 2006. Regulatory roles of spnT, a novel gene located within transposon TnTIR. Biochem. Biophy. Res. Comm. 348: 1038–1046.

Wei, J., Y.H. Tsai, Y.T. Horng et al. 2006. A mobile quorum-sensing system in Serratia marcescens. J. Bacteriol. 188(4): 1518–1525.

Wei, Y., H.C. Lai, S.U. Chen et al. 2004. Biosurfactant production by Serratia marcescens SS-1 and its isogenic strain SMΔR defective in SpnR, a quorum-sensing LuxR family protein. Biotechnol. Lett. 26: 799–802.

Whiteley, M., K.M. Lee and E.P. Greenberg. 1999. Identification of genes controlled by quorum sensing in Pseudomonas aeruginosa. PNAS 96(24): 13904–13909.

Whitfield, C. and I.S. Roberts. 1999. Structure, assembly and regulation of expression of capsules in *Escherichia coli*. Mol. Microbiol. 31: 1307–1319.

Whitfield, C. and A. Paiment. 2003. Biosynthesis and assembly of group 1capsular polysaccharides in Escherichia coli and related extracellular polysaccharides in other bacteria. Carbohyd. Res. 338: 2491–2502.

Williams, P., M. Camara, A. Hardman, S. Swift, D. Milton, V. Hope, K. Winzer, B. Middleton, D. Pritchard and B.W. Bycroft. 2000. Quorum sensing and the population-dependent control of virulence. Philos. Trans. R Soc. London. B Biol. Sci. 355: 667–680.

Yakimov, M.M., K.M. Timmis, V. Wray and H. Fredrickson. 1995. Characterization of a new lipopeptide surfactant produced by thermotolerant and halotolerant subsurface Bacillus licheniformis BAS50. Appl. Environ. Microbiol. 61: 1706–1713.

Yakimov, M.M. and P.N. Golyshin. 1997. ComA-dependant transcriptional activation of lichenysin A synthetase promoter in Bacillus subtilis cells. Biotechnol. Prog. 13: 757–761.

Yakimov, M.M., L. Giuliano, K.N. Timmis and D.N. Golyshin. 2000. Recombinant acylheptapeptide lichenysin: High level of production by Bacillus subtilis cells. J. Mol. Microbiol. Biotechnol. 2: 217–224.

Yoneda, T., M. Yoshiaki, F. Kazuo and T. Toshi. 2006. Showa Denko KK (JP), Tokyo, Japan. Production Process of Surfactin, United States Patent. Application No. 7011969.

Zerkowski, J.A. and D. Solaiman. 2007. Polyhydroxy fatty acids derived from sophorolipids. J. Amer Oil Chemists Soc. 84(5): 463–471.

Zhang, J., A. Gorkovenko, R. Gross, L. Allen and D. Kaplan. 1997. Incorporation of 2-hydroxyl fatty acids by Acinetobacter calcoaceticus RAG-1 to tailor emulsan structure. Int. J. Bio Macromol. 20: 9–21.

Zosim, Z., E. Rosenberg and D.L. Gutnick. 1986. Changes in hydrocarbon emulsification specificity of the polymeric bioemulsifier emulsan: Effects of alkanols. Colloid. Polym. Sci. 264: 218–223.

Zuckerberg, A., A. Diver, Z. Peeri, D.L. Gutnick and E. Rosenberg. 1979. Emulsifier of Arthrobacter RAG-1: Chemical and physical properties. Appl. Environ. Microbiol. 37: 414–420.

4

Delving through Quorum Sensing and CRISPRi Strategies for Enhanced Surfactin Production

Shireen Adeeb Mujtaba Ali,[1] *R Z Sayyed*,[2] *M S Reddy*,[3] *Hesham El Enshasy*[4,5] and *Bee Hameeda*[1,*]

1. Introduction

Surfactants are amphipathic compounds of organic (bio) and synthetic (chemical) origin. Of these, the bio-surfactants are the extracellular secondary metabolites produced by microorganisms and also cell membrane bound. Bio-surfactants are amphiphilic in nature with hydrophobic and hydrophilic moieties. Apparently, they are of different types, i.e., lipopeptides, glycolipids, polysaccharides, phospholipids, fatty acids, protein complexes, etc., and possess high surface activity, anti-tumour, anti-bacterial, anti-viral property. They are non-toxic, highly biodegradable, stable at different pH, temperature, salt concentrations and holds wide application in medicine, food industry and agriculture (Bognolo 1999). Basically, they are classified based on their source of origin, molecular weight, chemical and physical composition (Desai and Banat 1997). According to reports by Global market insights, biosurfactants market size exceeded USD 1.5 Billion, in 2019 and is estimated to grow at over 5.5% CAGR between 2020 and 2026 (Biosurfactants Market Trends 2020–2026). Growth Projections (gminsights.com). The consumers for these biosurfactants are from different countries such as Asia, Latin America and Africa. They have huge market

[1] Department of Microbiology University College of Science, Osmania University, Hyderabad Telangana India.
[2] Department of Microbiology, PSGVP Mandal's Arts, Science, and Commerce College, Shahada 425409, Maharashtra, India.
[3] Asian PGPR Society Society for Sustainable Agriculture, Auburn University, Auburn, AL, USA.
[4] Institute of Bioproduct Development (IBD), Universiti Teknologi Malaysia (UTM), Johor Bahru, Malaysia.
[5] City of Scientific Research and Technology Applications, New Burg Al Arab, Alexandria, Egypt.
* Corresponding author: drhami2009@gmail.com, drhami2009@osmania.ac.in

and it is important to produce them in large quantity. Among the biosurfactants, lipopeptide surfactin is the principal representative with remarkable biological activities. It is commonly produced by bacillus spp. and has surface-interface and membrane active properties. These properties make it a potential candidate for application in pharmaceutical, industrial and environmental applications (Meena and Kanwar 2014). Many attempts were made to increase production of surfactin, by increased supply of fatty acids and overexpression of surfactin synthase by strategies like promoter replacement, quorum sensing mechsnism (QSM), carbon catabolite repression and CRISPRi framework wherein amino acid precursors were made available in increased quantity, that ultimately inhibited the branch pathways of amino acid biosynthesis that enhance surfactin production (Wang et al. 2019).

Properties of Biosurfactants

In terms of feedstock abundance in nature for their use as raw material for the biosurfactant production, stability at different pH and salinity, biosurfactants have many advantages over synthetic or chemical-derived surfactants, which include low toxicity, bioavailability, biodegradability, environmentally safe and low cost (Fig. 1). They are therefore better alternatives, in different industrial sectors such as agriculture, food, medicine and cosmetics (Akbari et al. 2018).

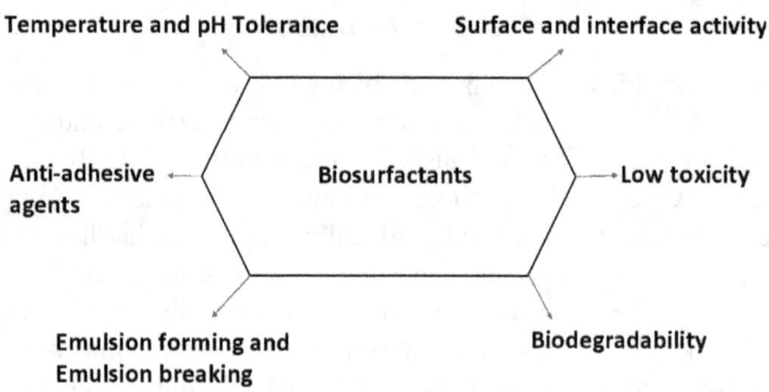

Figure 1. Different properties of bio-surfactants.

2. Lipopeptides

The most popular biosurfactants are lipopeptides for their reliable performance application and structural stability (Mnif and Ghribi 2015). Lipopeptides are most commonly produced by aerobic microorganisms, however *Bacillus licheniformis* JF-2 which is an anaerobic bacterium also produced lipopeptides (Javaheri et al. 1985). The members of the genus Bacillus are notable producers of broad-spectrum antimicrobial agents, i.e., cyclic lipopeptides (CLPs) (Borriss 2011, Falardeau et al. 2013). They are amphiphile macromolecules of non-ribosomal origin. Structurally they constitute an amino acid or alkyl/hydroxy fatty acid chain with a peptide component. Their synthesis requires non-ribosomal peptide synthetases (NRPSs), which is a multienzyme complex responsible for selection of amino acid sequence and condensation steps in peptide biosynthesis pathway (Sieber and Marahiel 2005).

Its producer strain is known to enjoy a selective advantage because of the NRPSs genes flexibility towards arbitrary evolution and innate rearrangements that leads to production of variant forms (Stein 2005).

The diversity in the form of lipopeptides is imparted to types and sequence of amino acids, peptide cyclization and length of fatty acid chain (Hamley 2015).

The members of genus Bacillus are Gram positive sporulating bacteria which magnificently produces CLPs. Following are the *Bacillus* spp. which are commonly reported for lipopeptide production (Fig. 2).

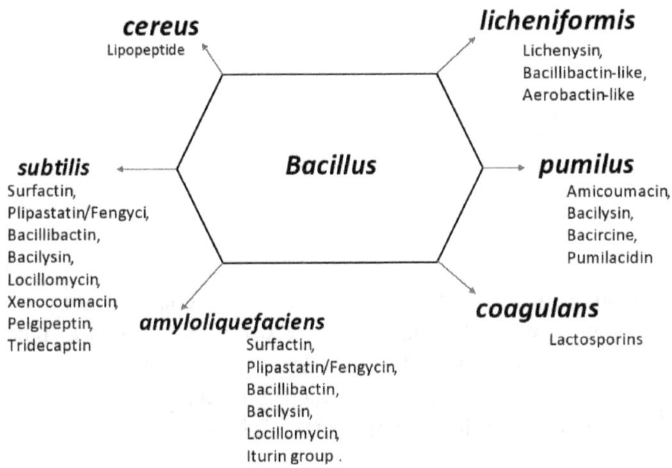

Figure 2. Different *Bacillus* spp. known for lipopeptide production.

3. Types and Structure

Lipopeptide consist of peptides that comprise cyclized amino acids (7–10 in number) via a lactone ring and varyin lengths of alkyl/β-hydroxy fatty acid chain (Baindara et al. 2013). They are extracellularly produced by bacillus spp. (Fig. 2) and divided into three families.

3.1 *Iturin*

Iturin structure is a heptapeptides that constitute 14–17 carbons and a beta-amino fatty acid chain. They form ion channels on the cell membrane, hence are well documented for their antifungal activity and not so profound antibacterial activity (Moyne et al. 2004) (Fig. 3).

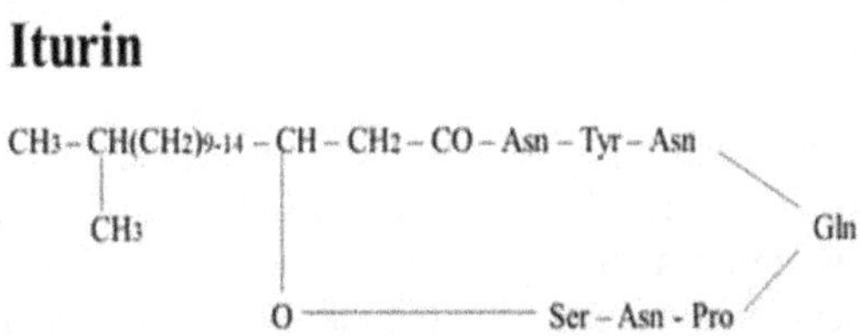

Figure 3. Structure of Iturin.

3.2 Fengycin

It is a decapeptide that constitutes 14–19 carbon and comprises of β-hydroxy fatty acid chain. They are reported for antifungal activity against filamentous fungi in specific (Wu et al. 2007, Li et al. 2013) (Fig. 4).

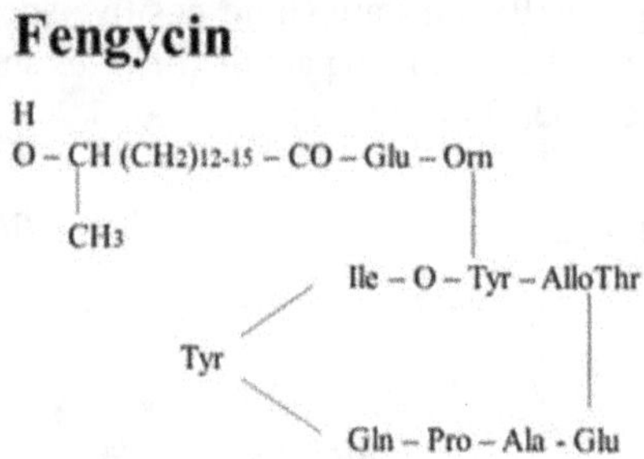

Figure 4. Structure of Fengycin.

3.3 Surfactin

It is a heptapeptides that constitute 13–15 carbons and comprises of β-hydroxy fatty acid chain (Peypoux et al. 1999). They have wide applications in pharmaceutics, agriculture and bioremediation (Fig. 5). The biofilm formation is enhanced in the presence of surfactin lipopeptide [(which is produced in presence of peptides responsible for quorum sensing and nutrient deficiency) (Bais et al. 2004)].

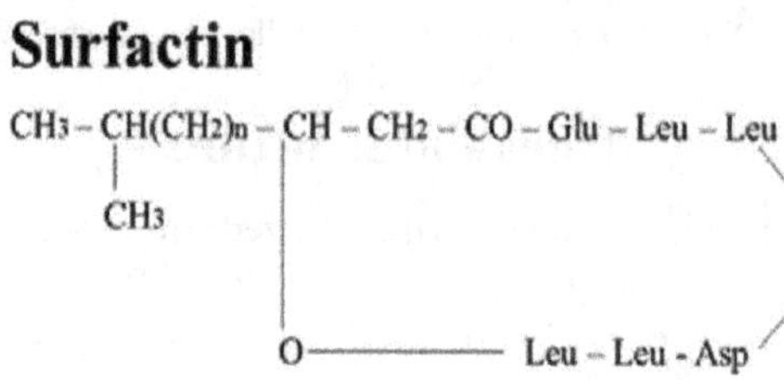

Figure 5. Structure of Surfactin.

The structural diversity of the CLP's is responsible for the existence of the variants which is because of composition, length, and type of, number of, and configuration of the lipid moiety and amino acids respectively (Raaijmakers et al. 2010) (Table 1). This structural variation governs the surface property and their bioactivities.

4. Biosynthesis, Production and Characterisation

4.1 Biosynthesis

The amphiphilic structure of a lipopeptide is made up of a hydrophobic and hydrophilic moiety (Table 2). The pathways for biosynthesis of each of the moiety, i.e., addition of an amino acid, acylation and ring closure (Peypoux et al. 1999) requires specific set of enzymes, mostly regulatory enzymes, hence the biosynthesis and regulation of lipopeptide have common features (Hommel and Ratledge 1993).

Table 1. List of variants in CLP's family (Mnif and Ghribi 2015).

Lipopeptide	Amino Acid Sequence Reference
Iturin Family	
Bacillomycin D	L-Asn-D-Tyr-D-Asn-L-Pro-L-Glu-D-Ser-L-Thr
Bacillomycin F	L-Asn-D-Tyr-D-Asn-L-Gln-L-Pro-D-Asn-L-Thr
Bacillomycin L	L-Asn-D-Tyr-D-Asn-L-Ser-L-Gln-D-Ser-L-Thr
Bacillomycin LC	L-Asn-D-Tyr-D-Asn-L-Ser-L-Glu-D-Ser-L-Thr
Iturin A	L-Asn-D-Tyr-D-Asn-L-Gln-L-Pro-D-Asn-L-Ser
Iturin AL	L-Asn-D-Tyr-D-Asn-L-Gln-L-Pro-D-Asn-L-Ser
Iturin C	L-Asn-D-Tyr-D-Asn-L-Gln-L-Pro-D-Asn-L-Ser
Mycosubtilin	L-Asn-D-Tyr-D-Asn-L-Gln-L-Pro-D-Ser-L-Asn
Fengycin Family	
Fengycin A**	L-Glu-D-Om-D-Tyr-D-aThr-L-Glu-D-Ala-L-Pro-L-Gln-L-Tyr-L-Ile
Fengycin B**	L-Glu-D-Om-D-Tyr-D-aThr-L-Glu-D-Val-L-Pro-L-Gln-L-Tyr-L-Ile
Plipastatin A	L-Glu-D-Om-L-Tyr-D-aThr-L-Glu-D-Ala-L-Pro-L-Gln-D-Tyr-L-Ile
Plipastatin B	L-Glu-D-Om-L-Tyr-D-aThr-L-Glu-D-Val-L-Pro-L-Gln-D-Tyr-L-Ile
Surfactin Family	
Esperin	L-Glu-L-Leu-D-Leu-L-Val-L-Asp-D-Leu-L-Leu-COOH
Lichenysin	L-XL_1-L-XL_2-D-Leu-L-XL_4-L-Asp-D-Leu-L-XL_7
Pumilacidin	L-Glu-L-Leu-D-Leu-L-Leu-L-Asp-D-Leu-L-XP_7
Surfactin	L-Glu-L-XS_2-D-Leu-L-XS_4-L-Asp-D-Leu-L-XS_7

Table 2. Biosynthetic pathways of hydrophobic and hydrophilic moieties synthesis.

Lipopeptide Structure		Biosynthetic Pathway	Reference
Hydrophobic moiety	Fatty acid chain, a hydroxy fatty acid (α-alkyl, β-hydroxy)	Hydrocarbon pathway	Hommel and Ratledge (1993)
Hydrophilic moiety	Cyclic peptides, carbohydrate, phosphate, carboxylic acid, amino acid, or alcohol	Carbohydrate pathway	

However, the proposed biosynthesis and linkage possibilities of the moieties are (Syldatk and Wagner 1987, Desai and Desai 1993).

- Biosynthesis of the hydrophilic and hydrophobic moieties takes place by two independent pathways.
- Pathways for synthesis of hydrophobic moiety is dependent on substrate but not hydrophilic moiety.
- Pathway for synthesis of hydrophilic moiety is dependent on substrate but not hydrophobic moiety.
- Synthesis of hydrophobic and hydrophilic moieties is dependent on substrate.

4.2 Production

Examination of the profile of bacterial metabolites that belongs to genus bacillus indicates that compounds from various lipopeptide groups, as well as multiple structural analogues of one specific lipopeptide, may be developed simultaneously by a single strain. For instance, *B. subtilis* can generate up to 12 surfactin analogues that vary in peptide residues and/or their duration and fatty acid chain branch (Maksimov et al. 2020).

Lipopeptides are produced by bacteria, yeast and actinomycetes on water miscible and immiscible substrates by solid state fermentation or batch fermentation processes. The non-ribosomal synthesis of bacterial lipopeptides by protein NRPS complexes and the structure quality of rare amino acids (ornithine, etc.) and D-amino acids (Arima et al. 1968, Tran et al. 2007) is a characteristic feature of bacterial lipopeptides. A cohort of macromolecules are involved in lipopeptide production whose nutritional nature governs the type of lipopeptide produced (Li et al. 2008), hence major limitations and bottlenecks need to be improved for the production of lipopeptides (Table 3). The various factors responsible for biosurfactant (lipopeptide) production are given below (Fig. 6).

Table 3. List of considerations to ponder in lipopeptide production.

Considerations to Overcome	For
Bottlenecks	Yield increase and cost reduction
Improvements	Industrial production processes
Address issues	Recognise high yield strains
Lipopeptide production and detection	Phenotypic and genetic analysis of producer strain and structural analysis of the lipopeptide

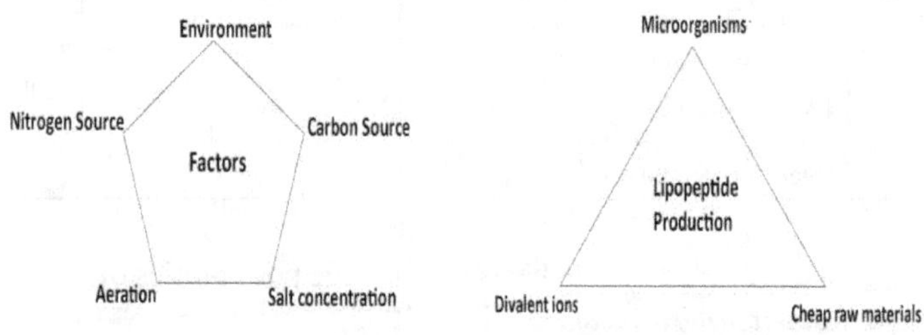

Figure 6. Factors responsible for biosurfactant (lipopeptide) production.

The cost-effectiveness of biosurfactant (lipopeptide) production by use of cheap raw materials makes the processes comparable with chemical processes. In certain cases, 50 percent of the final production cost of a bioprocess accounts for the carbon source and nitrogen sources, therefore several cheap raw materials (high in carbon and nitrogen content) and divalent ions are considered (Rufino et al. 2007). Given below are the commonly used cheap agro based raw materials and divalent ions in lipopeptide production (Table 4).

Table 4. List of commonly used raw materials and divalent ions.

Cheap Raw Material	Divalent Ions
Rice bran	Zinc sulphate ($ZnSO_4$)
Soybean	Iron trichloride ($FeCl_3$)
Potato peels	Manganese sulphate ($MnSO_4$)
Molasses	Magnesium sulphate ($MgSO_4$)
Sunflower oil cake	Potassium nitrate (KNO_3)

Kinetics of Fermentative Production of Lipopeptides

Kinetic modelling is considered as a valuable tool for fermentation process, since this type of models serves to forecast and reflect the growth of microorganism and formation of product (Dodić et al. 2012, Yalcin and Özbas 2005).

Kinetic models are classified into two types they are: (Zhu et al. 2013)

- **Structured Models:** are generally complicated and considers metabolic pathways and
- **Unstructured models:** are simple and most frequently used.

A two-temperature-stage method for the improvement of lipopeptide development in solid-state fermentation (SSF) has recently been suggested. It was found that the production of lipopeptide can be effectively enhanced at 37°C temperature (Zhu et al. 2014). Due to its cost effectiveness (low expenditure and operations), SSF in lipopeptide production from flask to fermentor has gained popularity for scale-up and is commercially relevant over the past few years. In SSF, it is important to research the relationship between the key state variables and quantitatively describe the fermentation behaviour at the fermentation phase. Biosurfactant production kinetics as explained by Desai and Banat (1997) can be considered to increase the yield of the bacillus lipopeptides.

The kinetics of fermentative production are as follows-

- ***Bacterial growth*** - Here, production of lipopeptide parallels with substrate utilization and bacterial growth.
- ***Production of lipopeptide under growth limiting conditions*** - Here, production of lipopeptide is increased when one of the medium components (carbon, nitrogen, and divalent ion sources) is deficient.
- ***Lipopeptide production by an immobilised or resting cell*** - This strategy fairly helps in reduction of production cost and time. As the cells can be maintained in a particular growth phase for longer period of time, the production can be initiated immediately. In addition, the cell mass/inoculum size, viability and aseptic condition (of no contamination by co cultures) can also be maintained.
- ***Lipopeptide production by supplementing a precursor*** - The valuable qualitative and quantitative changes in the lipopeptide can be attained upon supplementation of the medium by precursors (amino acids and fatty acids).

4.3 *Extraction of Lipopeptides*

Extraction of lipopeptides from the fermented beer is an important step, which defines its purity and further industrial applications. Several material and methods are proposed and reported in this context like adsorption of surfactin by use of various resins like AGI-4 and XAD-7, acid precipitation and foam fractionation by solvents like hexane, ethyl acetate (Chen et al. 2008a, 2008b), ultrafiltration (Sivapathasekaran et al. 2011), automated collection and use of response surface methodology for fengycin (Glazyrina et al. 2008, Wei et al. 2010).

4.4 *The Bio-physicochemical-characterization of Lipopeptides*

Below are the most common methods employed for identification and characterisation of lipopeptide producer strain and type of lipopeptides (Table 5) (Chen et al. 2008a).

Table 5. Different techniques involved in physico-chemical characterization of Lipopeptides.

Techniques	Purpose
Polymerase Chain Reaction (PCR)	Molecular analysis to identify the presence of specific gene
Chromatography: a) Paper chromatography, Thin layer chromatography b) High Performance Liquid chromatography, Reversed Phase Liquid chromatography	To establish and identify the presence of functional groups (Amino and fatty acid chains), and to collect purified compounds for further characterization and applications
Fourier-transform infrared spectroscopy (FTIR)	To identify and characterize an unknown compound based on the presence of a functional group
Mass spectroscopy (MS) Matrix-assisted laser desorption ionization–time of flight (MALDI-TOF) MS Nuclear Magnetic Resonance (NMR)	To characterise the lipopeptide based on molecular mass and magnetic resonance

5. Surfactin Biosynthesis

Surfactin is synthesized via a condensation reaction that involves fatty acids and amino acids as precursors. Strains of *Bacillus subtilis* have an ability to naturally produce surfactin, however a major bottleneck, i.e., low production has affected its broad production and commercial application (Wang et al. 2019). Molecular biological attempts have been made to overcome this hurdle. Standard attempts to overexpress the key genes in a heterologous microorganism (e.g., *E. coli*) were dropped, as the shift into the cloning plasmid of 32 kb gDNA (*srf*A operon and active *sfp*) is very difficult. Nonetheless, one attempt was performed with a bacterial artificial chromosome (BAC), but no surfactin was made (Lee et al. 2007). Therefore, several genetic experiments with the goal to improve the yield of surfactin in the *Bacillus* spp. themselves were designed. Genome shuffle, for example, was applied to a *B. amyloliquefaciens* strain with a 3.4-fold rise in surfactin (Zhao et al. 2012).

The biosynthetic pathway of surfactin production is divided into three steps (Wang et al. 2019), i.e.,

(1) formation of fatty acyl-CoA by activating fatty acids via fatty acyl-CoA ligase.

(2) biosynthesis of amino acids—L-Valine, L-Leucine, L-Aspartate, L-Glutamate.

(3) assembly of seven amino acids on fatty acyl-CoA via surfactin synthetase.

Here, surfactin synthase enzyme production is cell density regulated, is encoded by the *srfA* operon and phosphopantetheinyl transferase (PPTase) *Sfp.*Attempts reported to enhance surfactin production with focus on third step (as mentioned above) are given in Table 6.

Table 6. Attempted strategies reported to enhance surfactin production.

Strategies for Enhanced Production of Surfactin	References
Replacement of the original constitutive promoter of *srf*A (PsrfA) with stronger inducible promoters (Pg3 or P*spac*)	Jiao et al. (2017), Huigang et al. (2009)
Upregulation of the QS system (ComQXPA) or downregulation of the –ve factors	Wang et al. (2019)
The global regulation factor (codY), which had –ve impact on expression of srfA upon elimination resulting in enhanced production of surfactin	Dhali et al. (2017)
Increased supply of fatty acid precursors in modified synthetic medium could overexpress the genes of fatty acid biosynthesis pathways and enhance surfactin production (~4.9 g/L in a flask)	Wang et al. (2019)
Increased supply of amino acids and decreased metabolic flux in amino acid branch pathways might lead to enhanced surfactin production	Dokyun et al. (2013), Xu et al. (2013), Wu et al. (2015)

5.1 *Promoter Replacement Strategy for Surfactin Production*

Many studies are reported on decouple of the expression of the surfactin (*srfA*) operon in the presence of extracellular signal molecules for example replacement of regular promoter P*srf*A by Pspac by Sun et al. (2009) wherein the pMUTIN4 plasmid was employed to integrate a promoter region in *srfA* operon, nevertheless this type of plasmid usage wasn't fruitful. However, high surfactin yields were recorded when P*sac* promoter was induced with isopropyl-β-D-thiogalactopyranosid. Likewise, a study on marker less strategy for promoter replacement of P*srf*A with P*rep*U by Coutte et al. (2010) unfortunately didn't show high surfactin yield. The outcome of these studies indicates that P*srf*A has strong transcription initiation and optimisation for high surfactin production.

5.2 *Under QS Influence*

QS is the cell-cell communication between the microbial cells which uses signal molecules called autoinducers (Kalamara et al. 2018). It is one of the strategies employed by microorganisms for several purposes like to endure nutrient limitation (e.g., glucose deprivation), formation of biofilms, increased virulence, etc. The type of auto-inducers differs in a Gram positive (+ve) and Gram-negative (–ve) bacterium, i.e., acyl homoserine lactones (AHL) (in Gram –ve bacteria) and peptides (in Gram +ve bacteria). Impressively, production of a lipopeptide surfactin is observed in QS mechanism (QSM), that is when glucose nutrient is deficient in the environment which helps the bacillus to survive. In the course of glucose deprivation, the expression of *srfA* gene that encodes a non-ribosomal lipopeptide surfactin takes place. The expression of *srfA* gene is caused by the phosphorylation of autoinducers, i.e., transfer of phosphate moiety to ComA (response regulator) from auto-phosphorylised ComP (histidine kinase) when autoinducer molecule pre-ComX is modified to ComQ upon reaches a threshold concentration (Chen et al. 2020). Surfactin production indirectly activates regulators like *deg*Q by phosphorylation (DegQ) and leads to production of exoproteases and extracellular enzymes (Kalamara et al. 2018) (Fig. 7).

The produced biosurfactant surfactin forms pores on cell membrane, causes K^+ ion leakage which activates histidine kinase KinC finally guide to phosphorylation of Spo0A. The Spo0A is an important QS regulator in *B. subtilis* (López et al. 2009, Banse et al. 2011, Bendori et al. 2015), which controls biofilm formation (Kalamara et al. 2018).

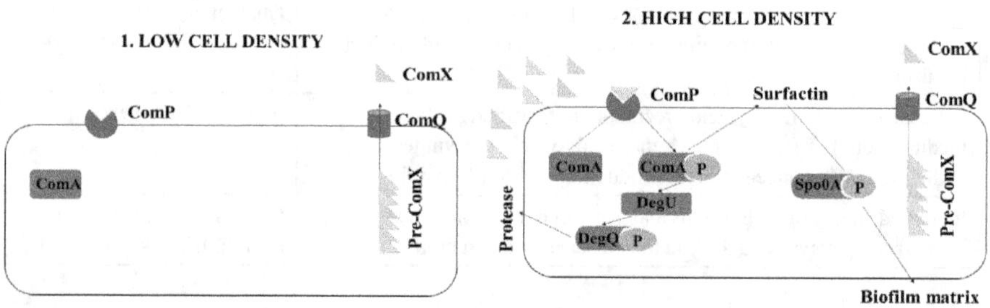

Figure 7. Production of Surfactin under QS influence (image adapted from Kalamara et al. 2018).

5.3 *Carbon Catabolite Repression (CCR)*

A vast variety of carbon sources are used by *Bacillus* spp., but a well-known phenomenon is that the most popular sugars (e.g., glucose, fructose, or malate) are given first preference than other sources of carbon. The mechanism for use of carbohydrates is complex, with many unique and global regulatory policies strictly regulated by the repression of carbon catabolites (CCR). By CCR, in the presence of non-referred inputs of large concentrations, *Bacillus* spp. first utilises preferred carbohydrates that are useful for effective competition with other microorganisms in the natural world (Marciniak et al. 2012, Kim and Burne 2017).

In *B. subtilis* CCR is controlled predominantly by the global transcription factor CcpA (catabolite regulation protein A) (Yang et al. 2017). Histidine-containing phosphotransfer protein (HPr) is phosphorylated by HPr kinase (HPrK) at Ser46 by bifunctional HPr kinase/phosphorylase HPrK/P at high concentrations of fructose-1,6-bisphosphate and ATP or pyrophosphate in the presence of preferred carbon sources like glucose, and then intercalates with CcpA which can bind to a preserved cis-acting sequence termed catabolite reaction (Schumacher et al. 2011, Ishii et al. 2013, Charbonnier et al. 2017, Kim and Burne 2017).

5.4 *Relationship between Carbon Catabolite Repression (CCR) ⟶ QS ⟶ Surfactin Production*

Bacillus spp. uses the desired carbon source (e.g., glucose) from a variety of compounds to allow rapid development. In a given environment, glucose content could also be an important nutritional factor that governs cell density. A study hypothesised that CCR would be relieved to use non-preferred sugars after glucose exhaustion to help initiate QSM, such as biofilm formation and sporulation (Fujita 2009). A research showed that surfactin is also active in mitigate CCR in *B. amyloliquefaciens*, as a signal peptide to influence QS which is directly related to CCR in order to address the restriction of carbon sources in the natural environment (Chen et al. 2020). Previous reports by Gonzalez-Pastor, 2011 reveals that surfactin deficiency can cause severe developmental defects in *B. subtilis* that may lead to cell death as well as addition of surfactin to *ΔcomA* mutant can partly alleviate QSM in the strain. Thus, from the reported literature it can be concluded that CCR is important for cellular development in Bacillus spp. and production of surfactin. Likewise, surfactin production regulates QSM by acting as signal peptides and the deficiency of surfactin is detrimental for cell growth.

5.5 *CRISPRi Technology*

A possible platform for selective gene regulation is provided by the CRISPR (clustered frequently interspaced short palindromic repeats) system (Barrangou et al. 2007). Approximately 40 percent of bacteria and 90 percent of archaea have CRISPR/CRISPR associated (Cas) structures to withstand foreign DNA components (Makarova et al. 2011).

A variety of compounds like drugs, antibiotics, enzymes, organic acids and biofuels are produced by microorganisms and their yield can be increased by manipulate metabolic pathways of the host cells' which includes operator specific regulators of transcription that binds on DNA, which repress or derepress the transcription of target genetic codes on the operons. A transcriptional regulatory tool like RNA interference (RNAi) is a good example of modification of metabolic pathways that inhibits one or more gene expression (Schultenkämper et al. 2020).

It is a sequence-specific gene edit technique with role in up (CRISPRa) and down (CRISPRi) regulation of gene expression (Lau 2014, Jinek et al. 2012, Qi et al. 2013, Larson et al. 2013). All it requires is a CRISPR-associated protein 9 and a single-guide RNA (sgRNA) to target a DNA sequence (Qi et al. 2013). The dCas9 protein consists of CRISPR-associated protein 9, with two amino acids exchange (D10A and H841A) in its endonuclease domains (RuvC and HNH) and hence lacks endonuclease activity (Qi et al. 2013) (Fig. 8).

Jennifer Doudna (founder) and Emmaneulle Charpentier (co-founder) are the two celebrated names in CRISPR technology who were awarded Nobel prize

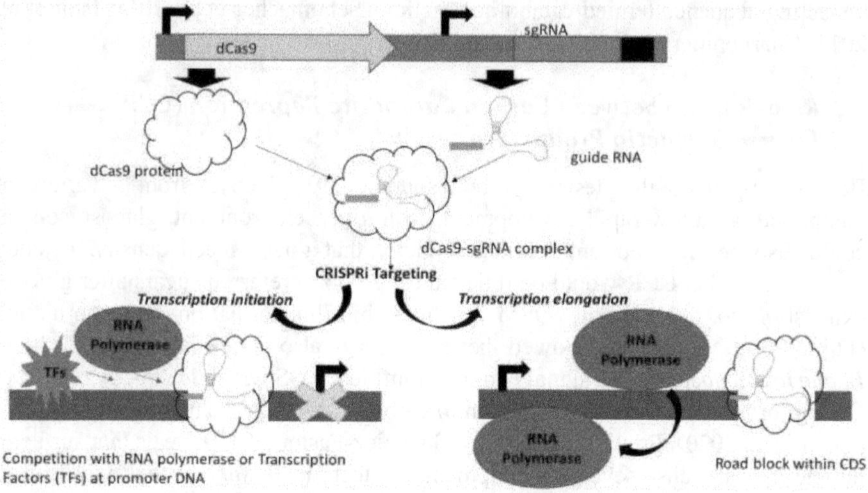

Figure 8. Diagrammatic illustration of CRISPRi technique by dCas9 enzyme and sgRNA complex which (1) competes with RNA Polymerase and Transcription factors (TFs) during transcription initiation and (2) interferes and blocks transcription elongation (image adapted from Schultenkämper et al. 2020).

in October 2020. The CRISPR interference is a quick, easy and fast modification method and finds application in microbial biotechnology in strain development and fix metabolic pathways (Larson et al. 2013).

5.6 Building the CRISPRi Framework to Study Surfactin Biosynthetic Pathway for its Enhanced Production

In a study reported by Wang et al. (2019) surfactin production in *B. subtilis* 168 was enhanced by increased supply of amino acids. Firstly, the surfactin biosynthetic pathway was incorporated in BS168NU by exposure of exogenous *sfp* gene (BS168NU-S). Secondly, an effective CRISPRi system was constructed, by RT-qPCR measured transcriptional repression. Thirdly, the transcription of twenty genes chosen from the amino acid branched metabolic pathways were individually interfered with CRISPRi. With the *ccpA* gene as the internal control and BS168NU-Sd as the calibrator, the $2^{-\Delta\Delta CT}$ (Livak et al. 2001) approach was employed to quantify the relative transcription levels of the target gene. The outcome proved that gene expression in *B. subtilis* 168 was effectively blocked by the CRISPRi framework. The repression efficiencies of various genes, however, differed.

Of the 20 recombinant strains obtained, 16 recombinant strains acquired the improved production of surfactin. In specific, the strains in which *yrpC, racE* or *murC* genes were inhibited raised the production of surfactin to 0.54, 0.41 or 0.42 g/L respectively. Then, in combination, the 3 genes were further repressed. The results revealed that strains that co-inhibited *yrp*C and *rac*E had the highest development (surfactin production), which were directly linked to the metabolism of L-glutamate, the acylation of which was the first step in the assembly of surfactin. This study revealed that engineered the metabolic pathway of amino acids is can be an effective technique to improve surfactin production.

6. Applications

Human intervention has imbalanced both human and environmental health and even the latest technologies available have negative impact, hence the use of microorganisms and their metabolites is an eco-friendly approach that will lead to sustainability. The annual revenue of >US $1 billion is recorded for lipopeptides with approval for use in more than 70 countries (Meena and Kanwar 2015). Lipopeptides majorly are reported for their ability to cause apoptosis of fungi and bacteria through leakage in the cell membrane (Koumoutsi et al. 2004, Zeriouh et al. 2014). Exponential rise in pesticidal resistance towards synthetic pesticides, has led to a global call for a healthy plant nutrition and sustainable agriculture and search for new pest control strategies and microbial bioactive agents (Penha et al. 2020). Lipopeptides are immunomodulators with properties like anti-tumour, anti-adhesive, thrombocytolytic activity, etc. Increase in drug resistance and incessant side effects of synthetic drugs is one of the major reasons which has urged for the exploration of new green bioactive compounds, be used for clinical applications, food preservation and dairy products (Mandal et al. 2013). Hence, they are considered as potent candidates in agriculture and biomedical research. In addition, due to high-cost remediation technologies against the accumulated organic and inorganic contaminants from natural sources and industrial plants in soil, water and air, requires low-cost bioactive agents/microorganisms for bioremediation of environment (Bustamante et al. 2012). Lipopeptides are also used in bioremediation for their properties like metal sequestration, viscosity reduction, hydrocarbon solubilisation, etc. (Mnif and Ghribi 2015). Due to their applications in many fields, there are several patents on lipopeptide biosurfactants, as of 2019 there are 27 patents on biosurfactants and predicted global worth of 729 billion USD by 2025 (Ugalmule and Swain 2019). The Table 7 lists out the information on lipopeptide applications.

7. Conclusion

Surfactin, iturin and fengycin, the LPs produced by Bacillus, are potent antagonistic agents that are capable to act directly and indirectly against pathogens to influence, destroy or trigger systemic resistance and biofilm formation. They display cytotoxic, antimicrobial, antiviral, insecticidal, antitumor, and enzyme-inhibitory activities. These characteristics of Bacillus LPs make them economically important weapons against plant pathogensand an incredible alternative for biocontrol systems, which are required by both tightened regulatory constraints and quickly evolved disease resistance. To enhance the production of surfactin, strategies like genome shuffle, promoter replacement, carbon catabolite repression, genome and metabolic pathway edit strategieslike CRISPRi which is still in its infancy can be explored further to meet the global demand of biosurfactants.

Table 7. Applications of different lipopetides.

Field	**Lipopeptide**	**Under the Title**	**Reference**
Agriculture	Fengycin, Iturin and Surfactin	Biological control of peach brown rot (*Monilinia* spp.) by *Bacillus subtilis* CPA-8 is based on production of fengycin-like lipopeptides.	Yanez-Mendizábal et al. (2012b)
		• The plant-associated *Bacillus amyloliquefaciens* strains MEP 218 and ARP 23 capable of producing the cyclic lipopeptides iturin or surfactin and fengycin are effective in biocontrol of sclerotinia stem rot disease. • Putative use of a *Bacillus subtilis* L194 strain for biocontrol of *Phoma medicaginis* in *Medicago truncatula* seedlings. • Biocontrol of Fusarium wilt disease in muskmelon with *Bacillus subtilis* Y-IVI. • Biological activity of the lipopeptide-producing *Bacillus amyloliquefaciens* PGPBacCA1 on common bean *Phaseolus vulgaris* L. pathogens. • Screening Bacillus species as biological control agents of *Gaeumannomyces graminis var. Tritici* on wheat. • Iturinic Lipopeptide Diversity in the *Bacillus subtilis* Species Group—Important Antifungals for Plant Disease Biocontrol Applications.	Alvarez et al. (2012) Slimene et al. (2012) Zhao et al. (2013) Torres et al. (2017) Yang et al. (2018) Dunlap et al. (2019)
Medical	Iturin and Surfactin	• Cloning, sequencing, and characterization of the genetic region relevant to biosynthesis of the lipopeptides iturin A and surfactin in *Bacillus subtilis*. • Micellization of surfactin and its effect on the aggregate conformation of amyloid β (1-40). • Induction of apoptosis in human leukemia K562 cells by cyclic lipopeptide from *Bacillus subtilis natto* T-2. • Mechanism of inactivation of enveloped viruses by the biosurfactant surfactin from *Bacillus subtilis*. • Antimycoplasma properties and application in cell culture of surfactin, a lipopeptide antibiotic from *Bacillus subtilis*. • Antitumor activity of *Bacillus natto* 5. Isolation and characterization of surfactin in the culture medium of *Bacillus natto* KMD 2311. • Antibacterial activity of Bacillus species-derived surfactin on *Brachyspira hyodysenteriae* and *Clostridium perfringens*. • Lipopeptide antibiotic production by *Bacillus velezensis* KLP2016.	Yao et al. (2003) Han et al. (2008) Wang et al. (2007) Vollenbroich et al. (1997a) Vollenbroich et al. (1997b) Kameda et al. (1974) Horng et al. (2019) Meena et al. (2018)

Bioremediation	Surfactin	• Production of surfactin isoforms by *Bacillus subtilis* BS-37 and its applicability to enhanced oil recovery under laboratory conditions.	Liu et al. (2015)
	Not specified	• Biotechnology in petroleum recovery: The microbial EOR.	Sen (2008)
	Not specified	• Characterization and Enhanced Degradation Potentials of Biosurfactant-Producing Bacteria Isolated from a Marine Environment.	Wu et al. (2019)
	Not specified	• Bioremediation of petroleum contaminated soils by lipopeptide producing *Bacillus subtilis* SE1.	Nimrat et al. (2019)
	Not specified	• Bioremediation of 2,4-Diaminotoluene in Aqueous Solution Enhanced by Lipopeptide Biosurfactant Production from Bacterial Strains.	Carolin et al. (2020)

Acknowledgements

The authors (HB, SA) are thankful to SERB EMR for the financial support (SERB/F/942/2017-18). Author (SA) acknowledge the research fellowship under the SERB EMR grant.

References

Akbari, S., H.N. Abdurahman, M.R. Yunus, F. Fayaz and R.O. Alara. 2018. Biosurfactants—a new frontier for social and environmental safety: A mini review. Biotechnology Research and Innovation 2: 1, 81–90. DOI: 10.1016/j.biori.2018.09.001.

Alvarez, F., M. Castro, A. Príncipe et al. 2012. The plant-associated *Bacillus amyloliquefaciens* strains MEP 218 and ARP 23 capable of producing the cyclic lipopeptidesiturin or surfactin and fengycin are effective in biocontrol of sclerotinia stem rot disease. J. Appl. Microbiol. 112: 159–174. DOI: 10.111 1/j.1365-2672.2011.05182.x.

Arima, K., A. Kakinuma and G. Tamura. 1968. Surfactin, a crystalline peptidelipid surfactant produced by *Bacillus subtilis*: Isolation, characterisation and its inhibition of fibrin clot formation. Biochem. Biophys. Res. Commun. 31: 3, 488–494. DOI: 10.1016/0006-291X(68)90503-2.

Baindara, P., S.M. Mandal, N. Chawla, P.K. Singh, A.K. Pinnaka and S. Korpole. 2013. Characterization of two antimicrobial peptides produced by a halotolerant *Bacillus subtilis* strain SK.DU.4 isolated from a rhizosphere soil sample. AMB Express 3: 2. DOI: 10.1186/2191-0855-3-2.

Bais, H.P., S.W. Park, T.L. Weir, R.M. Callaway and J.M. Vivanco. 2004. How plants communicate using the underground information superhighway. Trends Plant Sci. 9: 26–32. DOI: 10.1016/j.tplants.2003.11.008.

Banse, A.V., E.C. Hobbs and R. Losick. 2011. Phosphorylation of Spo0A by the histidine kinase KinD requires the lipoprotein Med in *Bacillus subtilis*. J. Bacteriol. 193: 3949–3955. DOI: 10.1128/JB.05199-11.

Barrangou, R., C. Fremaux, H. Deveau, M. Richards, P. Boyaval, S. Moineau, D.A. Romero and P. Horvath. 2007. CRISPR provides acquired resistance against viruses in prokaryotes. Science 315: 1709–1712. DOI: 10.1126/science.1138140.

Beerli, R.R. and C.F. Barbas. 2002. 3rd Engineering polydactyl zinc-finger transcription factors. Nat. Biotechnol. 20: 135–141. DOI: 10.1038/nbt0202-135.

Bendori, S.O., S. Pollak, D. Hizi and A. Eldar. 2015. The RapP-PhrP quorum-sensing system of *Bacillus subtilis* strain NCIB3610 affects biofilm formation through multiple targets, due to an atypical signal-insensitive allele of RapP. J. Bacteriol. 197: 592–602. DOI: 10.1128/JB.02382-14.

Bognolo, G. 1999. Biosurfactants as emulsifying agents for hydrocarbons. Colloids and Surfaces A: Physicochemical and Engineering Aspects 152: 41–52. DOI: 10.1016/S0927-7757(98)00684-0.

Borriss, R. 2011 Use of plant-associated Bacillus strains as biofertilizers and biocontrol agents in agriculture. pp. 41–76. *In*: Maheshwari, D.K. (ed.). Bacteria in Agrobiology: Plant Growth Responses. Berlin Heidelberg: Springer Verlag.

Bustamante, M., N. Durán and M. Diez. 2012. Biosurfactants are useful tools for the bioremediation of contaminated soil: A review. J. Soil Sci. Plant. Nut. 12: 667–687. DOI: 10.4067/S0718-95162012005000024.

Carolin, C.F., S.P. Kumar, J.G. Joshiba, R. Ramamurthy and J.S. Varjani. 2020. Bioremediation of 2,4-diaminotoluene in aqueous solution enhanced by lipopeptide biosurfactant production from bacterial strains. Journal of Environmental Engineering 146. DOI: 10.1061/(ASCE)EE.1943-7870.0001740.

Charbonnier, T., D. Le Coq, S. McGovern, M. Calabre, O. Delumeau, S. Aymerich et al. 2017. Molecular and physiological logics of the pyruvate-induced response of a novel transporter in *Bacillus subtilis*. mBio 8: e00976-17. DOI: 10.1128/mBio.00976-17.

Chen, B., J. Wen, X. Zhao, J. Ding and G. Qi. 2020. Surfactin: A quorum-sensing signal molecule to relieve CCR in *Bacillus amyloliquefaciens*. Front Microbiol. 11: 631. DOI: 10.3389/fmicb.2020.00631.

Chen, H.L., Y.S. Chen and R.S. Juang. 2008a. Flux, decline and membrane cleaning in cross-flow ultrafiltration of treated fermentation broths for surfactin recovery. Separation and Purification Technology 62: 47–55. DOI: 10.1016/j.seppur.2007.12.015.

Chen, H.L., Y.S. Chen and R.S. Juang. 2008b. Purification of surfactin in pretreated fermentation broths by adsorptive removal of impurities. Biochemical Engineering Journal 40: 452–459. DOI: 10.1016/j.bej.2008.01.020.

Cleto, S., J.V. Jensen, V.F. Wendisch and T.K. Lu. 2016. *Corynebacterium glutamicum* metabolic engineering with CRISPR Interference (CRISPRi). ACS Synth. Biol. 5: 375–385. DOI: 10.1021/acssynbio.5b00216.

Coutte, F., V. Leclère, M. Béchet, J.S. Guez, D. Lecouturier, M. Chollet-Imbert, P. Dhulster and P. Jacques. 2010. Effect of pps disruption and constitutive expression of *srfA* on surfactin productivity, spreading and antagonistic properties of *Bacillus subtilis* 168 derivatives. J. Appl. Microbiol. 109: 480–491. DOI: 10.1111/j.1365-2672.2010.04683.x.

Desai, J.D. and A.J. Desai. 1993. Production of biosurfactants. pp. 65–97. *In*: Kosaric, N. (ed.). Biosurfactants: Production, Properties, Applications. Marcel Dekker, Inc., New York, N.Y.

Desai, J.D. and I.M. Banat. 1997. Microbial production of surfactants and their commercial potential. Microbiology and Molecular Biology Reviews 61(1): 47–64. PMID: 9106364.

Dhali, D., F. Coutte, A.A. Arias, S. Auger, V. Bidnenko, G. Chataigné, M. Lalk, J. Niehren, S.J. De and C. Versari. 2017. Genetic engineering of the branched fatty acid metabolic pathway of *Bacillus subtilis* for the overproduction of surfactin C14 isoform. Biotechnol. J. 12: 1600574. DOI: 10.1002/biot.201600574.

Dodić, J.M., D.G. Vučurović, S.N. Dodić, J.A. Grahovac, S.D. Popov and N.M. Nedeljković. 2012. Kinetic modelling of batch ethanol production from sugar beet raw juice. Appl. Energy 99: 192–197. DOI: 10.1016/j.apenergy.2012.05.016.

Dokyun, N., M.Y. Seung, C. Hannah, P. Hyegwon, P.J. Hwan and L.S. Yup. 2013. Metabolic engineering of *Escherichia coli* using synthetic small regulatory RNAs. Nat. Biothechnol. 31: 170–4. DOI: 10.1038/nbt.2461.

Dunlap, C.A., M.J. Bowman and A.P. Rooney. 2019. Iturinic lipopeptide diversity in the *Bacillus subtilis* species group—Important antifungals for plant disease biocontrol applications. Front. Microbiol. 10: 1794. DOI: 10.3389/fmicb.2019.01794.

Falardeau, J., C. Wise, L. Novitsky and T.J. Avis. 2013. Ecological and mechanistic insights into the direct and indirect antimicrobial properties of *Bacillus subtilis* lipopeptides on plant pathogens. J. Chem. Ecol. 39: 869–878. DOI: 10.1007/s10886-013-0319-7.

Fujita, Y. 2009. Carbon catabolite control of the metabolic network in *Bacillus subtilis*. Biosci. Biotechnol. Biochem. 73(2): 245–59. DOI: 10.1271/bbb.80479.

Glazyrina, J., S. Junne, P. Thiesen, K. Lunkenheimer and P. Goetz. 2008. *In situ* removal and purification of biosurfactants by automated surface enrichment. Applied Microbiology and Biotechnology 81: 23–31. DOI: 10.1007/s00253-008-1620-1.

González-Pastor, J.E. 2011. Cannibalism: A social behavior in sporulating *Bacillus subtilis*. FEMS Microbiol. Rev. 35(3): 415–24. DOI: 10.1111/j.1574-6976.2010.00253.x.

Hamley, I.W. 2015. Lipopeptides: From self-assembly to bioactivity. Chem. Commun. 51: 8574–8583. DOI: 10.1039/C5CC01535A.

Han, Y., X. Huang, M. Cao and Y. Wang. 2008. Micellization of surfactin and its effect on the aggregate conformation of amyloid β (1-40). Journal of Physical Chemistry B 112: 47 15195–15201. DOI: 10.1021/jp805966x.

Hannon, G.J. 2002. RNA interference. Nature 418: 244–251. DOI: 10.1038/418244a.

Hommel, R.K. and C. Ratledge. 1993. Biosynthetic mechanisms of low molecular weight surfactants and their precursor molecules. pp. 3–63. *In*: Kosaric, N. (ed.). Biosurfactants: Production, Properties, Applications. Marcel Dekker, Inc., New York, N.Y.

Horng, Y., Y. Yu, A. Dybus et al. 2019. Antibacterial activity of Bacillus species-derived surfactin on *Brachyspirahyodysenteriae* and *Clostridium perfringens*. AMB Expr. 9: 188. DOI: 10.1186/s13568-019-0914-2.

Huigang, S., B. Xiaomei, L. Fengxia, L. Yaping, W. Yundailai and L. Zhaoxin. 2009. Enhancement of surfactin production of *Bacillus subtilis* fmbR by replacement of the native promoter with the Pspac promoter. Can. J. Microbiol. 55: 1003–6. DOI: 10.1139/W09-044.

Ishii, H., T. Tanaka and M. Ogura. 2013. The *Bacillus subtilis* response regulator gene degU is positively regulated by CcpA and by catabolite-repressed synthesis of ClpC. J. Bacteriol 195(2): 193–201. DOI: 10.1128/JB.01881-12.

Javaheri, M., G.E. Jenneman, M.J. McInnerey and R.J. Knapp. 1985. Anaerobic production of a biosurfactant by *Bacillus licheniformis*. Applied and Environmental Microbiology 50: 698–700. DOI: 10.1128/AEM.50.3.698-700.1985.

Jiao, S., X. Li, H. Yu, H. Yang, X. Li and Z. Shen. 2017. *In situ* enhancement of surfactin biosynthesis in *Bacillus subtilis* using novel artificial inducible promoters. Biotechnol. Bioeng. 114: 832. DOI: 10.1002/bit.26197.

Jinek, M., K. Chylinski, I. Fonfara, M. Hauer, J.A. Doudna and E.A. Charpentier. 2012. Programmable Dual-RNA guided DNA endonuclease in adaptive bacterial immunity. Science 337: 816– 821. DOI: 10.1126/science.1225829.

Kalamara, M., M. Spacapan, I. Mandic-Mulec and R.N. Stanley-Wall. 2018. Social behaviours by *Bacillus subtilis*: Quorum sensing, kin discrimination and beyond. Molecular Microbiology 110: 863–878. DOI: 10.1111/mmi.14127.

Kameda, Y., S. Oira, K. Matsui, S. Kanatomo and T. Hase. 1974. Antitumor activity of *Bacillus natto* 5. Isolation and characterization of surfactin in the culture medium of *Bacillus natto* KMD 2311. Chemical and Pharmaceutical Bulletin 22: 938–944. DOI: 10.1248/cpb.22.938.

Kim, J.N. and R.A. Burne. 2017. CcpA and CodY coordinate acetate metabolism in *Streptococcus mutans*. Appl. Environ. Microbiol. 83(7). DOI: 10.1128/AEM.03274-16.

Klug, A. 2010. The discovery of zinc fingers and their applications in gene regulation and genome manipulation. Annu. Rev. Biochem. 79: 213–231. DOI: 10.1146/annurev-biochem-010909-095056.

Koumoutsi, A., X.H. Chen, A. Henne, H. Liesegang, G. Hitzeroth, P. Franke et al. 2004. Structural and functional characterization of gene clusters directing non ribosomal synthesis of bioactive cyclic lipopeptides in *Bacillus amyloliquefaciens* Strain FZB42. J. Bacteriol. 186: 1084–1096. DOI: 10.1128/jb.186.4.1084-1096.2004.

Larson, M., L. Gilbert, X. Wang et al. 2013. CRISPR interference (CRISPRi) for sequence-specific control of gene expression. Nat. Protoc. 8: 2180–2196. DOI: 10.1038/nprot.2013.132.

Lau, E. 2014. Genetic screens: CRISPR screening from both ways. Nat. Rev. Genet. 15: 778–779. DOI: 10.1038/nrg3850.

Lee, Y.-K., B.-D. Yoon, J.-H. Yoon, S.-G. Lee, J.J. Song, J.-G. Kim, H-M. Oh and H.-S. Kim. 2007. Cloning of *srfA* operon from *Bacillus subtilis* C9 and its expression in *E. coli*. Appl. Microbiol. Biot. 75: 567–572. DOI: 10.1007/s00253-007-0845-8.

Li, L., K. Li, K. Wang, C. Chen, C. Gao, C. Ma and P. Xu. 2014. Efficient production of 2,3-butanediol from corn stover hydrolysate by using a thermophilic *Bacillus licheniformis* strain. Bioresour. Technol. 170: 256–261. DOI: 10.1016/j.biortech.2014.07.101.

Li, X.Y., Y.H. Wang and Y.Q. He. 2013. Diversity and active mechanism of fengycin-type cyclopeptides from Bacillus subtilis XF-1 against *Plasmodiophora brassicae*. J. Microbiol. Biotechnol. 23: 313–321. DOI: 10.4014/jmb.1208.08065.

Li, Yi-M., N.I.A. Haddad, S.-Z. Yang and B.-Z. Mu. 2008. Variants of lipopeptides produced by *Bacillus licheniformis* HSN221 in different medium components evaluated by a rapid method ESI-MS. Int. J. Pept. Res. Ther. 14: 229–235. DOI: 10.1007/s10989-008-9137-0.

Liu, Q., J. Lin, W. Wang, H. Huang and S. Li. 2015. Production of surfactin isoforms by *Bacillus subtilis* BS-37 and its applicability to enhanced oil recovery under laboratory conditions. Biochem. Eng. J. 93: 31–37. DOI: 10.1016/j.bej.2014.08.023.

Livak, J.K. and D.T. Schmittgen. 2001. Analysis of relative gene expression data using real-time quantitative PCR and the $2^{-\Delta\Delta CT}$ method. Methods 25: 4, 402–408. DOI: 10.1006/meth.2001.1262.

López, D., H. Vlamakis, R. Losick and R. Kolter. 2009. Paracrine signaling in a bacterium. Gene Dev. 23: 1631–1638. DOI: 10.1101/gad.1813709.

Makarova, K.S., D.H. Haft, R. Barrangou, S.J.J. Brouns, E. Charpentier, P. Horvath, S. Moineau, F.J.M. Mojica, Y.I. Wolf, A.F. Yakunin et al. 2011. Evolution and classification of the CRISPR-Cas systems. Nat. Rev. Microbiol. 9: 467–477. DOI: 10.1038/nrmicro2577.

Maksimov, I.V., B.P. Singh, E.A. Cherepanova et al. 2020. Prospects and applications of lipopeptide-producing bacteria for plant protection (review). Appl. Biochem. Microbiol. 56: 15–28. DOI: 10.1134/S0003683820010135.

Marciniak, B.C., M. Pabijaniak, A. de Jong, R. Dűhring, G. Seidel, W. Hillen and O.P. Kuipers. 2012. High- and low-affinity *cre* boxes for CcpA binding in *Bacillus subtilis* revealed by genome-wide analysis. BMC Genomics 13: 401. DOI: 10.1186/1471-2164-13-401.

Malviya, D., K.P. Sahu, B.D. Singh, S. Paul, A. Gupta, R.A. Gupta, S. Singh, M. Kumar, D. Paul, P.J. Rai, V.H. Singh and P.G. Brahmaprakash. 2020. Lesson from ecotoxicity: Revisiting the microbial lipopeptides for the management of emerging diseases for crop protection. Int. J. Environ. Res. Public Health 17: 1–27. DOI: 10.3390/ijerph17041434.

Meena, K.R. and S.S. Kanwar. 2015. Lipopeptides as the antifungal and antibacterial agents: Applications in food safety and therapeutics. BioMed Research International 2015: 1–9. Article ID 473050. DOI: 10.1155/2015/473050.

Meena, K.R., T. Tandon, A. Sharma and S.S. Kanwar. 2018. Lipopeptide antibiotic production by *Bacillus velezensis* KLP2016. Journal of Applied Pharmaceutical Science 8: 91–98. DOI: 10.7324/JAPS.2018.8313.

Mijakovic, I., S. Poncet, A. Galinier, V. Monedero, S. Fieulaine, J. Janin, S. Nessler, J.A. Marquez, K. Scheffzek, S. Hasenbein, W. Hengstenberg and J. Deutscher. 2002. Pyrophosphate-producing protein dephosphorylation by HPr kinase/phosphorylase: A relic of early life? J. Proc. Natl. Acad Sci. USA 99(21): 13442–7. DOI: 10.1073/pnas.212410399.

Mnif, I. and D. Ghribi. 2015. Review lipopeptides biosurfactants: Mean classes and new insights for industrial, biomedical, and environmental applications. Biopolymers (PeptSci) 104: 129–147. DOI: 10.1002/bip.22630.

Moyne, A.L., T.E. Cleveland and S. Tuzun. 2004. Molecular characterization and analysis of the operon encoding the antifungal lipopeptide bacillomycin D. FEMS Microbiol. Lett. 234: 43–49. DOI: 10.1016/j.femsle.2004.03.011.

Nimrat, S., S. Lookchan, T. Boonthai and V. Verapong. 2019. Bioremediation of petroleum contaminated soils by lipopeptide producing *Bacillus subtilis* SE1 Afr. J. Biotechnol. 18(23): 494–501. DOI: 10.5897/AJB2019.16822.

Penha, R.O., L.P.S. Vandenberghe, C. Faulds et al. 2020. Bacillus lipopeptides as powerful pest control agents for a more sustainable and healthy agriculture: Recent studies and innovations. Planta 251: 70. DOI: 10.1007/s00425-020-03357-7.

Peypoux, F., J. Bonmatin and J. Wallach. 1999. Recent trends in the biochemistry of surfactin. Appl. Microbiol. Biotechnol. 51: 553–563. DOI: 10.1007/s002530051432.

Poncet, S., I. Mijakovic, S. Nessler, V. Gueguen-Chaignon, V. Chaptal, A. Galinier, G. Boël, A. Mazé and J. Deutscher. 2004. HPr kinase/phosphorylase, a Walker motif A-containing bifunctional sensor enzyme controlling catabolite repression in Gram-positive bacteria. J. Biochim. Biophys Acta 1697(1-2): 123–35. DOI: 10.1016/j.bbapap.2003.11.018.

Qi, G., F. Zhu, P. Du, X. Yang, D. Qiu, Z. Yu, J. Chen and X. Zhao. 2010. Lipopeptide induces apoptosis in fungal cells by a mitochondria-dependent pathway. Peptides 31(11): 1978–86. DOI: 10.1016/j.peptides.2010.08.003.

Qi, L.S., M.H. Larson, L.A. Gilbert, J.A. Doudna, J.S. Weissman, A.P. Arkin and W.A. Lim. 2013. Repurposing CRISPR as an RNA-guided platform for sequence-specific control of gene expression. Cell 152: 1173–1183.

Raaijmakers, M.J., I. Bruijn, de, O. Nybroe and M. Ongena. 2010. Natural functions of lipopeptides from Bacillus and Pseudomonas: More than surfactants and antibiotics. FEMS Microbiol. Rev. 34: 1037–1062.

Rufino, R.D., L.A. Sarubbo and G.M. Campos-Takaki. 2007. Enhancement of stability of biosurfactant produced by *Candida lipolytica* using industrial residue as substrate. World J. Microbiol. Biotechnol. 23: 729–734. DOI: 10.1007/s11274-006-9278-2.

Schumacher, M.A., M. Sprehe, M. Bartholomae, W. Hillen and R.G. Brennan. 2011. Structures of carbon catabolite protein A-(HPr-Ser46-P) bound to diverse catabolite response element sites reveal the basis for high-affinity binding to degenerate DNA operators. Nucleic Acids Res. 39(7): 2931–42. DOI: 10.1093/nar/gkq1177.

Schultenkämper, K., L.F. Brito and V.F. Wendisch. 2020. Impact of CRISPR interference on strain development in biotechnology. Biotechnol. Appl. Biochem. DOI: 10.1002/bab.1901.

Sen, R. 2008. Biotechnology in petroleum recovery: The microbial EOR. Progress in Energy and Combustion Science 34: 714–724. DOI: 10.1016/j.pecs.2008.05.001.

Sieber, S.A. and M.A. Marahiel. 2005. Molecular mechanisms underlying non ribosomal peptide synthesis: Approaches to new antibiotics. Chem. Rev. 105: 715–738. DOI: 10.1021/cr0301191.

Sivapathasekaran, C., S. Mukherjee, R. Sen, B. Bhatacharya and R. Samanta. 2011. Single step concomitant concentration, purification and characterization of two families of lipopeptides of marine origin. Bioprocess and Biosystem Engineering 34: 339–346. DOI: 10.1007/s00449-010-0476-9.

Slimene, B.I., O. Tabbene, N. Djebali et al 2012. Putative use of a *Bacillus subtilis* L194 strain for biocontrol of Phomamedicaginis in Medicagotruncatula seedlings. Res. Microbiol. 163: 388–397. DOI: 10.1016/j.resmic.2012.03.004.

Stein, T. 2005. *Bacillus subtilis* antibiotics: Structures, syntheses and specific functions. Mol. Microbiol. 56: 845–857. DOI: 10.1111/j.1365-2958.2005.04587. x.

Sun, H., X. Bie, F. Lu, Y. Lu, Y. Wu and Z. Lu. 2009. Enhancement of surfactin production of *Bacillus subtilis* fmbR by replacement of the native promoter with the Pspac promoter. Can. J. Microbiol. 55: 1003–1006. DOI: 10.1139/W09-044.

Syldatk, C. and F. Wagner. 1987. Production of biosurfactants. pp. 89–120. *In*: Kosaric, N., W.L. Cairns and N.C.C. Gray (eds.). Biosurfactants and Biotechnology. Marcel Dekker, Inc., New York.

Torres, M.J., C. Pérez Brandan, D.C. Sabaté, G. Petroselli, R. Erra-Balsells and M.C. Audisio. 2017. Biological activity of the lipopeptide-producing *Bacillus amyloliquefaciens* PGPBacCA1 on common bean *Phaseolus vulgaris* L. pathogens. Biological Control 105: 93–99. DOI: 10.1016/j.biocontrol.2016.12.001.

Tran, H., A. Ficke, T. Asiimwe, M. Höfte and J.M. Raaijmakers. 2007. Role of the cyclic lipopeptide massetolide A in biological control of Phytophthora infestans and in colonization of tomato plants by Pseudomonas fluorescens. New Phytol. 175(4): 731–42. DOI: 10.1111/j.1469-8137.2007.02138.x. PMID: 17688588.

Ugalmugale, S. and R. Swain. 2019, November. Biotechnology Market Share, Growth Forecasts Report 2025. Retrieved from Global Market Insights: https://www.gminsights.com/industry-analysis/biotechnology-market.

Vlamakis, H., Y. Chai, P. Beauregard, R. Losick and R. Kolter. 2013. Sticking together: Building a biofilm the *Bacillus subtilis* way. Nat. Rev. Microbiol. 11: 157–168. DOI: 10.1038/nrmicro2960.

Vollenbroich, D., M.R. Kamp and G. Pauli. 1997a. Mechanism of inactivation of enveloped viruses by the biosurfactant surfactin from *Bacillus subtilis*. Biologicals 25: 289–297. DOI: 10.1006/oil.1997.0099.

Vollenbroich, D., G. Pauli, M. Özel and J. Vater. 1997b. Antimycoplasma properties and application in cell culture of surfactin, a lipopeptide antibiotic from *Bacillus subtilis*. Applied and Environmental Microbiology 63: 44–49. PMID: 8979337 PMCID: PMC168300.

Wang, C., Y. Cao, Y. Wang et al. 2019. Enhancing surfactin production by using systematic CRISPRi repression to screen amino acid biosynthesis genes in *Bacillus subtilis*. Microb. Cell Fact 18: 90. DOI:10.1186/s12934-019-1139-4.

Wang, L.C., B.T. Ng, F. Yuan, K.Z. Liu and F. Liu. 2007. Induction of apoptosis in human leukemia K562 cells by cyclic lipopeptide from *Bacillus subtilis* natto T-2. Peptides 28: 1344–1350. DOI: 10.1016/j.peptides.2007.06.014.

Wei, Y.-H., L.-C. Wang, W.-C. Chen and S.-Y. Chen. 2010. Production and characterization of fengycin by indigenous *Bacillus subtilis* F29–3 originating from a potato farm. International Journal of Molecular Science 11: 4526–4538. DOI: 10.3390/ijms11114526.

Wu, J., G. Du, J. Chen and J. Zhou. 2015. Enhancing flavonoid production by systematically tuning the central metabolic pathways based on a CRISPR interference system in *Escherichia coli*. Sci Rep. 5: 13477. DOI: 10.1038/srep13477.

Wu, C.Y., C.L. Chen, Y.H. Lee, Y.C. Cheng, Y.C. Wu and H.Y. Shu. 2007. Non ribosomal synthesis of fengycin on an enzyme complex formed by fengycin synthetases. J. Biol. Chem. 282: 5608–5616. DOI: 10.1074/jbc.M609726200.

Wu, Y., M. Xu, J. Xue, K. Shi and M. Gu. 2019. Characterization and enhanced degradation potentials of biosurfactant-producing bacteria isolated from a marine environment. ACS Omega 4: 1645–1651. DOI: 10.1021/acsomega.8b02653.

Xing, X., X. Zhao, J. Ding, D. Liu and G. Qi. 2018. Enteric-coated insulin microparticles delivered by lipopeptides of iturin and surfactin. Drug Deliv. 25(1): 23–34. DOI: 10.1080/10717544.2017.1413443.

Xu, P., Q. Gu, W. Wang, L. Wong, A.G. Bower, C.H. Collins and M.A. Koffas. 2013. Modular optimization of multi-gene pathways for fatty acids production in *E. coli*. Nat. Commun. 4: 1409. DOI: 10.1038/ncomms2425.

Xu, Z., J. Shao, B. Li, X. Yan, Q. Shen and R. Zhang. 2013. Contribution of bacillomycin D in *Bacillus amyloliquefaciens* SQR9 to antifungal activity and biofilm formation. Appl. Environ. Microbiol. 79(3): 808–15. DOI: 10.1128/AEM.02645-12.

Yalçin, S.K. and Z.Y. Özbaş. 2005. Determination of growth and glycerol production kinetics of a wine yeast strain *Saccharomyces cerevisiae* Kalecik 1 in different substrate media. World J. Microbiol. Biotechnol. 21: 1303–1310. DOI: 10.1007/s11274-005-2634-9.

Yánez-Mendizábal, V., H. Zeriouh, I. Viñas et al. 2012. Biological control of peach brown rot (Monilinia spp.) by *Bacillus subtilis* CPA-8 is based on production of fengycin-like lipopeptides. Eur. J. Plant Pathol. 132: 609–619. DOI: 10.1007/s10658-011-9905-0.

Yang, L., X. Han, F. Zhang et al. 2018. Screening Bacillus species as biological control agents of *Gaeumannomyces graminis var. Tritici* on wheat. Biol Control. 118: 1–9. DOI: 10.1016/j.biocontrol.2017.11.004.

Yang, Y., H.J. Wu, L. Lin et al. 2015. A plasmid-born Rap-Phr system regulates surfactin production, sporulation and genetic competence in the heterologous host, *Bacillus subtilis* OKB105. Appl. Microbiol. Biotechnol. 99(17): 7241–52. DOI: 10.1007/s00253-015-6604-3.

Yang, Y., H.J. Wu, L. Lin, Y. Yang, L. Zhang, H. Huang, C. Yang, S. Yang, Y. Gu et al. 2017. A flexible binding site architecture provides new insights into CcpA global regulation in Gram-positive bacteria. DOI: 10.1128/mBio.02004-16.

Yao, S., X. Gao, N. Fuchsbauer, W. Hillen, J. Vater and J. Wang. 2003. Cloning, sequencing, and characterization of the genetic region relevant to biosynthesis of the lipopeptidesiturin A and surfactin in *Bacillus subtilis*. Current Microbiology 47: 272–277. DOI: 10.1007/s00284-002-4008-y.

Zeriouh, H., A. de Vicente, A. Pérez-García and D. Romero. 2014. Surfactin triggers biofilm formation of *Bacillus subtilis* in melon phylloplane and contributes to the biocontrol activity. Environ. Microbiol. 16: 2196–2211. DOI: 10.1111/1462-2920.12271.

Zhang, F., L. Cong, S. Lodato, S. Kosuri, G.M. Church and P. Arlotta. 2011. Efficient construction of sequence-specific TAL effectors for modulating mammalian transcription. Nat. Biotechnol. 29: 149–153. DOI: 10.1038/nbt.1775.

Zhao, J., Y. Li, C. Zhang, Z. Yao, L. Zhang, X. Bie, F. Lu and Z. Lu. 2012. Genome shuffling of *Bacillus amyloliquefaciens* for improving antimicrobial lipopeptide production and an analysis of relative gene expression using FQ RT-PCR. J. Ind. Microbiol. Biot. 39: 889–896. DOI: 10.1007/s10295-012-1098-9.

Zhao, Q., W. Ran, H. Wang et al. 2013. Biocontrol of *Fusarium* wilt disease in muskmelon with *Bacillus subtilis* Y-IVI. Biocontrol 58: 283–292. DOI: 10.1007/s10526-012-9496-5.

Zhen, Z., Li. Rui, Y. Guanghui, R. Wei and S. Qirong. 2013. Enhancement of lipopeptides production in a two-temperature-stage process under SSF conditions and its bioprocess in the fermenter. Bioresource Technology 127: 209–215. DOI: 10.1016/j.biortech.2012.09.119.

Zhu, Z., L. Sun, X. Huang, W. Ran and Q. Shen. 2014. Comparison of the kinetics of lipopeptide production by *Bacillus amyloliquefaciens* XZ-173 in solid-state fermentation under isothermal and non-isothermal conditions. World J. Microbiol. Biotechnol. 30(5): 1615–23. DOI: 10.1007/s11274-013-1587-7.

5

Applications and Future Prospects of Biosurfactants

Anita V Handore,[1] *Sharad R Khandelwal*,[2,*] *Rajib Karmakar*,[3] *Divya L Gupta*[4] and *Vinita S Jagtap*[5]

1. Introduction

1.1 *Surfactants*

Surfactants are substances which create self-assembled molecular clusters, i.e., micelles in a solution (water or oil phase) and adsorb into the interface between a solution and a different phase (gases/solids) (Nakama 2017). They are classified into three groups, viz., anionic surfactants, nonionic surfactants and cationic surfactants (Azarmi and Ali 2015). Surfactants play the role of significant agents in dispersion, wetting, foaming, emulsifying, anti-foaming, cleaning and in different products including adhesives, emulsions, paints, inks, sanitizers, toothpastes, shampoos, detergents, firefighting foams, insecticides and lubricants. They are extensively used in various industries such as petrochemical, agricultural, textile, paper, food, cosmetics and pharmaceuticals. Generally, surfactants are mostly synthesized by using different chemicals, thereby showing various adverse effects not only on the environment but also on the health of society.

1.2 *BioSurfactants*

Bio surfactants synthesized by various microorganisms possess similar properties that reduce the surface and interfacial tensions by mechanisms similar to those of synthetic surfactants.

[1] Research and Development Department, Sigma Wineries Pvt. Ltd. Nashik, M.S. India-422112.

[2] H.A.L. College of Science & Commerce, Nashik, Maharashtra, India 422207.

[3] Department of Agricultural Chemicals, Bidhan Chandra Krishi Viswavidyalaya, Directorate of Research, Research Complex Building, Nadia,W. Bengal, India-741252.

[4] Amneal Pharmaceuticals 1 New England Ave, Piscataway NJ 08854, USA.

[5] Medical Oncology Molecular Laboratory, Tata Memorial Hospital Mumbai, Maharashtra, India-400012.

* Corresponding author: sharad_khandelwal13@yahoo.com

However, they show numerous benefits over synthetic surfactants such as high biodegradability, low toxicity, low irritancy and compatibility with the human skin and ecology. Moreover, they possesses special characteristics like selectivity and specific activity at extreme temperatures, pH and salinity among other conditions.

It is found that a number of microorganisms like bacteria, yeasts, and fungi show the ability to produce different classes of bio surfactants on the basis of their chemical composition and microbial origin. The major classes of biosurfactants are, Glycolipids, Lipopeptides and Lipoproteins, Fatty Acids, Phospholipids, and Neutral, Polymeric and Particulate bio Surfactants.

Since the last few decades, Bio surfactants are coming up as 'Green products' with an increasing consumer demand. Off all natural and safe products they show substantial potential in respect of implantation of sustainable industrial processes such as the use of cheap and renewable resources. In this context, biosurfactants have remarkable growth in the global market due to their promising and diverse applications in numerous fields as follows.

2. Applications of Bio Surfactants

2.1 Applications in Environment

Numerous technologies have been established on basis of the capability of bio surfactants to interact with hydrophobic substrates causing reduction of surface and interfacial tensions, emulsification, solubilisation, dispersion, desorption and wetting. Generally chemically synthesized surfactants are non-biodegradable and causes various hazardous having an impact on the Environment and society health. In this context the demand for bio surfactants has increased in various sectors due to their ecofriendly and health support potential. Bio surfactants have recently become one of the promising products of the bio economy with wide a variety of applications as follows:

2.1.1 BioSurfactants and Phytoremediation

At higher concentrations, all metals are toxic as they form free radicals and cause oxidative stress. In addition, they disturb the normal activity of some essential plant enzymes and pigments by replacing them. Thus, heavy metals are found to be the main culprits for plant growth. The capability of the plant can be increased for phytoremediation by using both metal resistant and bio surfactant producing bacteria. It is reported that bio surfactant producing *Bacillus* sp. J119 strain can increase the efficiency of the plant growth of rapeseed, sundangrass, tomato and maize and also the uptake of cadmium (Singh et al. 2008).

2.1.2 Bio Surfactants and Bio Remediation

An alternative and eco-friendly method of remediation technology is the use of bio surfactants and bio surfactant-producing microorganisms to control environmental pollution.

Table 1. Bio surfactants and their applications in environmental biotechnology.

Group	Class	Microorganism	Applications	References
Glycolipids	Rhamnolipids	*P. aeruginosa*, *Pseudomonas* sp.	Enhancement of the degradation and dispersion of different classes of hydrocarbons; emulsification of hydrocarbons and vegetable oils; removal of metals from soil	(Sifour et al. 2007, Whang et al. 2008, Herman et al. 1995 Maier et al. 2000)
	Trehalolipids	*M. tuberculosis*, *R. erythropolis*, *Arthrobacter* sp., *Nocardia* sp., *Corynebacterim* sp.	Improvement of the bioavailability of hydrocarbons	(Franzetti et al. 2010)
	Sophorolipids	*T. bombicola*, *T. petrophilum*, *T. apicola*	Recovery of hydrocarbons from dregs and muds; removal of heavy metals from sediments; enhancement of oil recovery	(Whang et al. 2008, Pesce 2002, Baviere et al. 1994)
Fatty acids, phospholipids and neutral lipids	Corynomycolic acid	*C. lepus*	Enhancement of bitumen recovery	(Gerson et al. 1978)
	Spiculisporic acid	*P. spiculisporum*	Removal of metal ions from aqueous solution; dispersion action for hydrophilic pigments; preparation of new emulsion-type organogels, superfine microcapsules (vesicles or liposomes), heavy metal sequestrants	(Ishigami et al. 1983, Ishigami et al. 2000, Hong et al. 1998)
	Phosphatidylethanolamine	*Acinetobacter* sp., *R. erythropolis*	Increasing the tolerance of bacteria to heavy metals	(Appanna et al. 1995)
Lipopeptides	Surfactin	*B. subtilis*	Enhancement of the biodegradation of hydrocarbons and chlorinated pesticides; removal of heavy metals from contaminated soil, sediments and water; increasing the effectiveness of phytoextraction	(Jennema et al. 1983, Awashti et al. 1999, Arima et al. 1968)
	Lichenysin	*B. licheniformis*	Enhancement of oil recovery	(Thomas et al. 1993)

Table 1 Contd. ...

...Table 1 Contd.

Group	Class	Microorganism	Applications	References
Polymeric biosurfactants	Emulsan	*A. calcoaceticus* RAG-1	Stabilization of the hydrocarbon-in-water emulsions	(Zosim et al. 1982)
	Alasan	*A. radioresistens* KA-53		(Toren et al. 2001)
	Biodispersan	*A. calcoaceticus* A2	Dispersion of limestone in water	(Rosenberg et al. 1988)
	Liposan	*C. lipolytica*	Stabilization of hydrocarbon-in-water emulsions	(Cirigliano et al. 1984)
	Mannoprotein	*S. cerevisiae*		(Cameron et al. 1988)

Applications of bio surfactants not only allow the recovery of oil from oily sludge but also allow the reuse or recycling of valuable hydrocarbons recovered from the oily sludge. Oily sludge generated from petroleum processing plants is another source of environmental pollution. Schaller et al. (2004) has explored the potential use of surfactin bio surfactant as an agent for enhanced oil recovery and presented results that favor their application and cost-effectiveness. It is reported that bio surfactants play a key role in oil mobilization by emulsification in microbial enhanced oil recovery (Sen et al. 2008). For a sustainable environment, there is a need to treat and stabilize oily sludges prior to disposal as it reduces the environmental impact. Moreover, the recovered oil can also serves as an energy source which has led to the renewed and intensified efforts to treat the oily sludge (Chirwa et al. 2013).

2.2 *Application in Agriculture*

These days bio surfactants can be widely exploited in agriculture for enhancement of biodegradation of pollutants to improve the agricultural soil quality by soil bioremediation and phytoremediation by the elimination of plant pathogens. They are used for indirect plant growth promotion due to their antimicrobial potential and also used to increase the plant microbe interaction beneficial for the plant.

2.2.1 *Improvement of Soil Quality*

It is reported that the desorption of hydrophobic pollutants which are tightly bound to soil particles can be accelerated by bio surfactants. Besides, they can degrade certain chemical insecticides accumulated in the agricultural soil. Biosurfactants can efficiently remove the insoluble organic pollutants from soil as compared to synthetic surfactants. The requirement of iron for increased production of biosurfactant by *Pseudomonas* sp. and enhancement of bioavailability of poly aromatic hydrocarbons (PAH's) has also been reviewed (Dhara et al. 2013).

There are several reports on potential properties of bio surfactants produced by *Pseudomonas* sp, *Bacillus* sp., and *Acinetobacter* sp. in respect of removal of heavy metals from contaminated soil and even acceleration of biodegradation of pesticides

(Pacwa Plociniczak et al. 2011, Kassab and Roane 2006). Biosurfactants such as rhamnolipids and surfactins are known to remove heavy metals such as Ni, Cd, Mg, Mn, Ca, Ba, Li, Cu, and Zn (ions) from the soil with a new method of foaming-surfactant technology (Mulligan and Wang 2004, Mulligan et al. 2001, Dhara et al. 2013).

2.2.2 *Plant Pathogen Elimination*

The antimicrobial activity of biosurfactants against various plant pathogens makes them a promising biocontrol molecule to achieve sustainable agriculture. They can facilitate the biocontrol mechanism of plant growth promoting microbes like parasitism, antibiosis, competition, induced systemic resistance, and hypovirulence (Singh et al. 2007). It is reported that some biosurfactants are known to have antagonist properties and used to enhance the antagonistic activities of microbes and microbial products in agriculture (Jazzar and Hammad 2003, Kim et al. 2004, Nihorimbere et al. 2011). Biosurfactant producing rhizospheric isolates of *Pseudomonas* spp. and *Bacillus* spp have exhibited biocontrol of soft rot causing *Pectobacterium* spp. and *Dickeya* spp. (Krzyzanowska et al. 2012). Rhamnolipids have demonstrated inhibition of zoospore forming plant pathogens which have acquired resistance to commercial chemical pesticides (Sha et al. 2011, Kim et al. 2011, Hultberg et al. 2008). Rhamnolipids can stimulate the plant immunity which is considered as an alternative strategy to reduce the infection by plant pathogens (Vatsa et al. 2010). It is also reported that the plant growth-promoting *P. putida* can produce biosurfactant causing lysis of zoospores of the oomycete pathogen. Besides, lipopeptide biosurfactants produced by *Bacillus* spp. inhibit growth of phytopathogenic fungi like *Fusarium* spp., *Aspergillus* spp., and *B. sorokiniana.* Moreover, biosurfactant produced *P. fluorescens* show significant antifungal properties which are found to inhibit the growth of fungal pathogens like *P. ultimum*, *F. oxysporum*, *P. cryptogea*, *V. microsclerotia*, etc. (Dhara et al. 2013).

2.2.3 *Plant Microbe Interaction*

Generally, Microbial factors such as the ability to form a biofilm on the root surface and release of quorum sensing molecules thereby promoting motility, is important for the establishment of association with any plant. It is reported that these molecules affect the motility of microorganisms, participate in signaling and differentiation as well as in biofilm formation (Ron and Rosenberg 2001, Berti et al. 2007, Van Hamme et al. 2006, Kearns and Losick 2003). Moreover, the biosurfactant (rhamnolipid) produced by *Pseudomonas* spp. can efficiently regulate the process of quorum sensing cell to cell communication (Dusane et al. 2010). The biosurfactants produced by rhizobacteria can improve the bioavailability of hydrophobic molecules which can serve as nutrients. Also, the biosurfactants produced by soil microbes provide wettability to soil and support proper distribution of chemical fertilizers in soil thus supporting plant growth promotion (Dhara et al. 2013).

2.2.4 *Bio Pesticides*

At present biosurfactants have been promisingly used as bio pesticides due to their low toxicity, high degree of biodegradability, optimal activity at extreme

Table 2. Bio surfactants used as bio pesticides.

Bacterial Species	Biosurfactants	Composition	Activity	Insect
B. amyloliquefaciens AG1	Lipopeptides	Surfactin, fengycin, iturin, and bacillomycin	Larvicidal	*Tuta absoluta*
	Polyketides	Bacillaene, macrolactin, and difficidin		
B. amyloliquefaciens AG1	Lipopeptides	Surfactin, fengycin, iturin, and bacillomycin	Larvicidal	*Spodoptera littoralis*
B. subtilis	Cyclic lipopeptides	Surfactin	Larvicidal	Mosquito
	Lipopeptide	Not given	Larvicidal	*Spodoptera littoralis*
B. amyloliquefaciens	Cyclic lipopeptides	Not given	Pupicidal	
B. amyloliquefaciens G1	Lipopeptides	Surfactin	Insecticidal	*Myzuspersicae*
B. thuringiensis	Not given	Cry3Aa toxin	Larvicidal	Colorado potato beetle
B. thuringiensis	Not given	Cry1Da toxin	Larvicidal	*Spodoptera littoralis*
B. subtilis	Cyclic lipopeptides	Surfactin	Larvicidal	*Aedes aegypti* L.

(Adapted from, Tariku et al. 2018).

environmental conditions, and environmental friendly nature. Among various biosurfactant producing microbes, Bacillus species are one of the prominent biosurfactant-producing bacteria producing a broad-spectrum of lipopeptides such as surfactin, iturin, bacillomycin, fengycin, and lichenysin (Mukherjee and Das 2005).

2.3 *Application in Petroleum Industry*

In the petroleum industry, biosurfactants have been applied effectively for the exploration of heavy oil, offering advantages over their synthetic counterparts throughout the entire petroleum processing chain (extraction, transportation and storage). These biomolecules are efficiently used in microbial-enhanced oil recovery, the cleaning of contaminated vessels and to facilitate the transportation of heavy crude oil by pipelines (De et al. 2014, Assadi et al. 2010, Luna et al. 2012).

2.4 *Application as Commercial Laundry Detergents*

Due to public awareness about the environmental and health hazards, ecofriendly, natural substitutes of chemical surfactants in laundry detergents are in demand. Bio surfactants like Cyclic Lipopeptides (CLP) are stable over a wide pH range (7.0–12.0) and no surface-activity loss is observed at high temperatures. These compounds exhibit a noticeable capability for emulsion formation with vegetable oils and show remarkable compatibility as well as stability with commercial laundry detergents favoring their inclusion in laundry detergents formulations (Das and Mukherjee 2007, Fakruddin 2012).

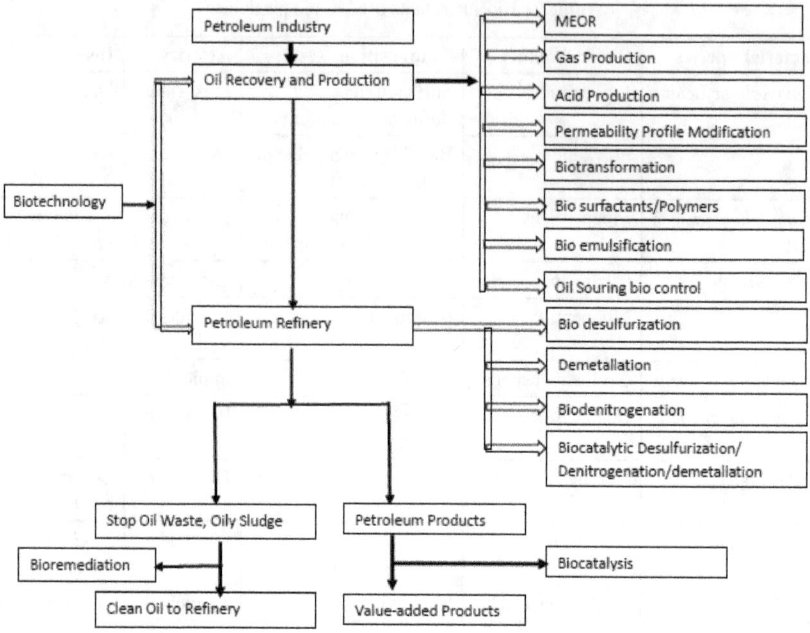

Figure 1. Potential applications of biotechnology in petroleum industry (Adapted and Redrawn from: De et al. 2016).

2.5 Application as Green Biocides

In industries, microbiologically induced corrosion, i.e., MIC is associated with formation of biofilms and frequently cause various problems in industries. It damages industrial equipment or infrastructures and leads to economic and environmental problems. Although, synthetic chemical biocides are normally used to solve these problems, most of them are ineffective against the biofilms also they are toxic and not degradable. However, bio surfactants are proved to be one of the significant eco-friendly anticorrosion compounds to inhibit bio corrosion (Graz and Varenyam 2020).

2.6 Application in Medicine and Pharmaceuticals

2.6.1 Anti-cancer Activity

Bio surfactants such as glycolipids and lipopeptides have emerged as possible broad-spectrum agents for cancer chemotherapy/biotherapy and as safe vehicles or ingredients in drug delivery formulations due to their structural novelty and diverse biophysical properties. It is reported that glycolipids and lipopeptides can selectively inhibit the proliferation of cancer cells and disrupt the cell membranes. Whereas, few microbial extracellular glycolipids are found to induce cell differentiation instead of cell proliferation in the human promyelocytic leukemia cell line. Moreover, exposure of PC 12 cells to MEL enhances the activity of acetylcholine esterase and interrupt the cell cycle at the G1 phase with resulting overgrowth of neurites and partial cellular differentiation, suggesting that MEL can induce neuronal differentiation

Table 3. Role of bio surfactants in bio corrosion process.

Strains	**Biosurfactants**	**Role**	**References**
B. species	Lipopeptides	Antagonistic effects against abroad spectrum of microorganisms	(Van et al. 2006, De et al. 2015)
B. amyloliquefaciens, B. subtilis	Peptide, lipopeptides	Antagonistic effects against many fungal pathogens	(Sarwar et al. 2018) (Sharma et al. 2018)
B. subtilis	Surfactin	Inhibit biofilm formed by *S. enterica, E. coli*, and *P. mirabilis*	(Mireles et al. 2001)
Pseudomonas sp. PS-17	Rhamnolipids	Inhibit corrosion of alloy	(Zin et al. 2018)
Pseudomonas spp.	Rhamnolipids	Algicidal, antiamoeba land zoo sporicidal properties	(Banat et al. 2010, Raval et al. 2017)
P. fluorescens	Biosurfactant	Inhibit corrosion of stainless steels	(Dagbert et al. 2006)
P. stutzeri F01	Biosurfactant	Antibacterial properties to corrosive bacterial strains	(Parthipan et al. 2018)

(Adapted from: Graż and Varenyam 2020).

Table 4. Effects of bio surfactants on various cancer cell lines.

Biosurfactants Class	**Biosurfactant Name**	**Source**	**Effect on Cell Line**
Lipopeptide	Surfacin	*B. subtilis*	Suppression of LoVo (colon carcinoma) cell line
Lipopeptide	Surfacin	*B. natto TK-1*	Killing of MCF-7 (human breast cancer) cell line
Lipopeptide	Iturin	*B. subtilis*	Inhibition of K562 leukaemia cells
Glycolipid	Mannosylerythritol lipid-A Mannosylerythritol lipid-B	*C. Antarctica* T-34	Induced HL60 (leukaemia cell line) differentiation
Sophorolipid	Sophorolipid	*C. bombicola* ATCC 22214	Increased in LN-229 differentiation
Sophorolipid	Di-acetylated lactonic C18:1	*W. domercqiae*	Apoptosis in H7402 (liver cancer) cells
Sophorolipid	Cetyl alcohol sophorolipid	*C. bombicola* ATCC 22214	Antiproliferation of HeLa cells
Sophorolipid	Various derivatives	*C. bombicola* ATCC 22214	Killing of human pancreatic cancercells
Sophorolipid	Various derivatives	*W. domercqiae*	Inhibition of oesophageal cancer cells
Sophorolipid	Various derivatives	*S. bombicola*	Killing of MDA-MB-231 breast cancer cells

(Adapted from: Naughton et al. 2018).

in PC 12 cells and provides the ground work for using microbial extracellular glycolipids as novel reagents to treat the cancer cells (Krishnaswamy et al. 2008). The biological activities of seven microbial extracellular glycolipids, including mannosylerythritol lipids (MEL)-A, mannosylerythritol lipids-B, polyol lipid, sophorose lipid, rhamnolipid, succinoyl trehalose lipid-3 and succinoyl trehalose lipid (STL)-1 have been studied (Isoda et al. 1999). All these glycolipids, except rhamnolipid, were found to induce cell differentiation instead of cell proliferation in the human promyelocytic leukaemia cell line HL60. STL and MEL markedly increased common differentiation characteristics in monocytes and granulocytes respectively. Bernheimer and kitomotoigard (1970) demonstrated that surfactin has various pharmacological applications such as inhibiting fibrin clot formation and hemolysis. Sheppard et al. (1991) depicted formation of ionic channels in lipid membranes by using surfactin. It has also been reported as having an antitumor activity against Ehrlich's as cite carcinoma cells (Kameda et al. 1974). Moreover, the anticancer activity of bio surfactant monoolein produced by dematiaccous fungus *E. dermatidis* confirmed the anti proliferative activity against cervical cancer (HeLa) and leukemia (U937) cell lines in a dose dependent manner. It was reported that there was no cytotoxicity with normal cells even at high concentrations (Chiewpattanaku et al. 2010).

2.6.2 *Antimicrobial Activity*

Several bio surfactants have strong antibacterial, antifungal and antivirus activities; they play a significant role of anti-adhesive agents to pathogens and make them useful for treating various diseases and efficiently used as therapeutic and probiotic agents (Gharaei 2011). It is reported that Rhamnolipids produced by *Pseudomonas aeruginosa* (Itoh et al. 1971), lipopeptides produced by *Bacillus subtilis* (Sandrin et al. 1990, Leenhouts et al. 1995) and *Bacillus licheniformis* (Jenny et al. 1991) and mannosylerythritol lipids from *Candida antarctica* (Kitamoto et al. 1993) have remarkable antimicrobial potential and the antibiotic effects as well as inhibition of HIV virus growth in white blood corpuscles have opened up new fields for their applications (Neu et al. 1990). It is stated that there are possible applications, viz., emulsifying aids for drug transport to the infection site, for supplementing pulmonary surfactants and as adjuvants for vaccines (Tayler et al. 1985). Pratt et al. (1989) have described a release of this amazing molecule by an oral *Streptococcus mitis* strain, causing reduction in adhesion of *Streptococcus mutans*. Also Velraeds et al. (1997) stated inhibition of adhesion of pathogenic enteric bacteria by biosurfactants produced by a *Lactobacillus* spp. as well as describes the dose related inhibition of the initial deposition rate of *Escherichia faecalis* and other bacteria adherent on both hydrophobic and hydrophilic substrata (Velraeds et al. 1997). They also speculated on other possible therapeutic agents through the development of antiadhesive biological coatings for catheter materials to delay the onset of bio film growth. It is found that the antimicrobial activities of glycolipids are effective against *Staphylococcus* sp. mainly on methicillin resistant *Staphylococcus aureus* (MRSA) (Das et al. 2008). Moreover, mannosylerythritol lipids (MEL-A and MEL-B) produced by *Candida antarctica* exhibit antimicrobial action against Gram-positive bacteria.

2.6.3 *Anti-adhesive Agents*

The adhesion of pathogenic microorganisms on medical utensils and surgical instruments is inhibited by microbial bio surfactants through combating colonization

(Rivardo et al. 2009). The pre-coating of bio surfactant solutions on vinyl urethral catheters demonstrated the decreased amount of biofilm formation by the urinary tract bacterial pathogens such as *E. coli, S. enterica, S. typhimurium* and *P. mirabilis* (Rodrigues et al. 2004). Numerous biosurfactants have been reported to demonstrate antimicrobial activity against many human pathogenic microorganisms. Besides, bio surfactants are found to exhibit anti-adhesive and anti-biofilm activities which make them applicable to reduce adhesion and colonization of pathogenic microorganisms (Rodrigues et al. 2006). According to Rodrigues et al. (2006) biosurfactants inhibit the adhesion of pathogenic organisms to solid surfaces or to the infection site. Again absorption of biosurfactants to the substratum alters the hydrophobicity of the surface and causes microbial adhesion and desorption processes (Desai and Banat 1997, Bai et al. 1997). Bio surfactants reduce the microbial population on prostheses and decrease the airflow resistance that occurs on voice prostheses after biofilm formation (Rodrigues et al. 2004). Bio surfactants can alter the physical and chemical conditions of the environment where biofilm formation is in progress and has direct action against pathogens (Mireles et al. 2001). It has been found that two lipopeptide biosurfactants produced by *B. subtilis* V9T14 and *B. licheniformis* VI9T2I1 have interesting anti-adhesive activities that inhibit the biofilm formation by two pathogenic strains mainly *S. aureus* ATCC 29213 and *E. coli* CFT073 (Rivardo et al. 2009). Serotypes of group II capsular polysaccharides produced by uropathogenic *E. coli* (UPFC strain CFT073) behave like surface active polymers and have good anti-adhesive properties. It inhibits mature bio film development of broad range Gram positive bacteria and Gram negative bacteria (Valle et al. 2006).

2.6.4 ImmunoModulatory Action

Bacterial lipopeptides constitute potent non-toxic, nonpyrogenic immunological adjuvants when mixed with conventional antigens. A remarkable enhancement in the humoral immune response was observed to poly-L-lysine coupled with low molecular mass antigens such as herbicolin A, iturin AL and microcystin in rabbits and chickens (Rodrigues et al. 2006). Sophorolipids are promising immune modulators, which decrease sepsis related mortality by modulation of adhesion molecules, nitric oxide and cytokine production in 36 hours in the vivo rat model of septic peritonitis. Also, they decrease the production of IgE by affecting plasma cell activity (Cameotra and Makkar 2004).

2.6.5 Antiviral Activity

Biosurfactants like Surfactin and its analogues shows antiviral activity. It is reported that due to physico-chemical interactions between virus envelopes and the surfactant, antiviral activity of Surfactin on enveloped viruses is found to be more prominent than on the non-enveloped viruses. Moreover, the biosurfactants have proved to inhibit growth of human immune deficiency virus in leucocytes (Desai and Banat 1997, Krishnaswamy et al. 2008). The sophorolipids from *C. bombicola* and its structural analogues like sophorolipid diacetate ethyl ester are potent against HIV, posess cytotoxic activities and act as virucidal agents (Krishnaswamy et al. 2008). It is described that the cell free virus of *P. parvovirus*, *Pseudo rabies* virus, Newcastle disease virus and bursal disease virus can be inactivated by the lipopeptides produced by *B. subtilis* which can also inhibit the replication and infectivity of the Newcastle disease virus and bursal disease virus but ineffective for pseudo rabies virus and porcine parvovirus (Huang et al. 2006). Also, Rhamnolipids and their complexes with alginate produced by *Pseudomonas* sp. have antiviral activity against Herpes simplex viruses type 1 and 2 (Remichkova et al. 2008).

2.6.6 Genetic Manipulation

Lipofection using cationic liposomes is considered to be a safe way to deliver foreign genes to the target cells without side effects among various known methods of gene transfection (Zhang et al. 2010). The use of liposomes made from biosurfactants is an important strategy for gene transfection. Kitamoto et al. (2002) demonstrated that liposome based biosurfactants show increasing efficiency of gene transfection in comparison with commercially available cationic liposome. In the last decade, for the liposome-based gene transfection some techniques and methodologies have been developed. Members of the *Candida antarctica* strain produce two kinds of mannosylerythritol lipids (MEL-A and MEL-B) that exhibit antimicrobial activity, particularly against Gram-positive bacteria; Ueno et al. in 2007 examined MEL-A-containing liposome for gene transfection by introducing biosurfactants in this field. Nakanishi et al. (2009) were able to produce nanovectors containing a biosurfactant that increases the efficacy of gene transfection *in vivo* and *in vitro*. Surfactin-mediated gold nanoparticles have opened a new and fascinating application in the field of drug, gene delivery, and targeted therapy (Reddy et al. 2009).

2.6.7 Cosmetics

Biosurfactants are widely used in formulations of various cosmetics due to their properties such as emulsification and de-emulsification, foaming, water binding capacity, spreading and wetting properties and effect on viscosity among others. These surfactants are used as emulsifiers, foaming agents, solubilizers, wetting agents, cleansers, antimicrobial agents, mediators of enzyme action in production of bath products, acne pads, anti dandruff products, contact lens solutions, baby products, mascara, lipsticks, toothpaste, dentine cleansers (Tugrul and Cansunar 2005, Tuleva et al. 2002). The low irritancy and high skin compatibility of many bio surfactants constitute a strong advantage over non-natural counterparts (Rodríguez et al. 2019). These biomolecules have been used as essential components to process various cosmetic products like shampoo, hair conditioners, soaps, shower gel, toothpastes, creams, moisturizers, cleansers, and many other skin care and healthcare products (Akbari et al. 2018).

2.7 Application in Food Industry

In this decade, alteration in the trend of consumers towards natural additives and the increasing health and environmental concerns have created a high demand for new "green" food additives (Marcia and Sumária 2018). So,the use of bio surfactants as food additives is in line with the growing consumer interest in natural, organic, vegan foods, requiring the development of different biomolecules exhibiting technological ability to reduce or eliminate use of synthetic surfactants (Jahan et al. 2020).

Apart from their innate surface-active property, bio surfactants exhibit remarkable emulsification ability, anti-adhesive and antioxidant potential, antimicrobial and anti-biofilm activities against food pathogens. In this context bio surfactants are significantly used in the food industry for the cleaning and/or treatment of contact surfaces, acting as antimicrobial and anti-biofilm agents, bioemulsifier, and can also be incorporated directly into formulations as an additive or ingredient (Ribeiro et al. 2020, Nitschke and Silva 2018, Khanna and Pattnaik 2019).

Generally, lipopeptides and glycolipids are prominently used due to their desirable properties for significant use in the food sector. It is described that these biomolecules show remarkable antioxidant properties (Zouari et al. 2016, Jamshidi et al. 2019), antibacterial and anti-adhesive activity against various microbial species such as *P. aeruginosa*, *E. coli*, *B. subtilis*, and *S. aureus*, etc. (De et al. 2018, Gaur et al. 2019), and low cytotoxicity (Balan et al. 2019). These compounds have been efficiently used to solubilize vegetable oils, stabilize fats, and to improve the organoleptic properties of bread. Bio surfactants are used in muffins instead of baking powder and eggs (Kiran et al. 2017); in cookies they replace synthetic additives (Zouari et al. 2016) and used as emulsifiers in salad dressings (Campos et al. 2019). In food processing industries, this biomolecule has been efficiently used to control the agglomeration of fat globules, to stabilize aerated systems, for improving the texture and shelf life of food products enriched with starch. Moreover it is used to improve the consistency and texture of fat-based products (Muthusamy et al. 2008). In bakery and ice cream manufacturing, bio surfactants are used to: control consistency, extend freshness, and solubilize flavor oils. During the cooking of oils and fats, these are used as fat stabilizers and anti-spattering agents. It is reported that addition of Rhamnolipids can improve the stability in dough; the texture, volume, and conservation of bakery products; and the properties of butter cream, croissants, and frozen confectionery products (Muthusamy et al. 2008, Danka et al. 2014).

3. Promising Prospects and Strategies for Biosurfactant Production

In spite of extensive research work in the last few decades on economizing the production of biosurfactants, there is still an economic challenge in respect of its commercial success in comparison with their synthetic counterparts. In such circumstances, globalizing and economizing the sustainable production and applications of biosurfactants following promising prospects and strategies can be implemented (Singh et al. 2018) as:

- The production of highly demanded biosurfactants can be made cost effective by changing strategies like the use of Fe-NPs and Fe-enriched immobilizing carriers to make lipopeptide production cost effective.
- Media optimization strategies in respect of limited Fe availability, optimized C: N ratios and the use of Lactones, i.e., growth enhancers helpful for enhanced production of rhamnolipid and SL yield as well as restraining the specificity of glycolipid congeners.

- During the continuous culturing and downstream processing of biosurfactants, a main constraint is the washout of cells from the bio reactor and the effect of changing reactor conditions and unwanted metabolites on cell growth apart from the foam-inducing property of free cells. Therefore, use of immobilized organism like *P. fluorescens, B. Subtilis, P. nitroreducens, P. Aeruginosa* is as alternative.
- Use of nano particles, solid-state fermentation, directed fermentation; foam fractionation, and fill and draw mode of operation could prove to be other promising processes for the enhanced industrial production of various biosurfactants.
- Use of microbioreactors for optimization studies, *in situ* product removal by automated surface enrichment, employment of novel oxygenation processes and the use of novel techniques like pertraction.
- Use of unprocessed, fortified waste substrates and biosurfactant coproduction with another industrially economical product needs to be more critically studied.
- The solid waste disposal issue should be addressed by efficient conversion of low-cost solid industrial and agricultural waste into revenue generating value-added products.
- Identification of new producer organisms and the genetic manipulation of existing known producers should also be promoted for continuing research. Moreover, for the application of Biosurfactants in environmentally feasible technologies, biosurfactant production can be carried out using cheap carbon sources (Danyelle et al. 2016).
- Cost effective bio production of bio surfactants can be carried out by using Feedstock like biomass derived from cereals, Oilseeds among others. Besides, insects have been investigated as a potential feedstock for biosurfactants. Moreover, Algae is being considered for microbial synthesis of biosurfactants (Dublin 2020).

In context to these prospects and strategies for cost effective, ecofriendly biosurfactant production and promising diverse applications in various fields signifies the potential scope of bio surfactants in the global market.

4. Conclusion

Although, surfactants are extensively used in numerous applications, they show various adverse impacts on the environment and society health . On the other hand, biosurfactants exhibit numerous benefits over synthetic surfactants such as high biodegradability, low toxicity, low irritancy and compatibility with human skin and ecological acceptance. These biomolecules synthesized by microorganisms like bacteria, fungi and yeast possess identical characteristics to reduce the surface and interfacial tensions by similar mechanisms of synthetic surfactants. They exhibit numerous benefits over synthetic surfactants like high biodegradability, low toxicity, low irritancy and compatibility with human skin and ecological acceptance. They tends to show some special properties like selectivity, specific activity at extreme temperatures, pH and salinity. In spite of extensive research in respect of cost effective production of biosurfactants, there is still an economic challenge with respect to their commercial success in comparison with their synthetic counterparts. In such circumstances these potent biomolecules can come up as 'Green and economically viable safe products' as per increasing demand of consumers by implementing some promising prospects and strategies for economizing their sustainable production as well as promoting their diverse applications for remarkable growth in the global market.

Acknowledgements

The authors are grateful to Research and Development Department, Sigma Wineries Pvt. Ltd. Nashik. Special thanks to Mr. Yash Khandelwal, Mr. V.C. Handore and Mrs. Hira V. Handore, Mr. P. Padekar, Mr. Saurabh. S. Sonar, Mr. Abhijeet Jagtap, and Ms. Mrunal Ghayal for valuable technical assistance.

References

Akbari, S., N.A. Hamid, R.Y. Mohd, F. Fayaz, O.A. Ruth et al. 2018. Biosurfactants a new frontier for social and environmental safety: A mini review. Biotechnology Research and Innovation 2: 81–90.

Appanna, V.D., H. Finn and M. Pierre. 1995. St. Exocellular phosphatidylethanolamine production and multiple-metal tolerance in *Pseudomonas fluorescens*. FEMS Microbiol. 131: 53–56.

Arima, K., A. Kakinuma and G. Tamura. 1968. Surfactin, a cristalline peptid lipid surfactant produced by *Bacillus subtilis*: Isolation, characterization and its inhibition of fibrin clot formation. Biochem. Biophys. Res. Commun. 31: 488–494.

Assadi, M. and M.S. Tabatabaee. 2010. Biosurfactants and their use in upgrading petroleum vacuum distillation residue: A review. Int. J. Environ. Res. 4: 549–572.

Awashti, N., A. Kumar, R. Makkar and S. Cameotra. 1999. Enhanced biodegradation of endosulfan, a chlorinated pesticide in presence of a biosurfactant. J. Environ. Sci. Heal. B. 34: 793–803.

Azarmi Reyhaneh and Ali Ashjaran. 2015. Type and application of some common surfactants. J. Chem. Pharm. Res. 7(2): 632–640.

Bai, G., M.L. Brusseau and R.M. Miller. 1997. Biosurfactant-enhanced removal of residual hydrocarbon from soil. J. Contam. Hydrol. 25: 157–170.

Balan, S.S., G.C. Kumar and S. Jayalakshmi. 2019. Physicochemical, structural and biological evaluation of Cybersan (trigalactomargarate), a new glycolipid biosurfactant produced by a marine yeast, Cyberlindnerasaturnus strain SBPN-27. Process Biochem. 80: 171–180.

Banat, I.M., A. Franzetti, I. Gandolfi, G. Bestetti, M.G. Martinotti et al. 2010. Microbial biosurfactants production, applications and future potential. Appl. Microbiol. Biotechnol. 87: 427–444.

Baviere, M., D. Degouy and J. Lecourtier. 1994. Process for washing solid particles comprising a sophoroside solution. U.S. Patent 5: 326–407.

Bernheimer, A.W. and L.S. Avigard. 1970. Nature and properties of a cytolytic agent produced by *B. subtilis*. J. Gen. Microbiol. 61: 361–369.

Berti, A.D., N.J. Greve, Q.H. Christensen and M.G. Thomas. 2007. Identification of a biosynthetic gene cluster and the six associated lipopeptides involved in swarming motility of Pseudomonas syringaepv. tomato DC3000. J. Bacteriol. 189: 6312–6323.

Cameotra, S.S. and R.S. Makkar. 2004. Recent applications of biosurfactants as biological and immunological molecules. Current Opinion in Microbiology 7(3): 262–266.

Cameron, D.R., D.G. Cooper and R.J. Neufeld. 1988. The mannoprotein of *saccharomyces cerevisiae* is an effective bioemulsifier. Appl. Environ. Microbiol. 54: 1420–1425.

Campos, J.M., T.L.M. Stamford and L.A. Sarubbo. 2019. Characterization and application of a biosurfactant isolated from *Candida utilis* in salad dressings. Biodegradation 30: 313–324. Doi: 10.1007/s10532-019-09877-8.

Chiewpattanaku, P., P. Sirinet, D. Alain, M. Emmanuelle, S. Benjamas et al. 2010. Bioproduction and anticancer activity of biosurfactant produced by the dematiaceous fungus *Exophiala dermatitidis* SK80. J. Microbiol. Biotechnol. 20: 1664–1671.

Chirwa, E.M.N., T. Mampholo and O. Fayemiwo. 2013. Biosurfactants as demulsifying agents for oil recovery from oily sludge–performance evaluation. Water Sci. Technol. 67: 2875–2881.

Cirigliano, M.C. and G.M. Carman. 1984. Isolation of a bioemulsifier from *Candida lipolytica*. Applied Environmental Microbiology 48(4): 747–750.

Dagbert, C., T. Meylheuc and M.N. Bellon-Fontaine. 2006. Corrosion behavior of AISI 304 stainless steel in presence of biosurfactant produced by *Pseudomonas fluorescens*. Electrochem. Acta. 51: 5221–5227.

Danka Galabova, Anna Sotirova, Elena Karpenko and Oleksandr Karpenko. 2014. The Role of Colloidal Systems in Environmental Protection, 41–83.

Das, K. and A.K. Mukherjee. 2007. Crude petroleum-oil biodegradation efficiency of *Bacillus subtilis* and *Pseudomonas aeruginosa* strains isolated from petroleum oil contaminated soil from North-East India. Bioresource Technol. 98: 1339–1345.

Das, P., S. Mukherjee and R. Sen. 2008. Antimicrobial potential of alipopeptide biosurfactant derived from a marine Bacillus circulans. J. Appl. Microbiol. 104: 1675–1684.

de Cássia, F.S., R. Silva, D.G. Almeida, R.D. Rufino, J.M. Luna, V.A. Santos and L.A. Sarubbo. 2014. Applications of biosurfactants in the petroleum industry and the remediation of oil spills. Int. J. Mol. Sci. 15: 12523–12542.

De Freitas Ferreira, J., E.A. Vieira and M. Nitschke. 2018. The antibacterial activity of rhamnolipid biosurfactant is pH dependent. Food Res. Int. 116: 737–744.

De Rienzo, M.A., I.M. Banat, B. Dolman, J. Winterburn, P.J. Martin et al. 2015. Sophorolipid biosurfactants: Possible uses as antibacterial and antibiofilm agent. New Biotechnol. 32: 720–726.

Desai, J.D. and I.M. Banat. 1997. Microbial production of surfactants and their commercial potential. Microbiol. Mol. Biol. Rev. 61: 47–64.

Dhara, P., Sachdev, Swaranjit and S. Cameotra. 2013. Biosurfactants in agriculture. Appl. Microbiol. Biotechnol. 97: 1005–1016.

Dublin, Jan. 15, 2020 /PRNewswire/–The "Disruptive Innovations in Biosurfactants". https://www.researchandmarkets.com/r/nir1hv.

Dusane, D., P. Rahman, S. Zinjarde, V. Venugopalan, R. McLean et al. 2010. Quorum sensing; implication on rhamnolipid biosur-factant production. Biotech. Genetic Eng. Rev. 27: 159–184.

Fakruddin, Md. 2012. Biosurfactant: Production and application. J. Pet. Environ. Biotechnol. 3(4): 1–5.

Franzetti, A., I. Gandolfi, G. Bestetti, T.J. Smyth, I.M. Banat et al. 2010. Production and applications of trehalose lipid biosurfactants. Eur. J. Lipid. Sci. Tech. 112: 617–627.

Gaur, V.K., R.K. Regar, N. Dhiman, K. Gautam, J.K. Srivastava et al. 2019. Biosynthesis and characterization of sophorolipid biosurfactant by Candida spp.: application as food emulsifier and antibacterial agent. Bioresour. Technol. 285: 121314.

Gerson, O.F. and J.E. Zajic. 1978. Surfactant production from hydrocarbons by Corynebacterium lepus, sp. nov. and Pseudomonas asphaltenicus, sp. nov. Dev. Ind. Microbiol. 178(19): 577–599.

Gharaei-Fathabad, E. 2011. Biosurfactants in pharmaceutical industry: A Mini-Review. American Journal of Drug Discovering and Development 1: 58–69.

Grażyna, P. and A. Varenyam. 2020. Bio surfactants: Eco-Friendly and innovative biocides against bio corrosion. Int. J. Mol. Sci. 21: 2152.

Herman, D.C., J.F. Artiola and R.M. Miller. 1995. Removal of cadmium, lead, and zinc from soil by a rhamnolipid biosurfactant. Environ. Sci. Technol. 29: 2280–2285.

Hong, J.J., S.M. Yang, C.H. Lee, Y.K. Choi, T. Kajiuchi et al. 1998. Ultrafiltration of divalent metal cations from aqueous solution using polycarboxylic acid type biosurfactants. J. Colloid Interf. Sci. 202: 63–73.

Huang, X., Z. Lu, H. Zhao, X. Bie, F.X. Lü et al. 2006. Antiviralactivity of antimicrobial lipopeptide from *Bacillus subtilis* fmbj against pseudorabies virus, porcine parvovirus, new castle disease virus and infectious bursal disease virus *in vitro*. Int. J. Pept. Res. Ther. 12: 373–377.

Hultberg, M., K.J. Bergstrand, S. Khalil and B. Alsanius. 2008. Characterization of biosurfactant-producing strains of fluorescent pseudomonads in a soilless cultivation system. Antonie Van Leeuwenhoek 94(2): 329–334.

Ishigami, Y., S. Yamazaki and Y. Gama. 1983. Surface active properties of biosoap from spiculisporic acid. J. Colloid Interf. Sci. 94: 131–139.

Ishigami, Y., Y. Zhang and F. Ji. 2000. Spiculisporic acid. Functional development of biosurfactants. Chim. Oggi. 18: 32–34.

Isoda, H., H. Shinmoto, M. Matsumura and T. Nakahara. 1999. The neurite initiating effect of microbial extracellular glycolipids in PC12 cells. Cytotechnology 31: 163–170.

Itoh, S., H. Honda, F. Tomita and T. Suzuki. 1971. Rhamnolipid produced by Pseuidotmon as aeruginiosa grown on n-paraffin. J. Antibiot. 24: 855–859.

Jahan, R., A.M. Bodratti, M. Tsianou and P. Alexandridis. 2020. Biosurfactants, natural alternatives to synthetic surfactants: physicochemical properties and applications. Adv. Colloid. Int. Sci. 275: 1–22.

Jamshidi-Aidji, M., I. Dimkiæ, P. Ristivojeviæ, S. Stankoviæ, G.E. Morlock et al. 2019. Effect-directed screening of Bacillus lipopeptide extracts via hyphenated high-performance thin-layer chromatography. J. Chromatogr. A 1605: 460366.

Jazzar, C. and E.A. Hammad. 2003. The efficacy of enhanced aqueous extracts of melia azedarach leaves and fruits integrated with the Camptotylusreuteri releases against the sweet potato whitefly nymphs. Bull Insectol. 56: 269–275.

Jennema, G.E., M.J. McInerney, R.M. Knapp, J.B. Clark, J.M. Feero et al. 1983. A halotolerant, biosurfactants-producing Bacillus species potentially useful for enhanced oil recovery. Dev. Ind. Microbiol. 24: 485–492.

Jenny, K., O. Kappeli and A. Fiechter. 1991. Biosurfactants from Bacillus licheniformis, structural analysis and characterization. Appl. Microbiol. Biotechnol. 36: 5–13.

Kameda, Y., S. Oira, K. Matsui, S. Kanatomo, T. Hase et al. 1974. Antitumor activity of Bacillus natto. V. Isolation and characterization of surfactin in the culture medium of Bacillus natto KMD 2311. Chem. Pharm. Bull. 22: 938–944.

Kassab, D.M. and T.M. Roane. 2006. Differential responses of a mine tailings Pseudomonas isolate to cadmium and lead exposures. Biodegradation 17: 379–387.

Kearns, D.B. and R. Losick. 2003. Swarming motility in undomesticated Bacillus subtilis. Mol. Microbiol. 49: 581–590.

Khanna, S. and P. Pattnaik. 2019. Production and functional characterization of food compatible biosurfactants. Appl. Food Sci. J. 3: 1–4.

Kim, P.I., H. Bai, D. Bai, H. Chae, S. Chung et al. 2004. Purification and characterization of a lipopeptide produced by Bacillus thuringiensis CMB26. J. Appl. Microbiol. 97: 942–949.

Kim, S.K., Y.C. Kim, S. Lee, J.C. Kim, M.Y. Yun et al. 2011. Insecticidal activity of rhamnolipid isolated from Pseudomonas sp. EP-3 against green peach aphid (Myzuspersicae). J. Agric. Food Chem. 59: 934–938.

Kiran, G.S., S. Priyadharsini, A. Sajayan, G.B. Priyadharsini, N. Poulose et al. 2017. Production of lipopeptide biosurfactant by a marine Nesterenkonia sp. and its application in food industry. Front. Microbiol. 8: 1138.

Kitamoto, D., H. Yanagishita, T. Shinbo, T. Nakane, C. Kamisawa et al. 1993. Surface active properties and antimicrobial activities of mannosylerythritol lipids as biosurfactants produced by *Candida antarctica*. Journal of Biotechnology 29: 91–96.

Kitamoto, D., H. Isoda and T. Nakahara. 2002. Functions and potential applications of glycolipid biosurfactants—from energy-saving materials to gene delivery carriers. J. Biosci. Bioeng. 94: 187–201.

Krishnaswamy, M., G. Subbuchettiar, T.K. Ravi and S. Panchaksharam. 2008. Biosurfactants properties, commercial production and application. Current Science 94: 736–747.

Krzyzanowska, D.M., M. Potrykus, M. Golanowska, K. Polonis, A. Gwizdek-Wisniewska et al. 2012. Rhizosphere bacteria as potential biocontrol agents against soft rot caused by various Pectobacterium and Dickeya spp. strains. J. Plant Pathol. 94(2): 367–378.

Leenhouts, J.M., P.W.J. Van der Winingard, A.I.P.M. DeKroon and B. DerKruij. 1995. Anionic phospholipids can mediate membrane insertion of the anionic part of a bound peptide. FEBS Lett. 370: 361–369.

Luna, J.M., R.D. Rufino, G.M. Campos-Takakia and L.A. Sarubbo. 2012. Properties of the biosurfactant produced by Candida Sphaerica cultivated in low-cost substrates. Chem. Eng. Trans. 27: 67–72.

Maier, R.M. and G. Soberón-Chávez. 2000. Pseudomonas aeruginosa rhamnolipids: biosynthesis and potential applications. Appl. Microbiol. Biotechnol. 54: 625–633.

Marcia, N. and Sumária Sousa e Silva. 2018. Recent food applications of microbial surfactants. Critical Reviews in Food Science and Nutrition 58(4): 631–638.

Mireles, J.R., A. Toguchi and R.M. Harshey. 2001. Salmonella enterica serovar typhimurium swarming mutants with altered biofilm forming abilities: Surfactin inhibits biofilm formation. J. Bacteriol. 183: 5848.

Mukherjee, A.K. and K. Das. 2005. FEMS Microbiology Ecology. 54: 479–89.

Mukherjee, A.K. 2007. Potential application of cyclic lipopeptide biosurfactants produced by Bacillus subtilis strains in laundry detergent formulations. Lett. Appl. Microbiol. 45: 330–335.

Mulligan, C.N. and S. Wang. 2004. Remediation of a heavy metal contaminated soil by a rhamnolipid foam. pp. 544–551. *In*: Thomas, H.R. and R.N. Yangt (eds.). Geoenvironmental Engineering Integrated Management of Groundwater and Contaminated Land. London, Thomas Telford.

Mulligan, C.N., R.N. Yong and B.F. Gibbs. 2001. Heavy metal removal from sediments by biosurfactants. J. Hazard. Mater. 85: 111–125.

Mulligan, C.N., R.N. Yong and B.F. Gibbs. 2001. Remediation technologies for metal-contaminated soils and groundwater: An evaluation. Engineering Geology 60(1-4): 193–207.

Muthusamy, K., S. Gopalakrishnan, T.K. Ravi and P. Sivachidambaram. 2008. Biosurfactants: Properties, commercial production and application. Curr. Sci. 94: 736–747.

Nakama, Y. 2017. Cosmetic Science and Technology Surfactants, 231–244. Doi: 10.1016/B978-0-12-802005-0.00015-X.

Nakanishi, M., Y. Inoh, D. Kitamoto and T. Furuno. 2009. Nano vectors with a biosurfactant for gene transfection and drug delivery. J. Drug Delivery Sci. Technol. 19: 165–169.

Naughton, P.J., R. Marchant, V. Naughton and I.M. Banat. 2018. Microbial bio surfactants: Current trends and applications in agricultural and biomedical industries. Journal of Applied Microbiology 127: 12–28.

Neu, T.R., T. Hartner and K. Poralla. 1990. Surface active properties of viscosin, a peptidolipid antibiotic. Appl. Microbiol. Biotechnol. 32: 518–520.

Nihorimbere, V., M. Marc Ongena, M. Smargiassi and P. Thonart. 2011. Beneficial effect of the rhizosphere microbial community for plant growth and health. Biotechnol. Agron. Soc. Environ. 15: 327–337.

Nitschke, M. and S.S.E. Silva. 2018. Recent food applications of microbial surfactants. Food Sci. Nut. 58: 631–638.

Pacwa-Płociniczak, M., G. APłaza, Z. Piotrowska-Seget and S. Cameotra. 2011. Environmental applications of biosurfactants: Recent advances. Int. J. Mol. Sci. 12: 633–654.

Parthipan, P., D. Sabarinathan, S. Angaiah and A. Rajasekar. 2018. Glycolipid biosurfactant as an eco-friendlly microbial inhibitor for the corrosion of carbon steel in vulnerable corrosive bacterial strains. J. Mol. Lipids 261: 473–479.

Pesce, L. 2002. A biotechnological method for the regeneration of hydrocarbons from dregs and muds, on the base of biosurfactants. World Patent 02/062,495.

Pratt-Terpstar, I.H., A.H. Weerkamp and H.J. Busscher. 1989. Microbial factors in a thermodynamic approach of oral Streptococci adhesion to solid substrata. J. Colloid Interface Sci. 129: 568–574.

Raval, M.S., S.S. Gund, N.R. Shah and R.P. Desai. 2017. Isolation and characterization of biosurfactant producing bacteria and their application as an antibacterial agent. Int. J. Pharma Bio. Sci. 8: 302–310.

Reddy, A.S., C.Y. Chen, C.C. Chen, J.S. Jean et al. 2009. Synthesis of gold nanoparticles via an environmentally benign route using a biosurfactant. J. Nanosci. Nanotechnol. 9: 6693–6699.

Remichkova, M., D. Galabova, I. Roeva, E. Karpenko, A. Shulga et al. 2008. Anti-herpesvirus activities of Pseudomonas sp. S-17 rhamnolipid and its complex with alginate. Z Naturforsch C 63: 75–81.

Ribeiro, B.G., J.M.C. Guerra and L.A. Sarubbo. 2020. Potential food application of a biosurfactant produced by Saccharomyces cerevisiae URM 6670. Front. Bioeng. Biotechnol. 8: 434.

Rivardo, F., R.J. Turner, G. Allegrone, H. Ceri, M.G. Martinotti et al. 2009. Anti-adhesion activity of two biosurfactants produced by Bacillus spp. prevents biofilm formation of human bacterial pathogens. Appl. Microbiol. Biotechnol. 83: 541–553.

Rodrigues, L., H. van der Mei, J.A. Teixeira and R. Oliveira. 2004. Biosurfactant from Lactococcus lactis 53 inhibits microbial adhesion on silicone rubber. Appl. Microbiol. Biotechnol. 66(3): 306–311.

Rodrigues, L., H. van der Mei, I.M. Banat, J. Teixeira, R. Oliveira et al. 2006. Inhibition of microbial adhesion to silicone rubber treated with biosurfactant from Streptococcus thermophilus A. FEMS Immunol. Med. Microbiol. 46(1): 107–112.

Rodríguez-López, L., M. Rincón-Fontán, X. Vecino, J.M. Cruz, A.B. Moldes et al. 2019. Preservative and irritant capacity of biosurfactants from different sources: A comparative study. J. Pharm. Sci. 108: 2296–2304.

Ron, E.Z. and E. Rosenberg. 2001. Natural roles of biosurfactants. Environ. Microbiol. Apr. 3(4): 229–36. Doi: 10.1046/j.1462-2920.2001.00190.x. PMID: 11359508.

Rosenberg, E., C. Rubinovitz, R. Legmann and E.Z. Ron. 1988. Purification and chemical properties of acinetobactercalcoaceticus A2 biodispersan. Appl. Environ. Microbiol. 54: 323–326.

Sandrin, C., F. Peypoux and G. Michel. 1990. Coproduction of surfactin and iturin A lipopeptides with surfactant and antifungal properties by *Bacillus subtilis*. Biotechnol. Appl. Biochem. 12: 370–375.

Santos, D.K., R.D. Rufino, J.M. Luna, V.A. Santos and L.A. Sarubbo. 2016. Biosurfactants: Multifunctional biomolecules of the 21st century. Int. J. Mol. Sci. 17(401): 1–31.

Sarwar, A., G. Brader, E. Corretto, G. Aleti, M. Abaidullah et al. 2018. Qualitative analysis of biosurfactants from Bacillus species exhibiting antifungal activity. PLoS ONE 13(6): e0198107.

Schaller, K.D., S.L. Fox, D.F. Bruhn, K.S. Noah, G.A. Bala et al. 2004. Characterization of surfactin from *Bacillus subtilis* for application as an agent for enhanced oil recovery. Appl. Biochem. Biotechno. 115: 827–836.

Sen, R. 2008. Biotechnology in petroleum recovery: The microbial EOR. Prog. Energy Combust. Sci. 34: 714–724.

Sha, R., L. Jiang, Q. Meng, G. Zhang, Z. Song et al. 2011. Producing cell-free culture broth of rhamnolipids as a cost-effective fungicide against plant pathogens. J. Basic Microbiol. 52: 458–466.

Sharma, R., J. Singh and N. Verma. 2018. Production, characterization and environmental applications of biosurfactants from *Bacillus amyloliquefaciens* and *Bacillus subtilis*. Biocatal. Agric. Biotechnol. 16: 132–139.

Sheppard, J.D., C. Jumarie, D.G. Cooper and R. Laprade. 1991. Ionic channels induced by surfactin in planar lipid bilayer membranes. Biochim. Biophys. Acta 1064: 13–23.

Sifour, M., M.H. Al-Jilawi and G.M. Aziz. 2007. Emulsification properties of biosurfactant produced from *Pseudomonas aeruginosa* RB 28. Pak. J. Biol. Sci. 10: 1331–1335.

Singh, A., J.D. Van Hamme and O.P. Ward. 2007. Surfactants in microbiology and biotechnology: Part 2: Application aspects. Biotechnol. Adv. 25: 99–121.

Singh, P. and Y. Patil. 2018. Biosurfactant production: emerging trends and promising strategies. Journal of Applied Microbiology 126(2): 1–13.

Singh, S., S. Hyun Kang, A. Mulchandani and W. Chen. 2008. Bioremediation: environmental clean-up through pathway Engineering. Current Opinion in Biotechnology 19: 437–444.

Tariku, T.E., H.J. Yong, K. Maryam and S.H. Yeon. 2018. Biosurfactants: Production and potential application in insect pest management. Trends in Entomology 14: 79–87.

Tayler, R.T., R.T. Damn, J. Miller, K. Spratt, J. Schilling et al. 1985. Isolation and characterization of the human pulmonary surfactant apoprotein gene. Nature 317: 361–365.

Thomas, C.P., M.L. Duvall, E.P. Robertson, K.B. Barrett, G.A. Bala et al. 1993. Surfactant-based EOR mediated by naturally occurring microorganisms 11: 285–291.

Toren, A., S. Navon-Venezia, E.Z. Ron and E. Rosenberg. 2001. Emulsifying activities of purified alasan proteins from acine- to bacterradioresistens KA53. Appl. Environ. Microbiol. 67: 1102–1106.

Tugrul, T. and E. Cansunar. 2005. Detecting surfactant-producing microorganisms by the drop-collapse test. World Journal of Microbiology and Biotechnology 21(6): 851–853.

Tuleva, B.K., G.R. Ivanov and N.E. Christova. 2002. Biosurfactant production by a new Pseudomonas putida strain. Z. Naturforsch C 57(3-4): 356–360.

Ueno, Y., N. Hirashima, Y. Inoh, T. Furuno, M. Nakanishi et al. 2007. Characterization of biosurfactant-containing liposomes and their efficiency for gene transfection. Biol. Pharm. Bull. 30: 169–172.

Valle, J., S. DaRe, N. Henry, T. Fontaine, D. Balestrino et al. 2006. Broad-spectrum biofilm inhibition by a secreted bacterial polysaccharide. Proc. Natl. Acad. Sci. USA 103: 12558–12563.

Van Hamme, J.D., A. Singh and O.P. Ward. 2006. Physiological aspects. Part 1 in a series of papers devoted to surfactants in microbiology and biotechnology. Biotechnol. Adv. 24: 604–620.

Vatsa, P., L. Sanchez, C. Clement, F. Baillieul, S. Dorey et al. 2010. Rhamnolipid biosurfactants as new players in animal and plant defense against microbes. Int. J. Mol. Sci. 11: 5095–5108.

Velraeds-Martine, M.C., H.C. Vander Mei, G. Reid and H.J. Busscher. 1997. Inhibition of initial adhesion of uropathogenic Enterococcus faecalis to solid substrate by an adsorbed biosurfactant layer from Lactobacillus acidophilus. Urology 49: 790–794.

Whang, L.M., P.W.G. Liu, C.C. Ma and S.S. Cheng. 2008. Application of biosurfactant, rhamnolipid, and surfactin, for enhanced biodegradation of diesel-contaminated water and soil. J. Hazard. Mater. 151: 155–163.

Zhang, Y., H. Li, J. Sun, J. Gao, W. Liu et al. 2010. Dc-chol/Dope cationic liposomes: A comparative study of the influence factors on plasmid pDNA and Si RNA gene delivery. Int. J. Pharm. 390: 198–207.

Zin, I.M., V.I. Pokhmurskii and S.A. Korniy. 2018. Corrosion inhibition of aluminium alloy by rhamnolipid biosurfactant derived from Pseudomonas sp. PS-17. Anti-Corros. Methods Mater. 65: 517–527.

Zosim, Z., D.L. Gutnick and E. Rosenberg. 1982. Properties of hydrocarbon-in-water emulsions stabilized by Acinetobacter RAG-1 emulsan. Biotechnol. Bioeng. 192(24): 281–292.

Zouari, R., D. Moalla-Rekik, Z. Sahnoun, T. Rebai, S. Ellouze-Chaabouni et al. 2016. Evaluation of dermal wound healing and in vitro antioxidant efficiency of *Bacillus subtilis* SPB1 biosurfactant. Biomed. Pharmacother. 84: 878–891.

6

Biosurfactants
The Ecofriendly Biomolecules of the Upcoming Era

Anita V Handore,[1,*] *Sharad R Khandelwal*,[2] *Rajib Karmakar*,[3] *Divya L Gupta*[4] and *Dilip V Handore*[1]

1. Introduction

1.1 Global Scenario of the Biosurfactant Market

These days the biosurfactant market has a noteworthy growth potential predominantly due to the increasing environmental concerns. The advancement in biotechnology and emergence of more rigorous laws have led biosurfactants to be considered as promising alternatives to chemical surfactants available in the market. It is reported that the global bio surfactant market was 3,44,068.40 tons in 2013 and expected to be 4,61,991.67 tons by 2020, growing at a CAGR of 4.3% from 2014 to 2020 (Grand View Research 2016). In 2016, market revenue generation was over $1.8 Billion and is expected to reach USD 2.6 Billion by 2023. Besides, the research projected the global bio surfactant market at over 5.52 billion by 2022, at a CAGR of 5.6% from 2017 to 2022 (https://www.marketsandmarkets.com/PressReleases/biosurfactant.asp). Europe is emerging and is expected to continue to grow as the biggest market (around 53%) followed by the United States mainly due to stricter regulatory guidelines in the region. However, increasing awareness and infrastructure in Asian countries is making them emerging consumers of bio surfactants (Singh and Patil 2018).

[1] Research and Development Department, Sigma Wineries Pvt. Ltd. Nashik, Maharashtra, India - 422112.
[2] H.A.L. College of Science & Commerce, Nashik, Maharashtra, India 422207.
[3] Department of Agricultural Chemicals, Bidhan Chandra Krishi Viswavidyalaya, Directorate of Research, Research Complex Building, Nadia, W. Bengal, India - 741252.
[4] Amneal Pharmaceuticals 1 New England Ave, Piscataway NJ 08854, USA.
* Corresponding author: avhandore@gmail.com

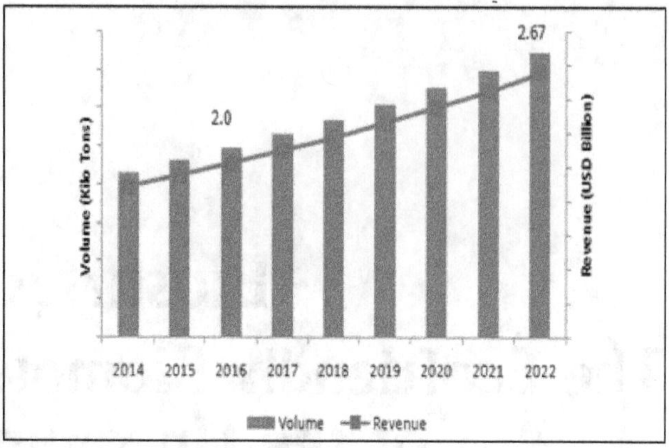

Figure 1. Biosurfactant Market 2016–2022 (USD Billion) (Kilo Tons) (Source: Zion Market Research 2017, https://www.zionmarketresearch.com/news/biosurfactants-market).

1.2 Overview of Biosurfactants

Since the last few decades, exploration of natural and renewable resources have been proven as a promising source for manufacturing natural products with high aggregate value in the world market. In this era, Bio surfactants are coming up as leading ecofriendly products with increasing demand. Besides exerting a strong positive impact on global problems, they exhibits significant potential with regard to implantation of sustainable industrial processes, like the use of renewable resources with a specific identity categorized as a "green" product. Bio surfactants are diverse groups of surface active molecules/chemical compounds synthesized by some microorganisms (Desai and Banat 1997).

Basically, microorganisms show the ability to produce bio surfactants for altering the surface or interfacial properties of cells or the local environment as a change in surface or interfacial tension is needed for erection of fruiting bodies, swarming of cells, gliding motility, and biofilm formation and development. These amphiphilic compounds are produced on living surfaces, mostly on the microbial cell surfaces, or excreted extracellularly. These biomolecules having both hydrophilic and hydrophobic domains confer the capability to accumulate between fluid phases and reduce surface and interfacial tensions (Karanth et al. 1999).

Due to their amphiphilic nature, these molecules have hydrophilic and lipophilic moieties which allow them to partition at water/air, oil/air, or oil/water interfaces. Due to such characteristics they can play a key role in emulsification, foam formation, detergency and dispersal, which are desirable qualities in various sectors. Most bio surfactants are either anionic or neutral and the hydrophilic moiety can be a carbohydrate, an amino acid, a phosphate group, or some other compounds. The hydrophobic moiety is mostly a long carbon chain fatty acid.

2. Significance of Biosurfactants to Microorganisms

Bio surfactants play fundamental and significant roles towards microorganisms. Their importance will help to understand the physiology of microbes to provide a logical framework for the finding a number of bio surfactants. Some natural roles of bio surfactants are as follow:

2.1 *Adhesion*

It is a physiological mechanism for growth and survival of cells in natural environments (Rosenberg 1986). Generally, microorganisms use their bio surfactants for regulating their cell surface properties for attaching or detaching from surfaces as per requirement. A special case of adhesion is the growth of bacteria on water insoluble hydrocarbons and is one of the primary processes affecting bacterial transport, which determines the bacterial fate in the subsurface.

2.2 *Emulsification*

It is found that various hydrocarbon-degrading microbes produce extracellular emulsifying agents, the inference being that emulsification plays a role in growth on water immiscible substrates (Rosenberg et al. 1993, 1997, 1999). For the growth of microbes on hydrocarbons, the interfacial surface area between water and oil can be a limiting factor and emulsification is a natural indirect process brought about by extracellular agents and there are certain conceptual difficulties in understanding how emulsification can provide an advantage for the microorganism producing the emulsifier.

There is correlation between emulsifier production and growth on hydrocarbons. The majority of *Acinetobacter* strains produce high-molecular-mass bio emulsifiers. Generally, the emulsifier producing organisms are able to grow on water insoluble substrates while, the mutants, which are unable to produce emulsifiers show poor growth on hydrocarbons.

2.3 *Bioavailability and Desorption*

One of the major reasons for the prolonged persistence of high-molecular-weight hydrophobic compounds is their low water solubility, which increases their sorption to surfaces and limits their availability to biodegrading microorganisms. Bio surfactants can enhance growth on bound substrates by desorbing them from surfaces or by increasing their apparent water solubility (Deziel et al. 1996). Surfactants which lower interfacial tension are particularly effective in mobilizing bound hydrophobic molecules making them available for biodegradation. Another important characteristic of biosurfactants is that above the critical micelle concentration, they form micelles (stable aggregates of 10 to 200 molecules) which, bring about a sudden variation in the relation between the concentration and the surface tension of the solution that can increase the solubility of Hydrophobic Organic Compounds (HOCs) (Makkar et al. 2003, Edwards et al. 1991).

Desorption plays a significant role in the growth of microbes. After a certain period of growth, conditions become unfavorable for further development of microorganisms, e.g., due toxin accumulation and impaired transport of necessary nutrients in crowded conditions. Desorption is advantageous at this stage for the cells and a need arises for a new habitat. The mechanisms for detachment seem to be important for all attached microorganisms to facilitate dispersal and colonization of new surfaces. One of the natural roles of an emulsifier/bio surfactant may be in regulating desorption of the producing strain from hydrophobic surfaces (Jordan et al. 1999).

2.4 *Defense Strategy*

Biosurfactants could be an evolutionary defense strategy for microbes as evidenced by high mycocidal activity of the mycocidal complex secreted by *C. humicola.* Similar analogies can be made for the lipopeptide biosurfactant producing strains of *B. subtilis.* The lipopeptide might have a strong influence on the survival of *B. subtilis* in its natural habitat, soil and the rhizosphere (De Souza et al. 2003, Nielsen et al. 2003).

3. Merits of Biosurfactants

Bio surfactants have several advantages over synthetic surfactants, including high biodegradability, low toxicity, low irritancy and compatibility with human skin (Banat et al. 2000, Cameotra and Makkar 2004). Therefore they are superior to the synthetic ones. The most significant advantage of a microbial surfactant over a chemical surfactant is its ecological acceptance (Desai and Banat 1997, Karsa et al. 1999). Some more advantages of biosurfactants over synthetic ones include selectivity, specific activity at extreme temperatures, pH, salinity and more. Some of the advantages of biosurfactants are discussed below:

- **Biodegradability:** Biosurfactants are easily degraded by natural processes of microorganisms (Mohan 2006).
- **Low toxicity:** Biosurfactants do not exert any adverse impact on the biotic ecosystem due to their low toxicity (Desai and Banat 1997).
- **Biocompatibility and digestibility:** Biosurfactants are biocompatible in nature and tolerated by living organisms (Rosenberg and Ron 1999).
- **Physical factors:** Most of the biosurfactants are not influenced by environmental factors like temperature, pH and ionic strength tolerances (Krishnaswamy 2008).
- **Availability of raw materials:** Bio surfactants can be produced using cheap raw materials which are abundantly available. Moreover, hydrocarbons, carbohydrates and/or lipids, might be used as carbon sources separately or in combination with each other (Fracchia et al. 2014).
- **Acceptable production economics:** Bio surfactants can also be produced from industrial wastes and by products for bulk production as per application, which is economically acceptable (Kosaric 1992).
- **Use in environmental control:** Synthetic chemical surfactants impose environmental problems and hence, Bio surfactants can be efficiently used in handling industrial emulsions, control of oil spills, biodegradation and detoxification of industrial effluents and in bioremediation of contaminated soil (Kosaric 1992).
- **Specificity:** Due to complex organic molecular structures with specific functional groups, bio surfactants are normally found to be specific in their action and efficiently applied for detoxification of specific pollutants, de-emulsification of

industrial emulsions, specific cosmetic, pharmaceutical and food applications (Kosaric 1992).

- **Surface and interface activity:** It is reported that a good surfactant can lower the surface tension of water from 75 to 35 mN/m and the interfacial tension between water/hexadecane from 40 to 1 mN/M. Surfactin shows potential to reduce the surface tension of water up to 25 mN/M and interfacial tension of water/hexadecane to < 1 mN/M (Krishnaswamy 2008, Mulligan 2005).

4. Classification of Biosurfactants

Bio surfactants are mainly categorized on the basis of their chemical composition and microbial origin. In general, their structure includes a hydrophilic moiety consisting of amino acids or peptides anions or cations; mono-or di- polysaccharides; and a hydrophobic moiety consisting of unsaturated, saturated, or fatty acids. Accordingly, the major class of biosurfactants are as follow

4.1 *Glycolipids*

These are carbohydrates in combination with long-chain aliphatic acids or hydroxyaliphatic acids. Among the glycolipids, the well known examples are rhamnolipids, trehalolipids, and sophorolipids and mannosylerythritol lipids. Except these, some other types like glycoglycerolipids (Nakata et al. 2000), sugar-based bioemulsifiers (Kim et al. 2000, Van et al. 2000), mannosylerythritol lipid A and many different hexose lipids (Golyshin et al. 1999).

4.1.1 *Rhamnolipids*

In these types, one or two molecules of rhamnose are linked to one or two molecules of β-hydroxydecanoic acid. Whereas, the –OH group of one of the acids is involved in glycosidic linkage with the reducing end of the rhamnose disaccharide, the –OH group of the second acid is involved in ester formation (Karanth et al. 1999) (Fig. 2). Certain species of *P. aeruginosa* are found to produce large amounts of such biosurfactants. L-Rhamnosyl- L-rhamnosyl-b-hydroxyde- canoyl-b-hydroxydecanoate and L-rhamnosyl- b-hydroxydecanoyl-b-hydroxydecanoate, referred to as rhamnolipid 1 and 2, respectively, are the principal glycolipids produced by *P. aeruginosa* (Gautam

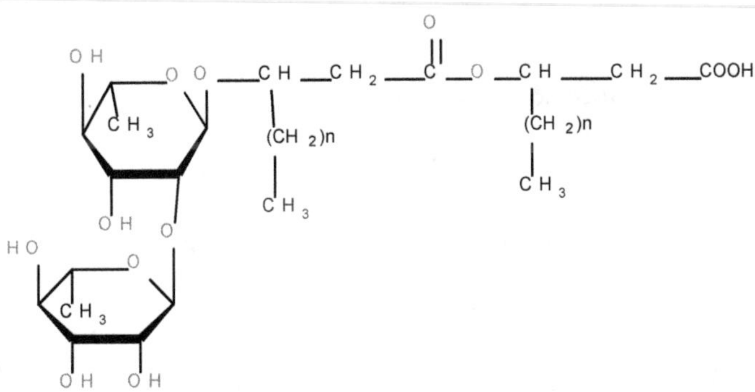

Figure 2. Structure of L-Rhamnosyl-L-rhamnosyl-b-hydroxydecanoly-b-hydroxydecanoate.

and Tyagi 2006). Due to excellent surface activity, physicochemical properties of Rhamnolipids have been significantly used (Pornsunthorntawee et al. 2009).

4.1.2 *Sophorolipids*

These are mixtures of at least six to nine different hydrophobic sophorosides. They have dimeric carbohydrate sophoroses attached to long chain hydroxy fatty acids (Fig. 3) and are mainly produced by yeasts such as *T. bombicola*, *T. apicola* (Tullock et al. 1967) and *W. domericqiae* (Chen et al. 2006). In *T. petrophilum*, and *T. apicola*, there are dimeric carbohydrate sophoroses linked to a long-chain hydroxy fatty acid. It is reported that *C. bogoriensis* produces glycolipids in which sophorose is found to be linked with docosanoic acid diacetate. Sophorolipids from *T. petrophilum* are produced on water-insoluble substrates such as alkanes and vegetable oils (Cooper and Paddock 1983).

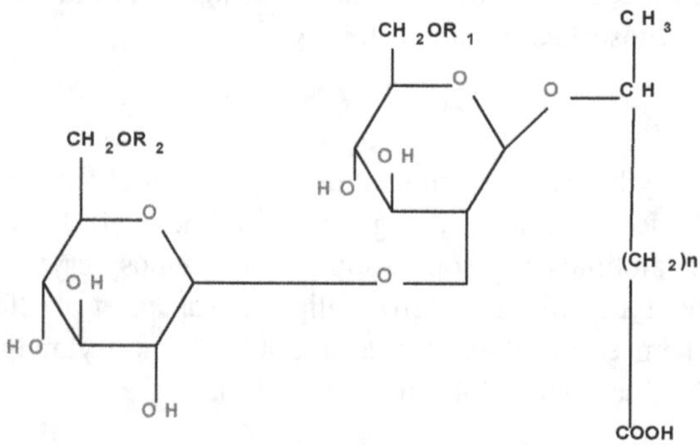

Figure 3. Structure of Sophorolipid.

4.1.3 *Trehalolipids*

Numerous structural types of microbial trehalolipid biosurfactants have been reported. It is found that Trehalolipids from different organisms vary with regard to size and structure of mycolic acid, number of carbon atoms, and degree of unsaturation (Cooper et al. 1989). It is found that Disaccharide trehalose linked at C-6 and C-6' to mycolic acids (Fig. 4) is accompanied with most species of

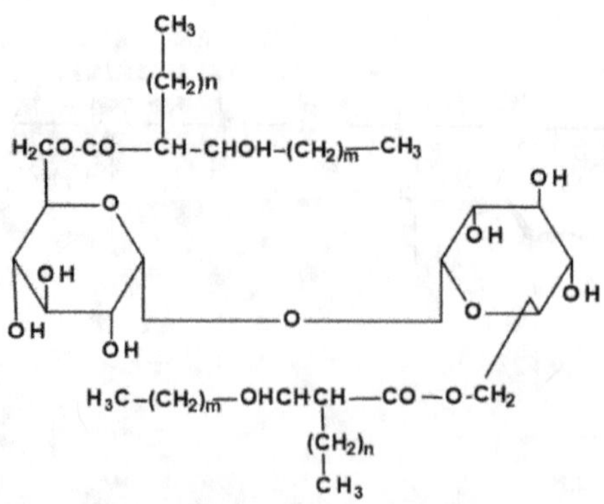

Figure 4. Structure of Trehalolipid.

Mycobacterium, *Nocardia*, and *Corynebacterium*. Mycolicacids are long chain, a-branched-b-hydroxy fatty acids. Trehalose dimycolate produced by *R. erythropolis* has been extensively studied. This organism can also synthesize a novel anionic trehalose lipid (Gautam and Tyagi 2006).

4.1.4 *Mannosylerythritol Lipids*

This biosurfactant consists of mannosylerythritol synthesized by yeasts like *Candida antarctica* (Crich et al. 2002, Kitamoto et al. 1993) and *Candida* sp. SY 16. The fatty acid component of the biosurfactant was determined to be hexanoic, dodecanoic, tetradecanoic or tetradecenoic acid (Kim et al. 1999) (Fig. 5). It is reported that Fukuoka et al. (2007) have characterized the surface active properties of a new glycolipid biosurfactant, mono acylated mannosylerythritol lipid produced by *P. antarctica* and *P. rugulosa.* These are generally produced by *Pseudozyma* sp. as a major component while *Ustilago* sp. produces them as a minor component. Recently they have gained attention due to their environmental compatibility, mild production conditions, structural diversity, self-assembling properties and versatile biochemical functions (Arutchelvi et al. 2008).

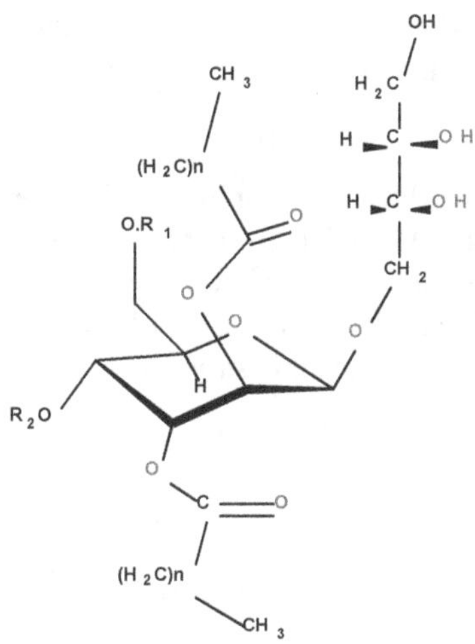

Figure 5. Structure of Mannosylerythritol lipids (Adapted from: Sajna et al. 2013).

4.2 *Lipopeptides and Lipoproteins*

Lipopeptides are amphiphilic molecules which incorporate one or more lipid chains attached to a peptide head group. This self-assembly is observed on the basis of hydrophile/lipophile balance of the molecules as well as interactions between the peptide units (Ian et al. 2015).

4.2.1 *Surfactin*

Surfactin, is one of the most effective biosurfactants exhibiting superior surface activity and belongs to a group of cyclic lipohepta peptides containing beta-hydroxyl fatty acids and D–/L-amino acid residues (Tang et al. 2007, Haddad et al. 2008). It is composed of a seven amino-acid ring structure coupled to a fatty-acid chain via a

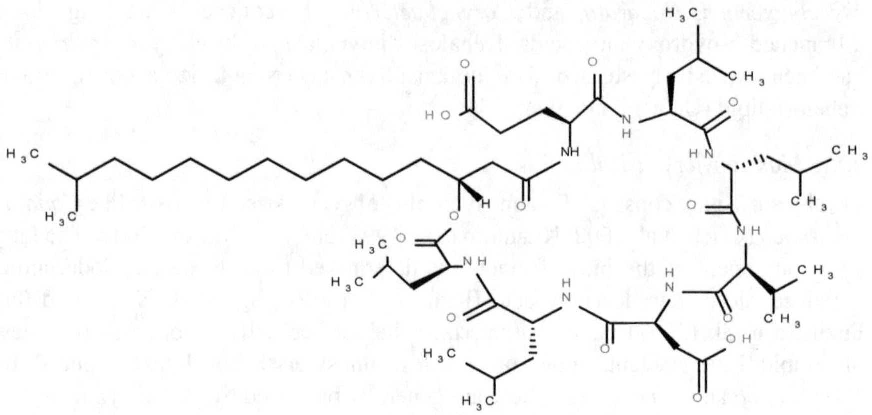

Figure 6. The structures of Surfactin (Adapted from, https://www.researchgate.net/publication/233879638_Classification_of_Bacillus_Beneficial_Substances_Related_to_Plants_Humans_and_Animals).

lactone linkage (Arima et al. 1968) (Fig. 6). It was first reported in *B. subtilis* ATCC-21332.28 (Peypoux et al. 1999). Surfactin can lower the surface tension from 72 to 27.9 mN/m30 and has a critical micelle concentration of 0.017 g/l (Sen et al. 2005).

4.2.2 *Iturin*

Iturins are a special class of pore forming lipopeptides which can be extracted from the culture media of various strains of *B. subtilis.* These amphiphilic compounds are characterized by a peptide ring of seven amino acid residues including an invariable D-Tyr2, with the constant chiral sequence LDDLLDL closed by a C14-C17 aliphatic beta-amino acid (Fig. 7). They exhibit strong antifungal activities against a wide

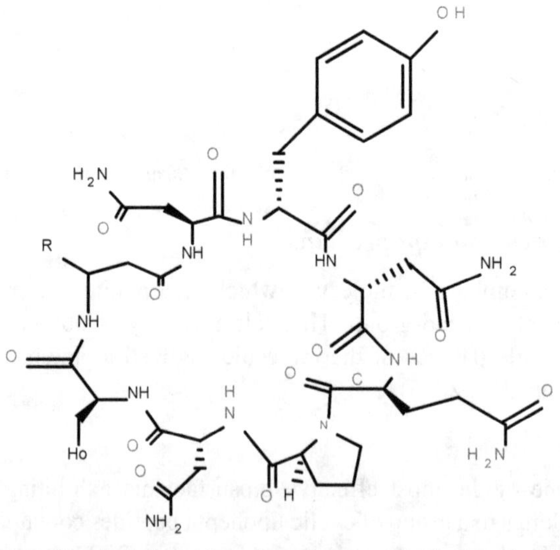

Figure 7. The structures of Iturin (Adapted from: https://www.researchgate.net/publication/233879638_Classification_of_Bacillus_Beneficial_Substances_Related_to_Plants_Humans_and_Animals).

variety of pathogenic yeasts and fungi but their antibacterial activities are restricted to some bacteria such as *M. luteus.* The biological activity of the iturin lipopeptides is modulated by the primary structure of the peptide cycle as illustrated by the methylation of the D-Tyr2 residue which dramatically decreases the activity by the inversion of the two adjacent Ser6-Asn7 residues which make mycosubtilin more active than iturin A.

The antifungal activity is related to the interaction of the iturin lipopeptides with the cytoplasmic membrane of target cells, the K+ permeability of which is greatly increased. The ability of Iturin compounds to increase the membrane cell permeability is due to the formation of ion-conducting pores, the characteristics of which depend both on the lipid composition of the membrane and on the structure of the peptide cycle (Maget-Dana et al. 1994).

4.2.3 Fengycin

Fengycin is a lipodecapeptide containing a β-hydroxy fatty acid in its side chain and comprises of C15 to C17 variants which have a characteristic Ala-Valdimorphy at position 6 in the peptide ring (Fig. 8). Fengycin is a novel antifungal lipopeptide antibiotic produced by *B. subtilis* f-29-3 (Vater et al. 2002, Schneider et al. 1999).

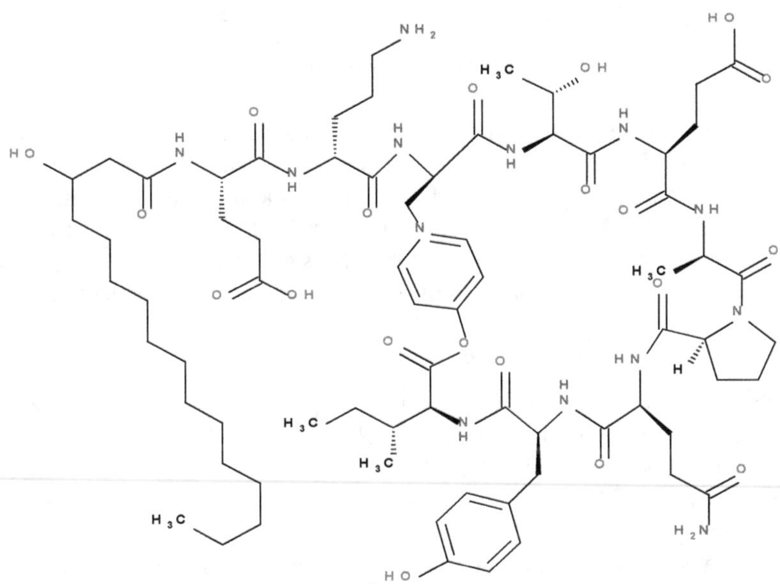

Figure 8. The structures of Fengycin (Adapted from: https://www.researchgate.net/publication/233879638_Classification_of_Bacillus_Beneficial_Substances_Related_to_Plants_Humans_and_Animals).

4.2.4 Lichenysin

Lichenysin exhibits structural similarities and physio-chemical properties like surfactin and produced by *B. licheniformis* (Fig. 9). This organism shows several other surface active agents which act synergistically and exhibit excellent temperature, pH and salt stability (McInerney et al. 1990). Lichenysin A produced by *B. licheniformis*

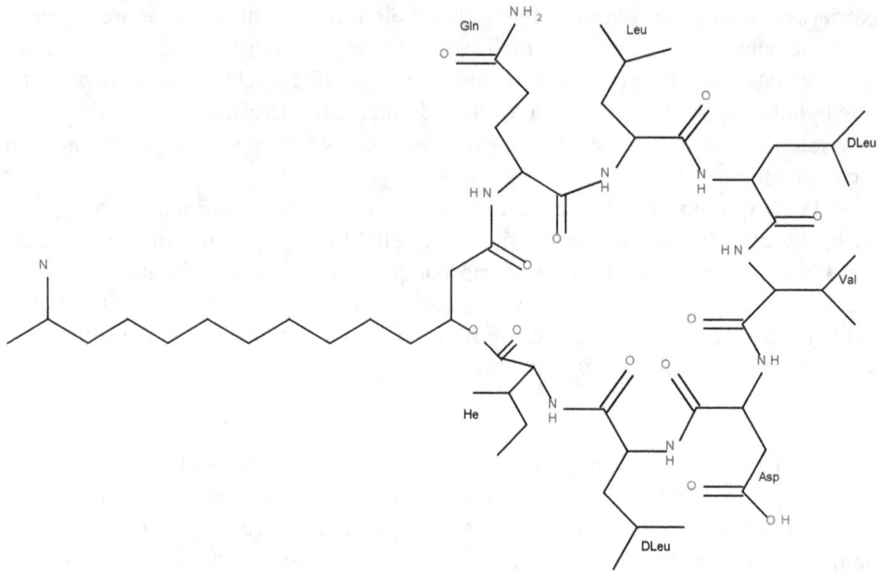

Figure 9. The structures of Linchenysin A (Adapted from: https://www.google.com/search?q=lichenysin+biosurfactant&sxsrf=ALeKk00UeRPjzETZVEu3P2I7hckH8_9KnA:1603250010621&source=lnms&tbm=isch&sa=X&ved=2ahUKEwio1_na28TsAhXyzzgGHSatDocQ_AUoAXoECAsQAw&biw=1366&bih=625#imgrc=uPI9E8R6BqBGOM).

strain BAS50, is characterized to contain a long chain beta-hydroxy fatty acid molecule (Yakimov et al. 1996). Lichenysin is reported to best able over a wide range of pH values, temperatures and NaCl concentrations and promotes dispersion of colloidal 3-silicon carbide and aluminum nitride slurries much more efficiently than chemical agents (Horowitz et al. 1990). It has also been reported that lichenysin is a more efficient cation chelator compared with surfactin (Grangemard et al. 2001).

4.3 *Fatty Acids, Phospholipids, and Neutral Lipids*

The fatty acid bio surfactants are saturated fatty acids in the range of C12 to C14 and complex fatty acids containing hydroxyl groups and alkyl branches (McDonald et al. 1981, Kretschmer et al. 1982). A number of bacteria and yeasts produce large amounts of fatty acids and phospholipid surfactants during growth on n-alkanes. It is reported that Phosphatidylethanolamine produced by *R. erythropolis* grown on n-alkane causes a lowering of interfacial tension between water and hexadecane to less than 1 mN/m and a critical micelle concentration (CMC) of 30 mg/l (Kretschmer et al. 1982). Moreover, Arthobacter strain AK-19 (Wayman et al. 1984), and *P. aeruginosa* 44T1 Robert et al. 1989) accumulated up to 40–80% (w/w) of such lipids when cultivated on hexadecane and olive oil respectively.

4.3.1 Corynomycolic Acid

Corynomycolic acids (R1-CH (OH)-CH (R2)-COOH), are a group of surface active compounds with a varying number of carbon atoms (Fig. 10). The substrate in the

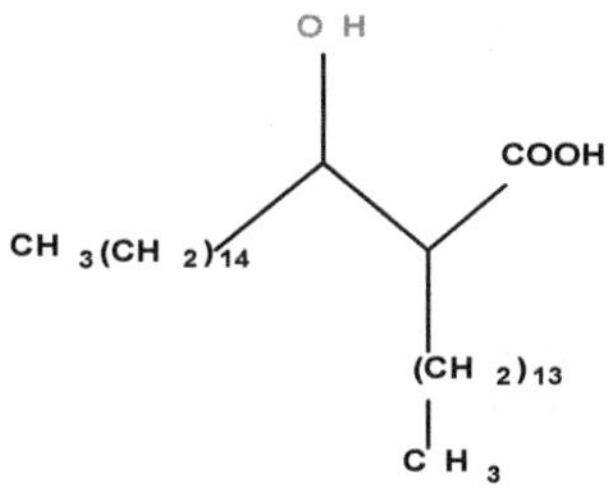

Figure 10. The structures of Corynomycolic acid.

growth media has a lot of influence in synthesizing biosurfactans with varying chain lengths. A mixture of corynomycolic acids with excellent surfactant properties has been isolated from *C. lepus*. It causes significant lowering of surface tension in aqueous solution and also lowers the interfacial tension between water and hexadecane at all Ph values between 2 and 10 (Cooper et al. 1981).

4.4 *Polymeric Biosurfactants*

Most polymeric biosurfactants have a backbone of three or four repeating sugars with fatty acids attached to the sugars (Rosenberg and Ron 1997). The best-studied polymeric biosurfactants are emulsan, liposan, mannoprotein, and other polysaccharide-protein complexes.

4.4.1 *Emulsan*

Emulsan is a complex extracellular acylated polysacharide synthesized by the *Acinetobacter calcoaceticus* with an average molecular weight of about 1000 KD (Kim et al., 1997). It is characterized as a polyanionic amphipathic heteropolysaccharide. The heteropolysaccharide backbone consists of repeating units of trisaccharide of N-acetyl-D-galac-tosamine, N-acetyl galactosamine uronic acid and an unidentified N-acetyl amino sugar (Zukerberg et al. 1979) (Fig. 11). Removal of the protein

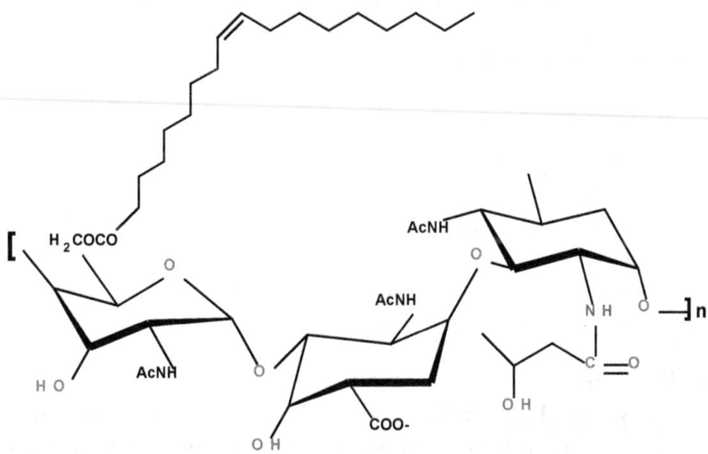

Figure 11. The structures of Emulsan.

fraction yields a product, apoemulsan, which exhibits much lower emulsifying activity on hydrophobic substrates such as n-hexadecane. One of the key proteins associated with the emulsan complex is a cell surface esterase (Bach et al. 2003).

4.4.2 *Biodispersan*

A. calcoaceticus A-2 produces an extracellular, non-dialyzable surface-active dispersing substance called biodispersan. The surface active component of biodispersan is an anionic heteropolysaccharide, with an average molecular weight of 51,400 and contains four reducing sugars, i.e., glucosamine, 6-methylaminohexose, galactosamine uronic acids and an unidentified amino sugar. It was suggested that mutants of strain *A. calcoaceticus* A-2 which are defective in protein secretion can be potentially used for biodispersan production (Rosenberg et al. 1988, Elkeles et al. 1994).

4.4.3 *Alasan*

Alasan is an anionic alanine-containing heteropolysaccharide protein biosurfactant produced by *A. radioresistens* KA-53 (Navonvenezia et al. 1995). Alasan produced by *A. radioresistens* KA-53 was reported to solubilise and degrade polyaromatic hydrocarbons (Barkay et al. 1999). The surface active component of alasan is a 35.77 kD protein called AlnA. This surface-active protein AlnA has a high amino acid sequence homology to *E. coli* outer membrane protein A (OmpA), but however OmpA does not possess any emulsifying activity (Toren et al. 2002). Three alasan proteins were purified from *A. radioresistens* KA-53 having molecular masses of 16, 31 and 45 kD and it was demonstrated that the 45-kD protein had the highest specific emulsifying activity, 11% higher than the intact alasan complex. The 16- and 31-kD proteins gave relatively low emulsifying activities, but they were significantly higher than that of apo-alasan (Toren et al. 2001).

4.4.4 *Liposan*

C. lipolytica produce an extracellular water soluble emulsifier, i.e., Liposan is composed of 83% (w/v) carbohydrate and 17% (w/v) protein. The carbohydrate portion is a heteropolysaccharide consisting of glucose, galactose, galactosamine and galacturonic acid (Cirigliano et al. 1984).

4.4.5 *Emulsifying Biopolymer from Fungus*

A bulky amount of mannoprotein production by *S. cerevisiae* showing remarkable emulsifier activity toward several alkanes, oils, and organic solvents had been reported (Cameron et al. 1988). The purified emulsifier contained 44% mannose and 17% protein. A mannose-fatty acid complex from alkane grown *C. tropicalis* has been successfully isolated and it has been found that this complex stabilizes hexadecane in water emulsion (Kappeli et al. 1984).

4.4.6 *Emulsifying Protein*

An emulsifying peptido glycolipid containing 52 amino acids, 11 fatty acids and a sugar unit is produced by *P. aeruginosa* P-20. This bio emulsifier, is composed of 50% carbohydrate, 19.6% protein and 10% lipid produced by *P. fluorescens* (Koronelli et al. 1983, Desai et al. 1988).

4.5 Particulate Biosurfactants

These are of two types, extracellular vesicles and whole microbial cells. Extracellular membrane vesicles partition hydrocarbons to form micro-emulsions, which play key a role in hydrocarbon uptake by microbial cells. Sometimes the whole bacterial cell itself can work as a surfactant.

4.5.1 Vesicles

Generally, *Acinetobacter* sp. when grown on hexadecane accumulated extracellular vesicles of 20 to 50 mm diameter with a buoyant density of 1.158 g/cm^3. These vesicles appear to play a role in the uptake of alkanes by *Acinetobacter* sp. HO1-N. These vesicles with a diameter of 20–50 nm and a buoyant density of 1.158 cubic g/cm are composed of protein, phospholipids and lipopolysaccharides (Kappeli and Finnerty 1979). Like *Acinetobacter* sp., *P. marginalis* also forms vesicles which work as surfactants.

4.5.2 Whole Microbial Cells

It is reported that most of the hydrocarbon-degrading microorganisms, nonhydrocarbon degraders, and some species of *Cyanobacteria* including a few pathogens have a strong affinity for hydrocarbon water and air-water interfaces. In such cases, the microbial cell itself is a surfactant (Karanth et al. 1999).

5. Sources of Biosurfactants

It is found that a number of microorganisms like bacteria, yeasts, and fungi show the ability to produce different classes of bio surfactants during their growth on water soluble and water insoluble substrates (Sheppard and Mulligan 1987, Desai et al. 1988, Ron and Rosenberg 2001). These microorganisms may be categorized into different groups on the basis of their ability of alkane utilization and extracellular lipids synthesis as under:

I. Microorganisms which can produce biosurfactants during growth on alkanes are particular strains of *Arthrobacter* sp., *Corynebacterium* sp. and *Nocardia* sp.
II. Microorganisms which can produce biosurfactants on both alkanes and water-soluble compounds. Most of the producers fall into this category such as *P. aeruginosa* producing rhamnolipids.
III. Microorganisms which can produce biosurfactants during growth on water soluble compounds. such as Surfactin-producing *B. subtilis*. Some species of yeast like *Rhodotorula* sp. can produce a mixture of mannitol and pentitol esters of beta-D-hydroxypalmitic acid and beta-D-hydroxystearic acid during growth on a complex medium with glucose as carbon source.

6. Biosurfactant-Production by Microorganisms

Normally, microorganisms utilizes carbon sources and energy for their growth. It is found that a combination of such carbon sources with insoluble substrates accelerates the intracellular diffusion and production of various substances like bio surfactants (Deleu et al. 2004, Chakraborty et al. 2014). Microorganisms like some filamentous fungi bacteria and yeasts exhibit potential for bio surfactants production with different molecular structures and surface activities (Campos et al. 2013).

These days, interest of the scientific community has been increasing towards isolation of microorganisms exhibiting remarkable biosurfactant producing ability with low CMC, low toxicity and high emulsifying activity. It is reported that *Pseudomonas* spp. and *Bacillus* spp. are great bio surfactant producers (Silva et al. 2014). Since last few decades, isolation of bio surfactant producing organisms has been carried out from uncontaminated and undisturbed environments like natural soils and marine environments (Thavasi et al. 2011).

However, it is found that polluted environmental sites contaminated with oil, like wastewater treatment plants have shown diversity of bio surfactant-producing microorganisms in huge numbers (Bento et al. 2005, Ndlovu et al. 2016). Generally, the hydrophilic substrates are primarily used by microorganisms for cell metabolism and the synthesis of the polar moiety of a biosurfactant, whereas hydrophobic substrates are used exclusively for the production of the hydrocarbon portion of the biosurfactant (Desai et al. 1997, Weber et al. 1992).

6.1 Metabolic Pathways of Biosurfactant Production

It is reported that biosurfactant synthesis occurs through four routes (a) Carbohydrate and Lipid synthesis (b) Synthesis of the Carbohydrate half and the Lipid half depends on the length of the chain of the carbon substrate in the medium. (c) Synthesis of the lipid half and the synthesis of the carbon half depends on the substrate employed; and (d) Synthesis of the carbon and lipid halves, which are both dependent on the substrate.

Diverse metabolic pathways are involved in the synthesis of precursors for biosurfactant production and depend on the nature of the main carbon sources employed in the culture medium. For instance, when carbohydrates are the only carbon source for the production of a glycolipid, the carbon flow is regulated in such a way that both lipogenic pathways (lipid formation) and the formation of the hydrophilic moiety through the glycolytic pathway are suppressed by the microbial metabolism, as illustrated in Fig. 12 (Haritash et al. 2009).

However, reactions catalysed by pyruvate kinase and phosphofructokinase-1 are irreversible. Thus, other enzymes exclusive to the process of gluconeogensis are required to avoid such reactions. Figure 13 illustrates the main reactions through the formation of glucose 6-phosphate, which is the main precursor of polysaccharides and disaccharides formed for the production of the hydrophilic moiety of glycolipids (Tokumoto et al. 2009). Therefore, the length of the n-alkane chain used as the carbon source alters the biosynthesis of a surfactant.

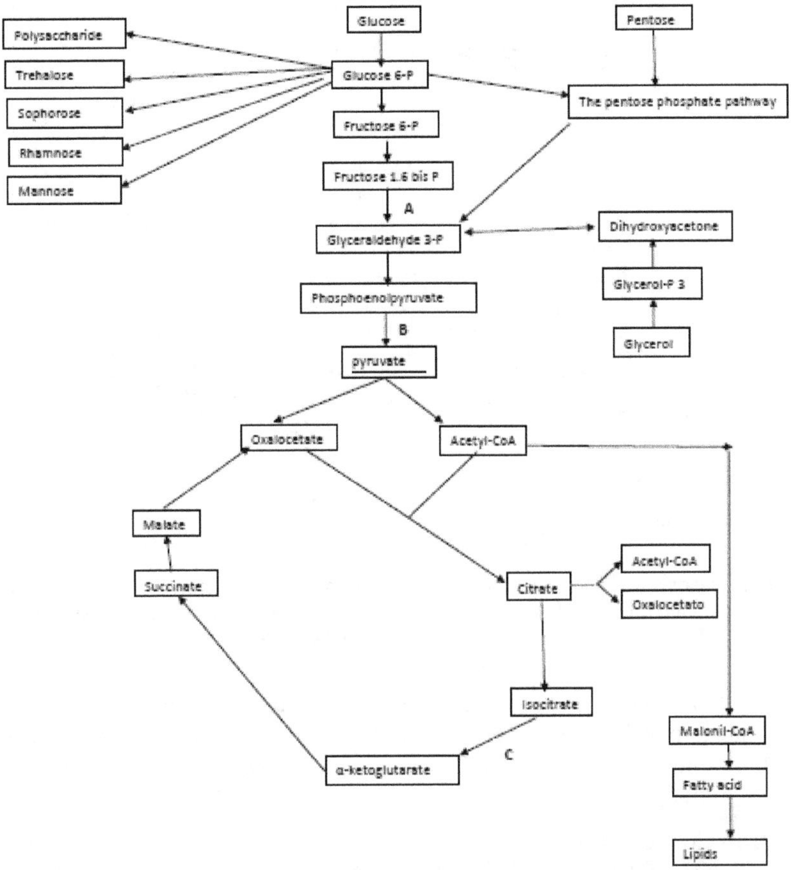

Figure 12. Intermediate metabolism associated with biosynthesis of biosurfactant precursors using carbohydrates as substrate. (Key Enzymes: A-Phosphofructokinase, B-Pyruvate Kinase, C-Isocitrate Dehydrogenase) (Adapted and Redrawn from: Danyelle Khadydja et al. 2016).

7. Screening Methods for Assessment of Biosurfactants

Usually, various screening methods have been employed for bio surfactant production assessment as per their interfacial or surface activity (Vanessa et al. 2010). Some methods are as follows:

I. Blood Agar Hemolysis Method

Isolated microorganisms should be screened on blood/agar plates containing 5% (v/v) blood and incubated at 28°C for 48 hours. Haemolytic activity should be detected as the presence of a definite clear zone (hemolysis) around the colony (Carrillo et al. 1996).

II. CTAB Agar Plate Method

This method should be used to detect anionic surfactants when isolates have been inoculated on modified minimal salt agar medium supplemented with CTAB

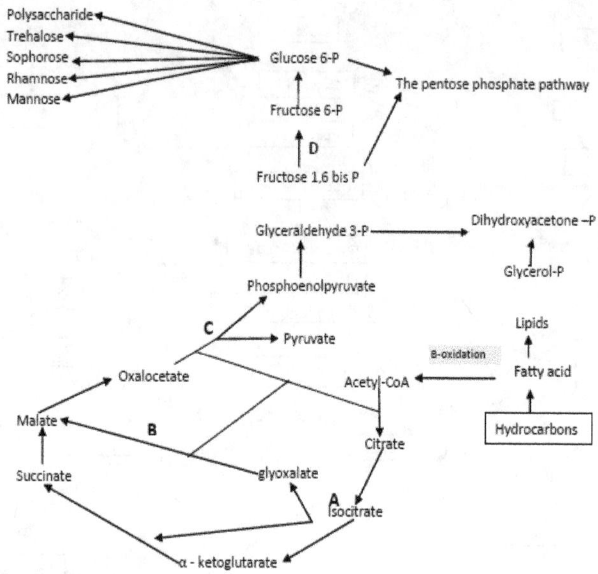

Figure 13. Intermediate metabolism associated with biosynthesis of precursors of biosurfactant using hydrocarbons as substrate. (Key Enzymes: A-Isocitrate Lyase, B-Malate Synthase, C-Phosphoenolpyruvate, D-Fructose-1) (Adapted and Redrawn from: Danyelle Khadydja et al.2016).

(0.2 g/L) along with methylene blue dye (0.005 g/L). In a positive result after incubation at 37°C for 72 hours, appearance of bluish halo around the colonies shows the presence of bio surfactant (Siegmund and Wagner 1991, Pendse and Arun 2018).

III. Drop Collapse Test

The drop collapse test should be performed using 96-well microlitre plates coated with 20 μl of oil and incubated at room temperature for 24 hours. One drop of cell-free extract is added and incubated for a minute. The collapsing drop indicates a positive test (Nayarisseri et al. 2018).

IV. Oil Displacement Method

In this method, 50 mL of distilled water should be added into the petri plate followed by addition of 2 mL of the crude oil in such way that it should uniformly spread on the water surface. Then, 500 μL of the culture supernatant should be added. The clear zones on the oil surface indicate a positive test (Reddy 2019).

V. Penetration Assay

ELISA 96 well micro plates can be used in this method; 200 μL of hydrophobic paste containing oil and silica gel should be added into the wells. 10 μL of crude oil, 90 μL of culture supernatant and 10 μL of safranin solutions should be subsequently added and the bio surfactant activity should be observed. If the hydrophilic liquid breaks through the oil film barrier into the paste, the test indicates the presence of Bio surfactant. During this silica enters the hydrophilic phase and the upper phase shows a colour change from clear red to cloudy white within 15 minutes (Kumar et al. 2017, Vanessa et al. 2010).

Table 1. Microorganisms involved in production of BioSurfactant.

Bio Surfactant	**Microorganisms Involved**	**Reference**
Rhamnolipids	*P. aeruginosa, Pseudomonas* sp., *B. glumae, B. plantarii, B. thailandensis*	Ndlovu et al. 2017
Trehalose lipids	*R. erythropolis, Arthrobacter* sp., *N. erythropolis, Corynebacterium* sp., *Mycobacterium* sp.	Muthusamy et al. 2008
Sophorolipids	*C. bombicola, C. antartica, T. petrophilum, C. botistae, C. apicola, C. riodocensis, C. stellata, C. bogoriensis*	Felse et al. 2007
Monnosylerythritol lipids	*C. antartica, Kurtzmanomyces* sp., *P. siamensis*	Kitamoto et al. 1993
Surfactin/Iturin	*B. subtilis, B. licheniformis, B. magiterium, B. amyloliquefaciens* ST34	Dey et al. 2015, Dey et al. 2017, Ndlovu et al. 2017
Subtilisin	*B. subtilis*	Sutyak et al. 2008
Emulsan	*A. calcoaceticus*	Choi et al. 1996
Alasan	*A. radioresistens*	Barkay et al. 1999
Lipopeptide	*B. subtilis* SPB1	Zouari et al. 2016
Mannosylerythritol lipids (MEL)	*Pseudozyma* spp.	Bae et al. 2018
Peptide lipids	*B. licheniformis*	Begley et al. 2009
Cellobiose lipids	*U. maydis*	Teichmann et al. 2007
Serrawettin	*S. marcescens*	Lai et al. 2009
Polyol lipids	*R. glutinis, R. graminis*	Amaral et al. 2006
Ornithine lipids	*Pseudomonas* sp., *T. thiooxidans, Agrobacterium* sp.	Desai et al. 1997
Viscosin	*P. fluorescens, L. mesenteriods*	Banat et al. 2010
Rhamnolipids	*P. aeruginosa, P. chlororaphis, S. rubidea*	Jadhav et al. 2011
Carbohydrate-lipid	*P. fluorescens, D. polmorphus*	Nerurkar et al. 2011
Protein PA	*P. aeruginosa*	Hisatsuka et al. 1971
Diglycosyl diglycerides	*L. fermentum*	Mulligan et al. 2001
Whole cell	*Cyanobacteria*	Levy et al. 1990
Fatty acids/neutral lipids	*C. michiganensis* subsp. *insidiosus*	Herman et al. 2002
Sophorolipids	*Candida bombicola, C. antartica, Torulopsis petrophilum C. botistae, C. apicola, C. riodocensis, C. stellata, C. bogoriensis*	Felse et al. 2007
Liposan	*C. tropicalis*	Cirigliano et al. 1984
Surfactin/Iturin	*B. subtilis, B. amyloliquefaciens*	Arguelles et al. 2009
Aminoacids lipids	*Bacillus* sp.	Cotter et al. 2005
Lichenysin	*Bacillus licheniformis, B. subtilis*	Yakimov et al. 1997
Phospholipids	*Acinetobacter* sp.	Kosaric et al. 2001
Vesicles & fimbriae	*A. calcoaceticus, P. marginilis, P. Maltophila*	Desai et al. 1997
Emulsan	*A. calcoaceticus*	Choi et al. 1996

VI. Emulsification Activity Test

Cell free supernatant (2 ml) should be added to distilled water (2-ml). The solution should be mixed with 1 ml substrate like soybean or diesel oils. After vigorous vortexing (2 min), the tubes should be allowed to stand for one hour for phase separation. The aqueous phase should be removed carefully and O.D. at 540 nm should be measured and compared with un-inoculated broth used as a negative control. Emulsification activity is defined as the measured optical density at 540 nm (Satpute et al. 2008).

VII. Emulsification Index Test

In the test tube having cell free supernatant (2 ml) crude oil, i.e., diesel oils/kerosene (2 ml) should be added and allowed to vortex (2 min) followed by a 24 hr incubation period (Cooper and Goldenberg 1987). The emulsion stability should be noted using the formula:

Emulsification Index = (Height of emulsion layer/Total height) × 100

VIII. Bacterial Adhesion to Hydrocarbon (BATH) Assay

Cell hydrophobicity should be measured by bacterial adherence to hydrocarbons as per the method of Rosenberg et al. (1980). The cell pellets should be washed and suspended in a buffer salt solution (g/L, 16.9 K_2HPO_4 and 7.3 KH_2PO_4) followed by dilution with the same buffer solution to an optical density (OD) of ~ 0.5 at 610 nm. Then the cell suspension (2 ml) should be taken into the test tubes, added with crude oil (100 µl) and vortexed for 3 min followed by incubation for 1 hour. After the phase separation, OD of the aqueous phase should be measured at 610 nm. Percentage of cells attached to crude oil is calculated by using the formula:

% of bacterial cell adherence = $(1\text{-}(OD_{\text{shaken with oil}}/OD_{\text{original}}) \times 100$

[Where, $OD_{\text{shaken with oil}}$:OD of the mixture containing cells and crude oil, OD_{original}:OD of the cell suspension in the buffer solution (before mixing with crude oil)]. Then, few drops of 2-(4-iodophenyl)-3-(4-nitrophenyl)-5-phenyltetrazolium chloride (INT) solution should be added to the above BATH assay solution and should observed under a light microscope. The INT turns red if it gets reduced inside the cells, indicating the viability of the cells adhered to the crude oil droplets (Nayarisseri et al. 2018).

IX. Hydrocarbon Overlay Agar

The Agar plate should be coated with 100 µl of crude oil and inoculated with the bacterial consortium followed by incubation at 30°C for 48–72 hours. The colony surrounded by the emulsified halo should be considered positive for bio surfactant production (Ron and Rosenberg 2001).

X. Salt Aggregation Assay

Sodium phosphate (0.002 M, pH 6.8) should be used to dilute a solution of 4 M $(NH_4)2SO_4$ in 0.002 M sodium phosphate (pH 6.8). Serial dilutions should be made giving $(NH_4)2SO_4$ concentration ranging from 4.0 to 0.2 M differing by 0.2 M per dilution (Lindahl et al. 1981). Microbial/bacterial suspension of 25 ml (approximately 10^{10} c.f.u.ml^{-1} in 0.002 M sodium phosphate buffer pH 6.8) should be mixed with an equal volume of salt solution in a 24 well tissue culture tray. The microbial/salt mixture should be gently shocked (2 min) at 25°C. Visual reading should be performed against a black background. Positive aggregation reaction will show clear solution and white aggregates (Pruthi and Cameotra 1997).

8. Methods for Biosurfactant Production

Yield of bio surfactant is influenced by the type of fermentation method. Batch, fed-batch and continuous batch methods are commonly used methods. The type of fermentation method used has a major influence on the bio surfactant yield. Common fermentation methods employed in bio surfactant research are applied for the production of bio surfactants (Sari et al. 2019).

A. Batch Process

In this process media and inoculum are simultaneously added to bioreactor, and the product is collected at the end. During the process, conditions of the bioreactor varity. Batch-type bioreactors can be used when a material available at certain times is suitable for a high solid (25%) content. For fibrous material, these types of reactors have proven to be more appropriate than continuous flow-type reactors. Due to presence of toxic materials, it is possible to stop the process and continue again (Zouari et al. 2014, Sari et al. 2019).

B. Continuous Process

In this process, streaming of substrate and collection of the product can be carried out continuously at any time after the maximum product concentration, or substrate limits have reached an almost constant value. Moreover, the inocula and substrates can be continuously added together which can extend the exponential phase. However, very less study has been reported due to hitches in regulating the substrate availability. These difficulties are due to the addition of new media which should reach a fixed volume at the time of maintaining output cultures in a constant cell physiology phase (Sari et al. 2019).

C. Fed-Batch Process

In this process, a gradual increase in the culture volume is possible as new media can be continuously added without eliminating the culture fluid in the fermenter, Also due to a consistent addition of nutrients, i.e., Carbon and Nitrogen sources, there is prevention of nutrient depletion till a product yield close to the possible maximum is obtained (Kronemberger et al. 2010). A sequential fed-batch fermentation process with fill-and-draw operations has been developed and recommended for improvement of Rhamnolipids production. In this high-performance fermentation process, cell components play a key role and provide an efficient strategy for high rhamnolipids production up to 150 g/L. Besides, high rhamnolipids productivity can be well-maintained (0.4 g/L h for 17 days). As this process has a great prospect for high-efficiency rhamnolipids production, the use of fed-batch type fermentation is recommended at industrial scales due to economic feasibility.

9. Factors Affecting Biosurfactant Production

Composition and emulsifying activity of the bio surfactant is not only influenced by the producer strain but also the culture conditions. Thereby, the quantity of bio surfactant produced as well as the type of polymer gets affected by the nature of the carbon and nitrogen sources, C:N ratio, nutritional limitations, physicochemical parameters such as temperature, aeration, divalent cations and pH (Salihu et al. 2009).

9.1 Carbon Sources

Growth and production of bio surfactants by microorganisms as well as quality and quantity of bio surfactant production are highly influenced by the nature and source of the carbon substrate and it is found to differ from species to species (Danyelle et al. 2016, Rahman and Gakpe 2008). Till date, diesel, crude oil, glucose, sucrose and glycerol have been reported as significant sources of carbon substrates for bio surfactant production (Desai and Banat 1997).

9.2 Nitrogen Sources

In the biosurfactant production medium, Nitrogen plays a vital role with respect to microbial growth as the protein and enzyme syntheses depend on it (Fakruddin 2012). It is reported that during the process of fermentation, C/N ratio has a remarkable influence w.r.t the buildup of metabolites, e.g., High C/N ratios (i.e., low nitrogen levels) limit bacterial growth and favor the cell metabolism towards the production of metabolites. Whereas, extreme nitrogen levels lead to the cellular material synthesis and limit the buildup of products (Robert et al. 1989). Since the last few decades, various organic and inorganic nitrogen compounds like sodium nitrate, ammonium sulphate, ammonium nitrate, urea peptone, yeast extract, meat extract and malt extracts have been efficiently used. Although, yeast extract is mostly used for bio surfactant production, its usage with respect to concentration may change according to the organism and culture medium (Fakruddin 2012).

9.3 Growth Conditions

Growth conditions (temperature, pH, agitation speed and oxygen) are very important with regard to yield and characteristics of the produced biosurfactant. Therefore, to obtain huge amounts of bio surfactants, optimizion of the bioprocess in respect with temperature, pH, aeration or agitation speed is always required. Most bio surfactant productions are reported to be performed in a temperature range of 25–30°C (Desai and Banat 1997). It is reported that remarkable bio surfactant production is observed if the pH is 8.0 (Zinjarde and Pant 2002). Besides, incubation time also exerts a significant effect on biosurfactant production. Generally, microorganisms are found to produce bio surfactants at different time intervals (Danyelle et al. 2016). Aeration and agitation have been found to facilitate oxygen transfer from the gas phase to the aqueous phase as well as transportation of the nutrient to microorganisms. In addition, it is associated to the physiological functions of the microbial emulsifier and thereby influences biosurfactant production. The best production value of the surfactant (45.5 g/l) can be obtained in case the air flow rate is 1 vvm and the dissolved oxygen concentration is maintained at 50% of saturation (Adamczak and Bednarski 2000). Generally, cellular activities of microorganisms are influenced by salt concentration of the medium and thereby the biosurfactant production (Fakruddin 2012).

10. Commercial Limitations of Biosurfactants

1. In spite of various remarkable advantages of bio surfactants there are a few commercial limitations too with respect to synthetic ones due to their economic feasibility and sometimes low yields.
2. There might be difficulty in obtaining pure bio surfactants, due to the downstream processing of diluted broths involved which need multiple consecutive steps (Pattanathu et al. 2008).
3. In various biotechnology processes, purification accounts for a huge part of the production cost.
4. The strains of bio surfactant producing bacteria or other microbes exhibiting high productivity are limited (Pattanathu et al. 2008).
5. Enhancement of production yield is hindered by strong foam formation. Subsequently, diluted media have to be applied and only immobilized systems offer enhanced productivity (Fiechter 1992).

Therefore, to address the cost aspects of bio surfactant production, there is a need for cost-effective substrates in the production process. In this regard, these days research into avenues to minimize cost and optimize the production process of bio surfactants is intertwined with their potential as agents of sustainability (Ibukun et al. 2018).

11. Cost Effective Production of Biosurfactants

Recently, application of industrial wastes for valuable compound production has been getting more importance to economize commercial production as well as sustainable and efficient management of unaccustomed waste generated from various processes (Patil and Rao 2015).

It is observed that improvement in fermentation technology, strain selection and use of cheaper, renewable substrates plays a vital role in enhancing the production processes of the Bio surfactants industry (Marchant and Banat 2012, Marchant et al. 2014). However, large scale production for most microbial surface active agents has not reached a satisfactory economical level due to low yields. Similarly, high cost inputs are essential for downstream processing to recover and purify microbial surfactants (Rodrigues et al. 2006, Smyth et al. 2010).

11.1 Cost Effective Renewable Substrates

Worldwide two basic strategies are adopted for cost-effectiveness associated with the use of cheap waste substrates for the preparation of fermentation media for reducing initial raw material costs and the development of efficient and magnificently optimized processes, including optimization of the culture conditions as well as cost-effective recovery processes for maximum bio surfactant production and recovery. It

Table 2. Various sources and generated green substrates (Low Cost/Renewable).

Source	Green Substrate (Low Cost/Renewable)	Reference
Agro-crop	Cassava, potato, sweet potato, soybean, sweet sugar beet, sorghum, etc.	Banat et al. 2014
Distillery industry	Industrial effluents, etc.	Banat et al. 2014
Slaughter house	Animal fat, etc.	Banat et al. 2014, Solaiman et al. 2003, Nitschke et al. 2010
Coffee processing residues	Coffee pulp, coffee husks, spent of free groundnut, etc.	Banat et al. 2014
Agro-industrial waste, Crop residues	Bran, beet molasses, Bagasse of sugarcane straw of wheat, cassava, cassava flour waste water, rice Straw of rice, hull of soy, corn, sugarcane molasses, etc.	Nitschke et al. 2004, Rashedi et al. 2005, Thavasi et al. 2011, Banat et al. 2014
Oil processing mills, soap stock, wastes from lubricating oil	Coconut cake, canola meal, olive oil mill waste water, palm oil mill waste, peanut cake, effluent, soybean cake, Glycerol and Byproducts, Refinery waste, etc.	Cooper and Paddock 1984, Robert et al. 1989, Mercadé et al. 1993, Banat et al. 2014, Nitschke et al. 2005
Food and starch processing industry	Frying edible oils and fats, olive oil, Soybean oil, potato peels, rape seed oil, sunflower, vegetable oils, ground-nut oil refinery residue, by-products, etc.	Mercadé et al. 1996, Banat et al. 2014, Makkar et al. 2011
Soya Milk and Tofu factory	Okara	Ohno et al. 1993, Ohno et al. 1995, Ohno et al. 1996
Dairy industry	Lactic whey, Curd whey, cheese whey, whey waste, effluent, etc.	Dubey and Juwarkar 2001, 2004, Makkar and Cameotra 2002, Dubey et al. 2005, Rodrigues and Teixeira 2008, Banat et al. 2014
Winery and other Fruit processing industries	Banana waste, Pomace of apple and grape, carrot industrial waste, pine apple	Portilla-Rivera et al. 2008, Banat et al. 2014
Forestry, agriculture and municipalities	Lignocellulosic materials	Makkar et al. 2011

is reported that globally millions of tons/yr of hazardous and non-hazardous wastes has been generated which require proper waste management and utilization (Saharan et al. 2011).

Moreover, increased public awareness of issues related to environmental pollution strongly influence the development of various technologies. In this regard, emphasis has been given on recovery, recycling and reuse of various industrial waste by products and effluents. But, the selection of waste products should ensure the proper balance of nutrients to allow microbial growth and consequent bio surfactant production. Normally, industrial waste having high content of carbohydrates or lipids is ideal for use as substrate.

It is reported that agro-industrial waste can be efficiently used for feasible bio surfactant production on an industrial scale after the optimization of different variables (Barros et al. 2007). The use of cheaper, renewable substrates from various industries such as the agricultural segment (sugars, molasses, plant oils, oil wastes, starchy substances, lactic whey), distillery wastes, animal fat, oil industries have been reported and reviewed thoroughly by several researchers (Makkar et al. 2011). Such cheap substrates not only act as nutrients for microbial growth but also act as an important source for isolation of promising bio surfactant producing microorganisms (Santos et al. 1986, Lee et al. 2008).

Selection of cheap raw materials is essential for the overall economic aspects of the commercial production of bio surfactants. The waste/raw materials with a high level of, nitrogen, carbohydrates and lipids are needed for use as substrates. Such wide spectrums of raw materials include various agricultural wastes and by-products which can lead to avoidance and reduction of environmental pollution. Several researchers have reviewed and reported the application of cheaper renewable substrates generated from different industries. Most of these substrates can act as nutrients for microbial growth as well as significant sources for isolation of bio surfactant producing microorganisms (Santos et al. 1986, Lee et al. 2008).

12. Promising Strategies for Biosurfactant Production

12.1 BioSurfactant Coproduction for Process Economization

Coproduction of bio surfactants with another economically important product in a single bioprocess would make the entire production set up highly economical (Singh and Patil 2018). The cost of the whole bio surfactant production process can be cut down by using cheap and renewable starting substrates. Makkar et al. (2011) have reported Bio surfactant coproduction with renewable substrates which can help to achieve more fruitful results by co-production of bio surfactants and other important metabolites, and enzymes like polyhydroxyalkanotes (PHA), lactic acid and proteases. Simultaneously PHAs have been applied for the manufacture of bottles, films and fibers as biodegradable packaging agents (Sudesh et al. 2000). Proteases are vital groups of industrial enzymes that satisfy the demand of approximately 60% of the world enzyme market. These enzymes have numerous industrial applications comprising detergents, food, leather, silk, waste management and pharmaceuticals (Gupta et al. 2002). Therefore, this approach can signify the cost effectiveness of simultaneous production of biosurfactants with other metabolites and will contribute to easing off the overall cost factor of the biosurfactant.

12.2 Effect of Substrate on Biosurfactant Production

Since the last few decades, several studies have been carried out for media optimization using promising bio surfactant producers of the *Pseudomonas, Bacillus* and *Candida* species. Generally, cell based bioprocesses include several interrelated parameters, and it is very costly, tedious and time consuming and to monitor each parameter separately. In this context, multiple data can be handled at the same time using statistical methods like Response Surface Methodology (RSM), Plackett-Burman Design (PBD) and Central Composite Design (CCD) to optimize the media for optimization of biosurfactant production processes (Singh and Patil 2018, Eswari

Table 3. Substrate variables for bio surfactant production.

Organism	Strategy and/Critical Variables Studied for Optimizing Media	Maximum Yield
B. subtilis N7	Plackett-Burman Design and Central Composite Design; Effect of Fe^{+2}, Mn^{+2}	0.706 g l^{-1}
B. subtilis HSO121	Plackett-Burman Design-Response Surface Methodology; $CaCl_2$, Maltose, L-arginine	47.58 g l^{-1} 38.06 fold increase
B. subtilis	Effect of Mn; Mn supplementation	Yield Not Documented 6.2 fold increase
B. megaterium	Effect of Fe; Intermittent Fe^{+2} feeding	4.2 ± 0.15 g l^{-1} 5.8 fold increase
P. aeruginosa OCD1	$ZnSO_4$ followed by intermittent $MnSO_4$ supplementation	0.980 g l^{-1}
P. aeruginosa ATCC9027	Response Surface Methodology & Batch *vs* fed Batch cultivation; Nitrate and magnesium salt	8.5 g l^{-1}
P. aeruginosa DN1	Palm oil as carbon source; C/N ratio 20	25.98 g l^{-1}
Streptomyces coelicoflavus NBRC 15539T	Plackett-Burman Design, Inoculum level Olive oil, $NaNO_3$	0.475 g l^{-1}
Rhodococcus erythropolis ATCC 4277	Use of $NaNO_3$, $MgSO_4$, Phosphate buffer (pH)	0.285 g l^{-1}

(Adapted from: Singh and Patil 2018).

Table 4. Upcoming and promising strategies for cost effective bio surfactant production.

Strategy	Remarks	Reference
Use of growth enhancers	Use of lactones enriches the production medium and enhances yield. Formation of a derivatized product makes recovery easier	Yeh et al. 2005, Santos et al. 2016c, Roelants et al. 2016
Use of Nanoparticles	Iron and manganese nanoparticles give enhanced biosurfactant yield probably by replenishing the critical metal ion requirement	Liu et al. 2013, Kiran et al. 2014b, Sahebnazar et al. 2018
Use of immobilized producer organism	Fe-nanoparticle enriched Immobilizing medium, especially alginate, would provide easy separation and additional yield enhancement with lesser byproducts. Activated charcoal acts as an enhancer as well as an immobilizing agent	Abouseoud et al. 2008, Gancel et al. 2009, Heyd et al. 2011, Onwosi and Odibo 2013
Use of Biofilm reactor, Pertraction and Rotating Discs Bioreactor for recovery	Immobilization and aeration in rotating discs takes care of foam production and subsequent loss of cells making recovery of the product easier thereby reducing production cost	Chtioui et al. 2010, Chtioui et al. 2012

(Adapted from: Singh and Patil 2018).

et al. 2016, Hassan et al. 2016, Santos et al. 2016). Apart from statistical methods, Artificial Intelligence (AI) based techniques like Artificial Neural Intelligence coupled with Genetic Algorithms (ANN-GA) is another approach that has been tried out for media optimization (Singh and Patil 2018, Sivapathasekaran and Sen 2013, Ahmad et al. 2016).

Additional strategies which can be applied for augmenting the yield of biosurfactants on a large-scale will include the use of micro bioreactors for process optimization. Use of automated surface enrichment, employment of a novel oxygenation process for *in situ* product removal as well as application of novel methods like pertraction (Chtioui et al. 2010).

13. Patents on Biosurfactants

A partial list of some of the patents granted during 1980 to 2019 on biosurfactant production, properties and applications is given in Table 5.

Table 5. Some patents on bio surfactants (Years 1980–2019).

Microorganisms	Title of Invention	Patent No.	Year	Inventor/Applicant
Arthrobacter spp. RAG-1 ATCC 31012*	Production of α-emulsans	US 4230801	1980	Gutnick, Rosenberg
Arthrobacter spp. RAG-1 ATCC 31012*	Production of α-emulsans	US 4234689	1980	Gutnick, Rosenberg, Shabtai
Acinetobacter spp. RAG-1 ATCC 31012	Cleaning oil-contaminated vessels with α-emulsans	US 4276094	1981	Gutnick and Rosenberg
Acinetobacter spp. RAG-1 ATCC 31012	Apo β-emulsans	US 4311829	1982	Gutnick, Rosenberg, Belsky, Zosim
Acinetobacter spp. RAG-1 ATCC 31012	Apo α-emulsans	US 4311830	1982	Gutnick, Rosenberg, Belsky, Zosim
Acinetobacter spp. RAG-1 ATCC 31012	Apo psi emulsans	US 4311831	1982	Gutnick, Rosenberg, Belsky, Zosim
Acinetobacter spp. RAG-1 ATCC 31012	Proemulsans	US 4311832	1982	Gutnick, Rosenberg, Belsky, Zosim
Acinetobacter spp. RAG-1 ATCC 31012	Psi emulsans	US 4380504	1983	Gutnick, Rosenberg, Belsky, Zosim
Acinetobacter spp. RAG-1 ATCC 31012	Emulsans	CA 1149302	1983	Gutnick, Belsky, Shabtai, Rosenberg, Zosim
Acinetobacter spp. RAG-1 ATCC 31012	Polyanionic heteropolysac charide biopolymers	US 4395353	1983	Gutnick, Rosenberg, Belsky, Zosim
Acinetobacter spp. RAG-1 ATCC 31012	α-Emulsans	US 4395354	1983	Gutnick, Rosenberg, Belsky, Zosim
B. licheniformis strain JF-2 (ATCC No. 39307)	Biosurfactant and enhanced oil recovery	US4522261	1985	Michael, McInerney Gary, Jenneman Roy, Knapp Donald, Menzie
A. calcoaceticus	Bathing agent	JP 63156714	1988	Osugi

Table 5 Contd. ...

...*Table 5 Contd.*

Microorganisms	Title of Invention	Patent No.	Year	Inventor/Applicant
A. calcoaceticus ATCC 31012, NRRL B-15616, NRRL B-15847, NRRL B-15848, NRRL B-15849, NRRL B-15850 and NRRL B-15860	Bioemulsified-containing personal care products for topical application to dermo-pathologic conditions of the skin and scalp	US 4870010	1989	Hayes
A. calcoaceticus NRRL B-15847, NRRL B-15848, NRRL B-15849, NRRL B-15850 and NRRL B-15860	Bioemulsifier production by *A. calcoaceticus* strains	US 4883757	1989	Gutnick, Nestaas, Rosenberg, Sar
A. calcoaceticus ATCC 31012, NRRL B-15616, NRRL B-15847, NRRL B-15848, NRRL B-15849, NRRL B-15850 and NRRL B-15860	Soaps and shampoos containing bioemulsifiers	CA 1266238	1990	Hayes and Holzner
A. calcoaceticus	Novel *A. calcoaceticus* and novel biosurfactant	EP 0401700	1990	Fukui, Negi, Tanaka
A. calcoaceticus 217 strain (FERM BP-2905)	Novel *A. calcoaceticus* and novel biosurfactant	JP 03130073	1991	Tanaka, Fukui, Negi
A. calcoaceticus ATCC 31012, NRRL B-15616, NRRL B-15847, NRRL B-15848, NRRL B-15849, NRRL B-15850 and NRRL B-15860	Personal care products containing bioemulsifiers	US 4999195	1991	Hayes
A. calcoaceticus ATCC 31012, NRRL B-15616, NRRL B-15847, NRRL B-15848, NRRL B-15849, NRRL B-15850 and NRRL B-15860	Personal care products containing bioemulsifiers	CA 1300512	1992	Hayes
A. calcoaceticus strains	Bioemulsifier production by *A. calcoaceticus* strains	CA 1316478	1993	Gutnick, Nestaas, Rosenberg, Sar

Table 5 Contd. ...

...Table 5 Contd.

Microorganisms	Title of Invention	Patent No.	Year	Inventor/Applicant
A. calcoaceticus ATCC 31012, NRRL B-15616, NRRL B-15847, NRRL B-15848, NRRL B-15849, NRRL B-15850 and NRRL B-15860	Hydrophobically Modified proteins	US 5212235	1993	Nestaas, Hrebenar, Lewis, Whitesides
A. radioresistens strain KA53	Novel bioemulsifiers	WO 9620611	1996	Ron and Rosenberg
A. calcoaceticus CL (KCTC 0081BP)	Microorganism producing biosurfactant	KR 9706157	1997	Hwang and Kim
Pseudomonas spp	Microbially Produced Rhamnolipids (biosurfactants) For The Control of Plant Pathogenic Zoosporic Fungi	US 5767090	1998	Michael, Stanghellini, Tucson, Arizon, Raina Margaret Miller, Tucson, Arizona, Scott Lynn Rasmussen, Tucson, Arizona, Do Hoon Kim, Tucson, Arizona, and Yimin Zhang, Tucson, Arizona
A. radioresistens strain KA53	Bioemulsifiers	US 5840547	1998	Ron, Rosenberg
B. subtilis, B. licheniformis, B. circulans, B. circulans, B. polymyxa, B. coagulans, B. macerans, Cl. Cellulolyticum, Cl. Aerotolerans, Cl. Acetobutylicum, B. azotofixans, B. macerans, B. polymyxa, Cl. Acetobutylicum, Cl. Pasturianum	Bacterial Preparation for agricultural use	US 5733355	1998	Susumu HibinoZenrou Minami
Pseudomonas sp. bacterium (NRRL B-18602)	Microbial production of a novel compound 7,10-dihydroxy-8-octadecenoic acid from oleic acid	US 5900496	1999	Ching T Hou

Table 5 Contd. ...

...Table 5 Contd.

Microorganisms	Title of Invention	Patent No.	Year	Inventor/Applicant
Acinetobacter spp. KRC-K4	Novel microorganism *Acinetobacter* spp. KRC-K4 and process for preparing emulsifier using the same	KR 170107	1999	Park Ae, Jung-Ill, Chang-Ho
Acinetobacter spp. KRC- K4	New lipopolysaccharide biosurfactant	JP 11269203	1999	Prosperi, Camilli, Crescenzi, Fascetti, Porcelli, Sacceddu
A. calcoaceticus CBS 962.97	Lipopolysaccharide biosurfactant	US 6063602	1999	Crescenzi, Sacceddu, Prosperi, Camilli, Fascetti, Porcelli
A. calcoaceticus	New lipopolysaccharide biosurfactant	JP 11269203	1999	Crescenzi, Sacceddu, Prosperi, Camilli, Fas cetti, Porcelli
Bacillus cell	Methods for producing polypeptides in mutants of bacillus cells	US 5958728	1999	Alan SlomaDavid SternbergLee AdamsStephen Brown
A. calcoaceticus	New lipopolysaccharide biosurfactant	EP 0924221	1999	Crescenzi, Sacceddu, Fascetti, Porcelli
A. calcoaceticus ATCC 31012, NRRL B-15847, NRRL B-15848, NRRL B-15849, NRRL B-15850 and NRRL B-15860	Bioemulsifier-stabilized hydrocarbosols	JP 1340969	2000	Murphy, Bolden Jr, Deal, Frances, Hayes, Futch Jr, Hrebenar
P. aeruginosa (USB-CS1)	Production of oily emulsions mediated by a microbial tensoactive agent	US 6060287	2000	Carlos Ali Rocha Dosinda Gonzalez, Maria Lourdes Iturralde Ulises Leonardo Lacoa Fernando Antonio Morales
Strain of *A. calcoaceticus*	Lipopolysaccharide biosurfactant	US6063602	2000	Giulio Prosperi Marcello Camilli Francesco Crescenzi Eugenio Fascetti Filippo Porcelli Pasquale Sacceddu
B. subtilis	Small peptides with antipathogenic activity, treated plants and methods for treating same	US 6183736B	2001	Anne-Laure Moyne Thomas, Clevel and Sadik Tuzun

Table 5 Contd. ...

...Table 5 Contd.

Microorganisms	Title of Invention	Patent No.	Year	Inventor/Applicant
A. lwoffi RAG-1, *A. calcoaceticus* BD4, BD413	Compositions containing bioemulsifiers and a method for their preparation	US 20020143071 WO 0248327 US 6512014	2002 2001 2003	Gutnick and Bach
A. radioresistens KA53	Novel bioemulsifiers	CA 2184897	NI	Rosenberg and Ron
A. calcoaceticus	New lipopolysaccharide biosurfactants	CA 2255157	NI	Prosperi, Porcelli, Fascetti, Crescenzi, Camilli, Sacceddu
Acinetobacter spp.	Helps to stabilize hydrocarbons, mineral oils, high viscosity hydrocarbons and/or high viscosity crude oil or emulsions and/or to disperse or remove unwanted hydrocarbons and oils from location(s)	US 6713062	2004	Merchant
C. albicans, C. rugosa, C. tropicalis, C. lipolytica, C. torulopsis	Microbial biosurfactants as agents for controlling pests	US 20050266036 A1	2005	Awada, Spendlove, Awada
P. aeruginosa	Biosurfactant production for development of biodegradable detergents	PI 1102592-1 A2	2011	Silvanito Alves Barbosa, Roberto Rodrigues De Souza
Pseudomonas spp.	Microbial biosurfactants as agents for controlling pests	US 7994138B2	2011	Salam, Awada Rex S Spendlove Mohamed Awada
Streptomyces sp.	Biosurfactant and production process	PI 1105951-6 A2	2011	Ana LF Porto, Eduardo, Santos, Leonie Sarubbo
C. guilliermondii	Production process of biosurfactant produced by *Candida guilliermondii* using agro-industrial waste	BR 102012023115	2012	Leonie Sarubbo, Valdemir, Santos, Raquel, Rufino, Juliana Luna
Strain of *Bacillus* sp.	*Bacillus* sp. Biosurfactants, Composition Including Same, Method For Obtaining Same, And Use Thereof	WO 2013/050700	2013	Coutte Francois, Jacques Philippe, Lecouturier Didier, Guez Jean-sebastien, Dhulster Pascal, Leclere Valerie, Bechet Max

Table 5 Contd. ...

...*Table 5 Contd.*

Microorganisms	Title of Invention	Patent No.	Year	Inventor/Applicant
C. bombicola ATCC 2214	Modified sophorolipid production by yeast strains and uses	EP 2580321 A1	2013	Soetaert, De MS, Saerens K, Roelants S, Van BI
B. subtilis, B. licheniformis, B. amyloliquefaciens, B. pumilus, B.s popilliae	Method of Producing Bosurfactants	US 20150037302	2015	Michael Paul Bralkowski Sarah Ashley Brooks Stephen M Hinton David Matthew Wright Shih-Hsin Yang
B. subtilis	Biosurfactant Production	EP 3502266 A1	2019	Bastiaens, Leen Simons, Queenie, Van Roy, Sandra, Ahmed, Safia Qazi, Muneer Ahmad, Naeem, Afshan Hina Umer, Aiman

* *A. calcoaceticus* RAG-1 was previously identified as *Arthrobacter* spp. RAG-1.

14. Future Potential of Biosurfactants

Generally chemically synthesized surfactants are non-biodegradable and cause a harmful impact on the Environment and society health. In this context consumer demand for eco-friendly and safe products has increased. Since the last few decades, biosurfactants have had remarkable growth due to their promising applications in petroleum, foods, beverages, cosmetics, detergents, textiles, paints, mining, cellulose, pharmaceutics and, oil recovery, cleanup of oil spills and the bioremediation of both soil and water. However, further, significant research and advances are needed to make Biosurfactant technology commercially viable.

Moreover, for the application of Biosurfactants in environmentally feasible technologies, promotion for their synthesis by using cheap carbon sources and renewable resources should be carried out (Danyelle Khadydja et al. 2016).

In this regard, use of cost effective production processes as well as identification of new producer organisms and the genetic manipulation of existing known producers should be promoted for continuing research. Besides, instead of using pure biosurfactants, crude fermentation broth and reuse of treated wastes will support sustainability of the environment. Therefore, biosurfactants can be proved to be ecofriendly biomolecules of the upcoming Era.

15. Conclusion

Since the last few decades, numerous microbial biosurfactants have been extensively investigated and their potential in diverse fields has been highlighted. However, very few had successful commercialization due to economic unfeasibility and issues with low yields. In this regard, innovative research into avenues to minimize cost and optimize the production process of bio surfactants has been intertwined with their diverse promising potential as agents of sustainability. Therefore, these amazing biomolecules are coming up as the leading ecofriendly products with increasing demand due to their significant potential with respect to implantation of sustainable industrial processes by using cost effective, renewable resources with the specific identity of "green" product in the upcoming era.

Acknowledgements

The authors are grateful to Research and Development Department, Sigma Wineries Pvt. Ltd. Nashik. Special thanks to Mr. Yash Khandelwal, Mr. VC Handore and Mrs. Hira V Handore, Mr. Saurabh Sonar, Mr. Abhijeet Jagtap, and Ms. Mrunal Ghayal for valuable assistance.

References

Abouseoud, M., A. Yataghene, A. Amrane and R. Maachi. 2008. Biosurfactant production by free and alginate entrapped cells of Pseudomonas fluorescens. J. Ind. Microbiol. Biotechnol. 35(11): 1303–1308.

Adamczak, M. and W. Bednarski. 2000. Influence of medium composition and aeration on the synthesis of biosurfactants produced by Candida antartica. Biotechnol. Lett. 22: 313–316.

Ahmad, Z., D. Crowley, N. Marina and S.K. Jha. 2016. Estimation of biosurfactant yield produced by Klebseilla sp. FKOD36 bacteria using artificial neural network approach. Measurement 81: 163–173.

Alan Sloma, David Sternberg, F. Lee and Adams Stephen. 1999. Brown Methods for producing polypeptides in mutants of bacillus cells. US5958728.

Amaral, P.F.F., J.M. da Silva, M. Lehocky, A.M.V. Barros-Timmons et al. 2006. Production and characterization of a bioemulsifier from Yarrowia lipolytica. Process Biochemistry 41(8): 1894–1898.

Ana, L.F. Porto, Eduardo F. Santos and Leonie A. Sarubbo. 2011. Biosurfactant and production process PI 1105951-6 A2.

Anne-Laure Moyne, Thomas E. Clevel and Sadik Tuzun. 2001. Small peptides with antipathogenic activity, treated plants and methods for treating same. US6183736B.

Arguelles-Arias, A., M. Ongena, B. Halimi, Y. Lara, A. Brans et al. 2009. *Bacillus amyloliquefaciens* GA1 as a source of potent antibiotics and other secondary metabolites for biocontrol of plant pathogens. Microbial Cell Factory 8: 63.

Arima, K., A. Kakinuma and G. Tamura. 1968. Surfactin, a cristalline peptidlipid surfactant produced by Bacillus subtilis: Isolation, characterization and its inhibition of fibrin clot formation. Biochem. Biophys. Res. Commun. 31: 488–494.

Arutchelvi, J.I., S. Bhaduri and P.V. Uppara. 2008. Doble *M. mannosylerythritol* lipids: A review. J. Ind. Microbiol. Biotechnol. 35(12): 1559–70.

Asselineau, C. and J. Asselineau. 1978. Trehalose containing glycolipids. Prog. Chem. Ftas Lipids 16: 59–99.

Awada, S., R. Spendlove and M. Awada. 2005. Microbial biosurfactants as agents for controlling pests. US 20050266036 A1.

Bach, H., Y. Berdichevsky and D. Gutnick. 2003. An exocellular protein from the oil-degrading microbe acinetobacter venetianus RAG-1 enhances the emulsifying activity of the polymeric bioemulsifier emulsan. Appl. Environ. Fticrobiol. 69: 2608–2615.

Bae, I.H., E.S. Lee, J.W. Yoo, S.H. Lee, J.Y. Ko et al. 2018. Mannosylerythritol lipids inhibit melanogenesis via suppressing ERK-CREB-MiTF-tyrosinase signalling in normal human melanocytes and a three-dimensional human skin equivalent. Exp. Dermatol., 2–5.

Banat, I.M., R.S. Makkar and S.S. Cameotra. 2000. Potential commercial applications of microbial surfactants. Applied Microbiology and Biotechnology 53(5): 495–508.

Banat, I.M., A. Franzetti, I. Gandolfi, G. Bestetti, M.G. Martinotti, R. Marchant et al. 2010. Microbial biosurfactants production, applications and future potential. Appl. Microbiol. Biotechnol. 87: 427–444.

Banat, I.M., S.K. Satpute, S.S. Cameotra, R. Patil, N.V. Nyayanit et al. 2014. Cost effective technologies and renewable substrates for biosurfactants' production. Front. Microbiol. Vol. 5. https://doi.org/10.3389/fmicb.2014.00697.

Barkay, T., S. Navon-Venezia, E. Ron and E. Rosenberg. 1999. Enhancement of solubilization and biodegradation of polyaromatic hydrocarbons by the emulsifier alasan. Applied Environmental Microbiology 65: 2697–2702.

Barros, F.F.C., C.P. Quadros, M.R. Maróstica and G.M. Pastore. 2007. Surfactina: Propriedades químicas, tecnológicase funcionais para aplicações em alimentos. Quím. Nova. 30: 1–14.

Bastiaens, Leen Simons, Queenie, Van Roy, Sandra Ahmed et al. 2019. Biosurfactant production. EP 3 502 266 A1.

Begley, M., P.D. Cotter, C. Hill and R. Paul Ross. 2009. Identification of a novel two-peptide antibiotic, lichenicidin, following rational genome mining for LanM proteins. Applied Environmental Microbiology 75: 5451–5460.

Bento, F.M., F.A. De oliveira Camargo, B.C. Okeke and W.T. Frankernberger. 2005. Diversity of biosurfactant producing microorganisms isolated from soils contaminated with diesel oil. Microbiological Research 160(3): 249–255.

Bodou, A.A., K.P. Drees and R.M. Maier. 2003. Distribution of biosurfactant-producing bacteria in undisturbed and contaminated arid southwestern soils. Applied and Environmental Microbiology 69(6): 3280–3287.

Cameotra, S.S. and R.S. Makkar. 2004. Recent applications of biosurfactants as biological and immunological molecules. Current Opinion in Microbiology 7(3): 262–266.

Cameron, D.R., D.G. Cooper and R.J. Neufeld. 1988. The mannoprotein of saccharomyces cerevisiae is an effective bioemulsifier. Appl. Environ. Fticrobiol. 54: 1420–1425.

Campos, J.M., T.L.M. Stamford, L.A. Sarubbo, J.M. Luna, R.D. Rufino et al. 2013. Microbial bio surfactants as additives for food industries. Biotechnol. Prog. 29: 1097–1108.

Carlos Ali Rocha Dosinda, Gonzalez Maria Lourdes, Iturralde Ulises Leonardo et al. 2000. Production of oily emulsions mediated by a microbial tenso-active agent. US6060287.

Carrillo, P.G., C. Mardaraz, S.I. Pitta-Alvarez and A.M. Giulietti. 1996. Isolation and selection of biosurfactant-producing bacteria. World J. Microbiol. Biotechnol. 12: 82–84.

Chakraborty, J. and S. Das. 2014. Biosurfactant-based bioremeditaion of toxic metals. pp. 167–201. *In*: Das, S. (ed.). Microbial Biodegradation and Bioremediation, 1st ed. Elsevier: Rourkela Odisha, India.

Chen, J., X. Song, H. Zhang and Y. Qu. 2006. Production, structure elucidation and anticancer properties of sophorolipid from wickerhamiella domercqiae. Enzyme Fticrob. Technol. 39: 501–506.

Ching, T. Hou. 1999. Microbial production of a novel compound 7,10-dihydroxy-8-octadecenoic acid from oleic acid. US5900496.

Choi, W.J., H.G. Choi and W.H. Lee. 1996. Effects of ethanol and phosphate on emulsan production by Acinetobacter calcoaceticus RAG-1. Journal of Biotechnology 45(3): 217–225.

Chtioui, O., K. Dimitrov, I.F. Gance and I. Nikov. 2010. Biosurfactants production by immobilized cells of Bacillus subtilis ATCC 21332 and their recovery by pertraction. Process Biochem. 45(11): 1795–1799.

Chtioui, O., K. Dimitrov, F. Gancel, P. Dhulster, I. Nikov et al. 2012. Rotating discs bioreactor, a new tool for lipopeptides production. Process Biochem. 47(12): 2020–2024.

Cirigliano, M.C. and G.M. Carman. 1984. Isolation of a bioemulsifier from Candida lipolytica. Appl. Environ. Microbiol. 48: 747–750.

Cirigliano, M.C. and G.M. Carman. 1985. Purification and Characterization of Liposan, a bioemulsifier from Candida lipolytica. Appl. Environ. Microbiol. 5: 846–850.

Cooper, D.G., J.E. Zajic and C. Denis. 1981. Surface-active properties of a biosurfactant from Corynebacterium lepus. J. Am. Oil Chem. Soc. 58: 77–80.

Cooper, D.G. and D.A. Paddock. 1983. Torulopsis Petrophilum and surface activity. Appl. Environ. Microbiol. 46: 1426–1429.

Cooper, D.G. and D.A. Paddock. 1984. Production of a biosurfactant from Torulopsis bombicola. Applied Environ. Microbiol. 47: 173–176.

Cooper, D.G. and B.G. Goldenberg. 1987. Surface active agents from two Bacillus species. Appl. Environ. Microbiol. 53: 224–9.

Cooper, D.G., S.N. Liss, R. Longay and J.E. Zajic. 1989. Surface activities of Mycobacterium and Pseudomonas. J. Ferment. Technol. 59: 97–101.

Cotter, P.D., C. Hill and R.P. Ross. 2005. Bacteriocins: Developing innate immunity for food. Nature Review Microbiology 3: 777–788.

Coutte Francois, Jacques Philippe, Lecouturier Didier, Guez Jean-sebastien, Dhulster Pascal et al. 2013. Bacillus sp. biosurfactants, composition including same, method for obtaining same, and use thereof. WO 2013/050700.

Crescenzi, F., P. Sacceddu, E. Fascetti and F. Porcelli. 1999. New lipopolysaccharide biosurfactant. EP0924221.

Crescenzi, F., P. Sacceddu, G. Prosperi, M. Camilli, E. Fascetti et al. 1999. New lipopolysaccharide biosurfactant. JP11269203.

Crescenzi, F., P. Sacceddu, G. Prosperi, M. Camilli, E. Fascetti, F. Porcelli et al. 1999. Lipopolysaccharide biosurfactant. US6063602.

Crich, D., A. de la ftora ft and R. Cruz. 2002. Synthesis of the mannosyl erythritol lipid ftEL A; confirmation of the configuration of the meso-erythritol moiety. Tetrahedron 58: 35–44.

Danyelle Khadydja, F., D. Santos Raquel, M. Rufino Juliana, A. Luna Valdemir et al. 2016. Biosurfactants: Multifunctional biomolecules of the 21st century. Int. J. Mol. Sci. 17(401): 1–31.

de Souza, J.T., de Boer, ft., de Waard P., Van Breek, T.A., J.A. Raajimakers et al. 2003. Biochemical, genetic and zoosporicidal properties of cyclic lipopeptide surfactants produced by Pseudomonas fluorescens. Appl. Environ. Fticrobiol. 69(12): 7161–7172.

Deleu, M. and M. Paquot.2004. From renewable vegetables resources to microorganisms: New trends in surfactants. R. Chimie. 7: 641–646.

Desai, A.J., Kft. Patel and J.D. Desai. 1988. Emulsifier production by pseudomonas fluorescence during the growth on hydrocarbon. Curr. Sci. 57: 500–501.

Desai, J.D. and I.M. Banat. 1997. Microbial production of surfactants and their commercial potential. Microbiol. Mol. Biol. Rev. 61: 47–64.

Dey, G., R. Bharti, G. Dhanarajan, S. Das, K.K. Dey et al. 2015. Marine lipopeptide Iturin A inhibits Akt mediated GSK3β and FoxO3a signaling and triggers apoptosis in breast cancer. Sci. Rep. 5: 1–14.

Dey, G., R. Bharti, A.K. Das, R. Sen, M. Mandal et al. 2017. Resensitization of Akt induced docetaxel resistance in breast cancer by 'Iturin A' a lipopeptide molecule from marine bacteria Bacillus megaterium. Sci. Rep. 7: 1–11.

Deziel, E., G. Paquette, R. Villemur, F. Lepine, J. Bisaillon et al. 1996. Biosurfactant production by a soil Pseudomonas strain G0072 owing on polycyclic aromatic hydrocarbons. Appl. Environ. Fticrobiol. 62(6): 1908–1912.

Dubey, K. and A. Juwarkar. 2001. Distillery and curd whey wastes as viable alternative sources for biosurfactant production. World J. Microbiol. Biotechnol. 17: 61–69.

Dubey, K.V., A.A. Juwarkar and S.K. Singh. 2005. Bio separations and downstream processing. Adsorption–desorption process using wood based activated carbon for recovery of biosurfactant from fermented) distillery waste water. Biotechnol. Prog. 21: 860–867.

Edwards, D.A., R.G. Luthy and Z. Liu. 1991. Solubilization of polycyclic aromatic hydrocarbons in micellar non-ionic surfactant solutions. Environ. Sci. Technol. 25(1): 127–133.

Elkeles, A., E. Rosenberg and E.Z. Ron. 1994. Production and secretion of the polysaccharide biodispersan of acinetobacter calcoaceticus A2 in protein secretion mutants. Appl. Environ. Fticrobiol. 60: 4642–4645.

Eswari, J.S., M. Anand and C. Venkateswarlu. 2016. Optimum culture medium composition for lipopeptide production by Bacillus subtilis using response surface model based ant colony optimization. Sadhana 4(11): 55–65.

Fakruddin, M.D. 2012. Biosurfactant: Production and application. J. Pet. Environ. Biotechnol. 3(4).

Felse, P.A., V. Shah, J. Chan and K.J. Rao. 2007. Sophorolipid biosynthesis by Candida bombicola from industrial fatty acid residues. Enzyme Microbiology Technology 40: 316–323.

Fiechter, A. 1992. Integrated systems for biosurfactant synthesis. Pure Applied Chem. 64: 1739–1743.

Fracchia, L., C. Ceresa, A. Franzetti, M. Cavallo, I. Gandolfi, J. Van Hamme et al. 2014. Industrial applications of biosurfactants. pp. 245–260. *In*: Kosaric, N. and F.V. Sukan (eds.). Biosurfactants: Production and Utilization—Processes, Technologies, and Economics, Boca Raton: CRC Press. Doi: 10.1201/b17599-15.

Fukui, T., T. Negi and Y. Tanaka. 1990. Novel *A. calcoaceticus* and novel biosurfactant. EP0401700.

Fukuoka, T., T. Morita, M. Konishi, T. Imura, D. Kitamoto et al. 2007. Characterization of new glycolipid biosurfactants, tri-acylated mannosylerythritol lipids, produced by pseudozyma yeasts. Biotechnol. Lett. 29: 1111–1118.

Gancel, F., L. Montastruc, T. Liu, L. Zhao, I. Nikov et al. 2009. Lipopeptide overproduction by cell immobilization on iron-enriched light polymer particles. Process Biochem. 44(9): 975–978.

Gautam, K.K. and V.K. Tyagi. 2006. Microbial surfactants: A review. J. Oleo Sci. 55(4): 155–166.

Giulio Prosperi, Marcello Camilli, Francesco Crescenzi, Eugenio Fascetti, Filippo Porcelli and Pasquale Sacceddu. 2000. Lipopolysaccharide biosurfactant. US 6063602.

Golyshin, P.M., H.L. Fredrickson, L. Giuliano, I.R. Rothme, K.N. Timmis et al. 1999. Effect of novel biosurfactants on biodegradation of polychlorinated biphenyls by pure and mixed bacterial cultures. New Microbiol. 22: 257–267.

Grand View Research. 2016. Biosurfactants Market by Product (Rhamnolipids, Sophorolipids, MES, APG, Sorbitan Esters, Sucrose Esters) Expected to Reach USD 2308.8 Million by 2020. Available online at: http://www.grandviewresearch.com/industry-analysis/biosurfactants-industry (Accessed April 15, 2016).

Grangemard, I., J. Wallach, R. Maget-Dana and F. Peypoux. 2001. Lichenysin: A more efficient cation chelator than surfactin. Appl. Biochem. Biotechnol. 90: 199–210.

Gupta, R., Q.K. Beg, S. Khan and B. Chahuan. 2002. An overview on fermentation, downstream processing and properties of microbial alkaline proteases. Appl. Microbiol. Biotechnol. 60: 381–395.

Gutnick, D.L. and E. Rosenberg. 1980. Production of α-emulsans. US 4230801.

Gutnick, D.L., E. Rosenberg and Y. Shabtai. 1980. Production of α-emulsans. US 4234689.

Gutnick, D.L. and E. Rosenberg. 1981. Cleaning oil-contaminated vessels with α-emulsans. US 4276094.

Gutnick, D.L., E. Rosenberg, I. Belsky and Z. Zosim. 1982. Apo α-emulsans. US 4311830.

Gutnick, D.L., E. Rosenberg, I. Belsky and Z. Zosim. 1982. Apo β-emulsans. US 4311829.

Gutnick, D.L., E. Rosenberg, I. Belsky and Z. Zosim. 1982. Apo psi emulsans. US 4311831.

Gutnick, D.L., E. Rosenberg, I. Belsky and Z. Zosim. 1982. Proemulsans. US 4311832.

Gutnick, D.L., E. Rosenberg, I. Belsky and Z. Zosim. 1983. Psi emulsans. US 4380504.

Gutnick, D.L., E. Rosenberg, I. Belsky and Z. Zosim. 1983. α-Emulsans. US 4395354.

Gutnick, D.L., E. Rosenberg, I. Belsky and Z. Zosim. 1983. Polyanionic heteropolysac charide biopolymers. US 4395353.

Gutnick, D.L., I. Belsky, Y. Shabtai, E. Rosenberg, Z. Zosim et al. 1983. Emulsans CA 1149302.

Gutnick, D.L.. E. Nestaas, E. Rosenberg and N. Sar. 1989. Bioemulsifier production by *A. calcoaceticus* strains. US4883757.

Gutnick, D.L.. E. Nestaas, E. Rosenberg and N. Sar. 1993. Bioemulsifier production by *A. calcoaceticus* strains. CA1316478.

Gutnick, D.L. and H.R. Bach. 2003. Compositions containing bioemulsifiers and a method for their preparation. US20020143071 WO0248327 US 6512014, 2002 2001 2003.

Haddad, N.I.A., X.Y. Liu, S.Z. Yang and B.Z. Mu. 2008. Surfactin isoforms from Bacillus subtilus HSO121: Separation and characterization. Protein Pept. Lett. 15: 265–269.

Haritash, A.K. and C.P. Kaushik. 2009. Biodegradation aspects of Polycyclic Aromatic Hydrocarbons (PAHs): A review. J. Hazard. Mater. 169: 1–15.

Hassan, M., T. Essam, A.S. Yassin and A. Salama. 2016. Optimization of rhamnolipid production by biodegrading bacterial isolates using Plackett-Burman design. Int. J. Biol. Macromol. 8: 573–579.

Hayes, M.E. 1989. Bioemulsified-containing personal care products for topical application to dermopathologic conditions of the skin and scalp. US 4870010.

Hayes, M.E. and G. Holzner. 1990. Soaps and shampoos containing bioemulsifiers. CA 1266238.

Hayes, M.E. 1991. Personal care products containing bioemulsifiers. US 4999195.

Hayes, M.E. 1992. Personal care products containing bioemulsifiers. CA 1300512.

Herman, D.C. and R.M. Maier. 2002. Biosynthesis and applications of glycolipid and lipopeptide biosurfactants. pp. 629–654. *In*: Kuo, T.M. and H.W. Gardner (eds.). Lipid Biotechnology. Marcel Dekker, New York (USA).

Heyd, M., M. Franzreb and S. Berensmeier. 2011. Continuous rhamnolipid production with integrated product removal by foam fractionation and magnetic separation of immobilized Pseudomonas aeruginosa. Biotechnol. Prog. 27(3): 706–716.

Hisatsuka, K., T. Nakahara, N. Sano and K. Yamada. 1971. Formation of rhamnolipid by Pseudomonas aeruginosa and its function in hydrocarbon fermentation. Agricultural Biology and Chemistry 35(5): 686–692.

Horowitz, S. and J.K. Currie. 1990. Novel dispersants of silicon carbide and aluminium nitrate. J. Dispersion Sci. Technol. 11: 637–659.

Hwang, K.A. and Y.S. Kim. 1997. Microorganism producing biosurfactant. KR 9706157.

Ian, W. Hamley. 2015. Lipopeptides: From self-assembly to bioactivity. Chem. Commun. 51: 8574–8583.

Ibukun O. Olasanmi and Ronald W. Thring. 2018. The role of biosurfactants in the continued drive for environmental sustainability. Sustainability 10(4817): 1–12.

Jadhav, M., S. Kalme, D. Tamboli and S. Govindwar. 2011. Rhamnolipid from Pseudomonas desmolyticum NCIM-2112 and its role in the degradation of Brown 3REL. Journal of Basic Microbiology 51: 1–12.

Jordan, R.N., E.P. Nichols and A.B. Cunningham. 1999. The role of (bio) surfactant sorption in promoting the bio-availability of nutrients localized at the solid-water interface. Water Sci. Technol. 39(7): 91–98.

Kappeli, O. and W.R. Finnerty. 1979. Partition of Alkane by an extracelular vesicle derived from hexadecane-grow acinetobacter. J. Bacteriol. 140: 707–712.

Kappeli, O., P. Walther, M. Mueller and A. Fiechter. 1984. Structure of cell surface of the yeast Candida tropicalis and its relation to hydrocarbon transport. Arch. Fticrobiol. 138: 279–282.

Karanth, N.G.K., P.G. Deo and N.K. Veenanadig. 1999. Microbial production of biosurfactants and their importance. Curr. Sci. 77: 116–123.

Karsa, D.R., R.M. Bailey, B. Shelmerdine and S.A. McCann. 1999. Overview: A decade of change in the surfactant industry. pp. 1–22. *In*: Karsa, D.R. (ed.). Industrial Applications of Surfactants. Royal Society of Chemistry, 4.

Kim, H., B. Yoon, C. Lee, H. Suh, H. Oh et al. 1997. Production and properties of a lipopeptide biosurfactant from *Bacillus subtilis* C9. Journal of Fermentation and Bioengineering 84(1): 41–46.

Kim, H.S., B.D. Yoon, D.H. Choung, H.M. Oh et al. 1999. Characterization of a biosurfactant, mannosylerythritol lipid produced from Candida sp. SY16. Appl. Microbiol. Biotechnol. 52(5): 713–21.

Kim, H.S., E.J. Lim, S.O. Lee, J.D. Lee, T.H. Lee et al. 2000. Purification and characterization of biosurfactants from nocardia sp. L-417. Biotechnol. Appl. Biochem. 31: 249–253.

Kiran, G.S., L.A. Nishanth, S. Priyadharshini, K. Anitha, J. Selvin et al. 2014. Effect of Fe nanoparticle on growth and glycolipid biosurfactant production under solid state culture by marine Nocardiopsis sp. MSA13A. BMC Biotechnol. 14(1): 1.

Kitamoto, D., H. Yanagishita, T. Shinbo, T. Nakane, C. Kamisawa et al. 1993. Surface active properties and antimicrobial activities of mannosylerythritol lipids as biosurfactants produced by Candida antarctica. Journal of Biotechnology 29: 91–96.

Koronelli, T.V., T.I. Komarova and Y.V. Denisov. 1983. Chemical composition and role of peptidoglycolipid of pseudomonas aeruginosa. Ftikrobiologiya 52: 767–770.

Kosaric, N. 1992. Biosurfactants in industry. Pure & Appl. Chern. 64(11): 1731–1737.

Kosaric, N. 2001. Biosurfactants and their application for soil bioremediation. Food Technology and Biotechnology 39: 295–304.

Kretschmer, A., H. Bock and F. Wagner. 1982. Chemical and physical characterization of interfacial-active lipids from rhodococcus erthropolis grown on n-alkane. Appl. Environ. Fticrobiol. 44: 864–870.

Krishnaswamy, M., G. Subbuchettiar, T.K. Ravi and S. Panchaksharam. 2008. Biosurfactants properties, commercial production and application. Current Science 94: 736–747.

Kronemberger, F.A., C.P. Borges and D.M. Freire. 2010. Fed-Batch Biosurfactant production in a bioreactor. International Review of Chemical Engineering 2: 513–518.

Kumar, P.N., T.H. Swapna, M.Y. Khan, G. Reddy, B. Hameeda et al. 2017. Statistical optimization of antifungal iturin A production from Bacillus amyloliquefaciens RHNK22 using agro-industrial wastes. Saudi J. of Biol. Sci. 24: 1722–1740.

Lai, C.C., Y.C. Huang, Y.H. Wei and J.S. Chang. 2009. Biosurfactant-enhanced removal of total petroleum hydrocarbons from contaminated soil. Journal of Hazardous Materials 167: 609–614.

Lee, S., S.J. Lee, S. Kim, I. Park, Y. Lee et al. 2008. Characterization of new biosurfactant produced by Klebsiella sp. Y6-1 isolated from waste soy bean oil. Bioresour. Technol. 99: 2288–2292.

Leonie, A. Sarubbo, Valdemir A. Santos, Raquel D. Rufino and Juliana M. Luna. 2012. Production process of biosurfactant produced by Candida guilliermondii using agro-industrial waste. BR 102012023115.

Levy, N., Y. Bar and S. Magdassi. 1990. Colloids Surfactants. 48: 337.

Lindahl, A., A. Faris, T. Wadstrom and S. Hjerten. 1981. A new test based on 'salting out' to measure relative hydrophobicity of bacterial cells. Biochim. Biophys. Acta 677: 471–476.

Liu, J., C. Vipulanandan, T.F. Cooper and G. Vipulanandan. 2013. Effects of Fe nanoparticles on bacterial growth and biosurfactant production. J. Nanopart. Res. 15(1): 1–13.

Maget-Dana, R. and F. Peypoux. 1994. Toxicology, Iturins, a special class of pore-forming lipopeptides: Biological and physicochemical properties. Toxicology 87(1-3): 151–174.

Makkar, R.S. and S.S. Cameotra. 2002. An update on the use of uncoventional substrates for biosurfactant production and their new applications. Appl. Micobiol. Biotechnol. 58: 428–434.

Makkar, R.S. and K.J. Rockne. 2003. Comparison of synthetic surfactants and biosurfactants in enhancing biodegradation of polycyclic aromatic hydrocarbons. Environ. Toxicol. Chem. 22(10): 2280–2292.

Makkar, R.S., S.S. Cameotra and I.M. Banat. 2011. Advances in utilization of renewable substrates for biosurfactant production. AMB Express. 1(5): 1–19.

Marchant, R. and I.M. Banat. 2012. Microbial biosurfactants: Challenges and opportunities for future exploitation. Trends Biotechnol. 30: 558–565.

Marchant, R., S. Funston, C. Uzoigwe, P.K.S.M. Rahman, I.M. Banat et al. 2014. Production of biosurfactants from nonpathogenic bacteria. pp. 73–82. *In*: Kosaric, N. and F.V. Sukan (eds.). Biosurfactants: Production and Utilization-Processes, Technologies, and Economics. Boca Raton: CRC Press.

McDonald, C.R., D.G. Cooper and J.E. Zajic. 1981. Surface-active lipids from nocardia erythropolis grown on hydrocarbons. Appl. Environ. Fticrobiol. 41: 117–123.

Mclnerney, M.J., M. Javaheri and P. David. 1990. Properties of the biosurfactant produced by Bacillus licheniformis strain JF-2. Journal of Industrial Microbiology 5: 95–102.

Mercadé, M.E., M.A. Manresa, M. Robert, M.J. Espuny, C. Andrés et al. 1993. Olive oil mill effluent (OOME): New substrate for biosurfactant production. Bioresour. Technol. 43: 1–6.

Mercadé, M.E., L. Monleon, C. de Andres, I. Rodon, E. Martinez et al. 1996. Screening and selection of surfactant-producing bacteria from waste lubricating oil. J. Appl. Bacteriol. 81: 161–168.

Merchant, J.L. 2004. Helps to stabilize hydrocarbons, mineral oils, high viscosity hydrocarbons and/or high viscosity crude oil or emulsions and/or to disperse or remove unwanted hydrocarbons and oils from location(s). US 6713062.

Michael Paul, Bralkowski Sarah, Ashley Brooks, Stephen M. Hinton, David Matthew, Wright Shih-Hsin Yang et al. 2015. Method of Producing Biosurfactants. US 20150037302.

Michael, E. Stanghellini, Raina, M. Miller, Scott Lynn Rasmussen, Do Hoon Kim and Yimin Zhang. 1998. Microbially produced rhamnolipids (biosurfactants) for the control of plant pathogenic Zoosporic Fungi. US 5767090.

Michael J. Mclnerney, Gary E. Jenneman, Roy M. Knapp, Donald E. Menzie, all of Norman, Okla. 1985. Biosurfactant and enhanced oil recovery. US 4522261.

Mohan, P.K., G. Nakhla and E.K. Yanful. 2006. Biokinetics of biodegradation of surfactants under aerobic, anoxic and anaerobic conditions. Water Res. 40: 533–540.

Mulligan, C.N., R.N. Yong and B.F. Gibbs. 2001. Heavy metal removal from sediments by biosurfactants. J. Hazard Mater. 85: 111–125.

Mulligan, C.N. 2005. Environmental applications for biosurfactants. Environ. Pollution 133: 183–198.

Murphy, P.L., P.L. Bolden Jr., J. Deal, I. Frances, M.E. Hayes, L.E. Futch Jr. and K.R. Hrebenar. 2000. Bioemulsifier-stabilized hydrocarbosols. JP 1340969.

Muthusamy, K., S. Gopalakrishnan, T.K. Ravi and P. Sivachidambaram. 2008. Biosurfactants: Properties, commercial production and application. Curr. Sci. 94: 736–747.

Nakata, K. 2000. Two glycolipids increase in the bioremediation of halogenated aromatic compounds. J. Biosci. Bioeng. 89: 577–581.

Navonvenezia, S., Z. Zosim, A. Gottieb, R. Legmann, S. Carmeli et al. 1995. Alasan, a new bioemulsifier from acinetobacter radioresistens. Appl. Environ. Fticrobiol. 61: 3240–3244.

Nayarisseri, A., P. Singh and S. Singh. 2018. Isolation and characterization of biosurfactant producing *Bacillus subtilis* strain ANSKLAB03. Bioinformation 14(6): 304–314.

Ndlovu, T., S. Khan and W. Khan. 2016. Distribution and diversity of biosurfactant-producing bacteria in a wastewater treatment plant. Environ. Sci. Pollut. Res. Int. 23(10): 9993–10004.

Ndlovu, T., M. Rautenbach, J.A. Vosloo, S. Khan, W. Khan et al. 2017. Characterisation and antimicrobial activity of biosurfactant extracts produced by Bacillus amyloliquefaciens and Pseudomonas aeruginosa isolated from a wastewater treatment plant. AMB Expr. 7: 108: 1–19.

Nerurkar, A.S.2010. Structural and molecular characteristics of lichenysin and its relationship with surface activity. Adv. Exp. Med. Biol. 672: 304–315.

Nestaas, E., K.R. Hrebenar, J.M. Lewis and G.M. Whitesides. 1993. Hydrophobically modified proteins. US 5212235.

Nielsen, T.H. and J. Sørensen. 2003. Production of cyclic lipopeptides by Pseudomonas fluorescens strains in bulk soil and in the sugar beet rhizosphere. Appl. Environ. Microbiol. 69: 861–868.

Nitschke, M., C. Ferraz and G.M. Pastore. 2004. Selection of microorganisms for biosurfactant production using agro industrial wastes. Braz. J. Microbiol. 35: 81–85.

Nitschke, M., S.G. Costa, R. Haddad, L.A.G. Gonc, N.C. Alves et al. 2005. Oil wastes as unconventional substrates for rhamnolipid biosurfactant production by Pseudomonas aeruginosa LBI. Biotechnol. Prog. 21: 1562–1566.

Nitschke, M., S.G. Costa and J. Contiero. 2010. Structure and applications of a rhamnolipid surfactant produced in soybean oil waste. Appl. Biochem. Biotechnol. 160: 2066–2074.

Ohno, A., T. Ano and M. Shoda. 1993. Effect of temperature change and aeration on the production of the antifungal peptide antibiotic iturin by Bacillus subtilis NB22 in liquid cultivation. Journal of Fermentation and Bioengineering 75: 463–465.

Ohno, A., T. Ano and M. Shoda. 1995. Production of a lipopeptide antibiotic, surfactin, by recombinant Bacillus subtilis in solid state fermentation. Biotechnol. Bioeng. 47: 209–214.

Ohno, A., T. Ano and M. Shoda. 1996. Use of soybean curd residue, okara, for the solid state substrate in the production of a lipopeptide antibiotic, iturin A, by Bacillus subtilis NB22. Process Biochem. 31: 801–806.

Onwosi, C.O. and F.J.C. Odibo. 2013. Rhamnolipid biosurfactant production by Pseudomonas nitroreducens immobilized on Ca2+ alginate beads and under resting cell condition. Annals Microbiol. 63(1): 161–165.

Osugi, T. 1988. Bathing agent. JP 63156714.

Park, Ae-R., K. Jung-Ill and J. Chang-Ho. 1999. Novel microorganism Acinetobacter spp. KRC-K4 and process for preparing bioemulsifier using the same. KR 170107.

Patil, Y. and P. Rao. 2015. Industrial waste management in the era of climate change—A smart sustainable model based on utilization of active and passive biomass. pp. 2079–2092. *In*: Walter Leal, F. (ed.). Handbook on Climate Change Adaptation. Springer-Verlag Berlin Heidelberg, Germany.

Pendse, A. and K. Arun. 2018. Use of various screening methods for isolation of potential biosurfactant producing microorganism from oil-contaminated soil samples. Journal of Pharmacy Research 12(4): 599–605.

Peypoux, F., Jft. Bonmatin and J. Wallach. 1999. Recent trends in the biochemistry of surfactin. Appl. Fticrobiol. Biotechnol. 51: 553–563.

Pornsunthorntawee, O., S. Chavadej and R. Rujiravanit. 2009. Solution properties and vesicle formation of rhamnolipid biosurfactants produced by Pseudomonas aeruginosa SP4. Colloids Surf. B 72: 6–15.

Portilla-Rivera, O., A. Torrado, J.M. DomiÌ·nguez and A.B. Moldes. 2008. A stability and emulsifying capacity of biosurfactants obtained from lignocellulosic sources using Lactobacillus pentosus. J. Agricu. Food Chem. 56: 8074–8080.

Prosperi, G., M. Camilli, F. Crescenzi, E. Fascetti, F. Porcelli et al. 1999. New lipopolysaccharide biosurfactant. JP 11269203.

Prosperi, G., F. Porcelli, E. Fascetti, F. Crescenzi, M. Camilli et al. 1999. New lipopolysaccharide biosurfactant. CA 2255157, NI.

Pruthi, V. and S. Cameotra. 1997. Rapid identification of biosurfactant producing bacterial strains using a cell surface hydrophobicity technique. Biotechnology Techniques 11(9): 671–674.

Rahman, P.K.S.M. and E. Gakpe. 2008. Production, characterisation and applications of biosurfactants—Review. Biotechnology 7(2): 360–370. https://doi.org/10.3923/biotech.2008.360.370.

Rashedi, H., M.M. Assadi, B. Bonakdarpour and E. Jamshidi. 2005. Environmental importance of rhamno lipid production from molassess carbon source. Int. J. Environ. Sci. Technol. 2: 59–62.

Reddy Obula Chittepu. 2019. Isolation and characterization of biosurfactant producing bacteria from groundnut oil cake dumping site for the control of foodborne pathogens. Grain & Oil Science and Technology 2(1): 15–20.

Robert, M., M.E. Mercade, M.P. Bosch, J.L. Parra, M.J. Espuny et al. 1989. Effect of carbon source on biosurfactant production by Pseudomonas aeruginosa 44T1. Biotechnol. Lett. 11: 871–874.

Rodrigues, L., H. van der Mei, I.M. Banat, J. Teixeira, R. Oliveira et al. 2006. Inhibition of microbial adhesion to silicone rubber treated with biosurfactant from Streptococcus thermophilus A. FEMS Immunol. Med. Microbiol. 46(1): 107–112.

Rodrigues, L.R. and J.A. Teixeira. 2008. Biosurfactants production from cheese whey. pp. 81–104. *In*: Cerd'an, M.E., M. Gonz'alez-Siso and M. Becerra (eds.). Advances in Cheese Whey Utilization. Trivandrum: India-Transworld Research Network, 8.

Roelants, S.L., K. Ciesielska, S.L. De Maeseneire, H. Moens, B. Everaert et al. 2016. Towards the industrialization of new biosurfactants: Biotechnological opportunities for the lactone esterase gene from Starmerella bombicola. Biotechnol. Bioeng. 113(3): 550–559.

Ron, E.Z. and E. Rosenberg. 1996. Novel bioemulsifiers. WO 9620611.

Ron, E.Z. and E. Rosenberg. 1998. Bioemulsifiers. US 5840547.

Ron, Rosenberg. 2001. Natural roles of biosurfactants. Environmental Microbiololgy 3: 229–236.

Rosenberg, E. and E.Z. Ron. 2006. Novel bioemulsifiers. CA 2184897, NI.

Rosenberg, E. 1986. The effect of pretreatments on surfactin production from potato process effluent by Bacillus subtilis. CRC Crit. Rev. Biotechnol. 3(109): 487–501.

Rosenberg, E., C. Rubinovitz, R. Legmann and E.Z. Ron. 1988. Purification and chemical properties of acinetobacter calcoaceticus A2 biodispersan. Appl. Environ. Fticrobiol. 54: 323–326.

Rosenberg, E. 1993. Exploiting microbial growth on hydrocarbons—new markets. Trends Biotechnol. 11(10): 419–424.

Rosenberg, E. and E.Z. Ron. 1997. Bioemulsans: Microbial polymeric emulsifiers. Curr. Opin. Biotechnol. 8: 313–316.

Rosenberg, E. and E.Z. Ron 1999. High- and low-molecular-mass microbial surfactants. Appl. Fticrobiol. Biotechnol. 52(2): 154–162.

Rosenberg, M., D. Gutnick and E. Rosenberg. 1980. Adherence of bacteria to hydrocarbons: A simple method for measuring cell-surface hydrophobicity. FEMS Microbiology Letters 9: 29–33.

Saharan, B.S., R.K. Sahu and D. Sharma. 2011. A review on biosurfactants: Fermentation, current developments and perspectives. Genetic Engineering and Biotechnology Journal GEBJ-29: 1–14.

Sahebnazar, Z., D. Mowla and G. Karimi. 2018. Enhancement of Pseudomonas aeruginosa growth and rhamnolipid production using iron-silica nanoparticles in low-cost medium. J. Nanostructures 8(1): 1–10.

Sajna, K.V., R.K. Sukumaran, H. Jayamurthy, K.K. Reddy, S. Kanjilal, R.B.N. Prasad and A. Pandeya. 2013. Studies on biosurfactants from Pseudozyma sp. NII 08165 and their potential application as laundry detergent additives. Biochem. Eng. J. 78: 85–92.

Salam, M., S. AwadaRex and Spendlove Mohamed Awada. 2011. Microbial biosurfactants as agents for controlling pests. US 7994138B2.

Salihu, A., I. Abdulkadir and M.N. Almustapha. 2009. An investigation for potential development of biosurfactants. Microbiol. Mol. Biol. Rev. 3: 111–117.

Santos, D.K.F., R.D. Rufino, J.M. Luna, V.A. Santos, L.A. Sarubbo et al. 2016. Biosurfactants: Multifunctional biomolecules of the 21st century. Int. J. Mol. Sci. 17: 401.

Santos Guerra, L., O. Kappeli and A. Fiechter. 1986. Dependence of Pseudomonas aeruginosa continuous culture biosurfactant production on nutritional and environmental factor. Applied Microbiol. Biotechnol. 24: 443–448.

Sari, C.N., R. Hertadi, M. Gozan and A.M. Roslan. 2019. Factors affecting the production of biosurfactants and their applications in Enhanced Oil Recovery (EOR). A review. IOP Conf. Ser.: Earth Environ. Sci. 353: 012048.

Satpute, S.K., B.D. Bhawsar, P.K. Dhakephalkar and B.A. Chopade. 2008. Assessment of different screening methods for selecting biosurfactant producing marine bacteria. Indian J. Mar. Sci. 37: 243–50.

Schneider, J., K. Taraz, H. Budzikiewicz, M.T. Deleu, P. Honart et al. 1999. The structure of two fengycins from bacillus subtilis S499. Z. Naturforsch. 54: 859–865.

Sen, R. and T. Swaminathan. 2005. Characterization of concentration and purification parameters and operating conditions for the small-scale recovery of surfactin. Process Biochem. 40: 2953–2958.

Sheppard, J.D. and C.N. Mulligan. 1987. The production of surfactin by Bacillus subtilis grown on peat hydrolysate. Appl. Microbiol. Biotechnol. 27: 110–116.

Siegmund, I. and F. Wagner. 1991. New method for detecting rhamnolipids excreted by Pseudomonas spp. during growth on mineral agar. Biotechnol. Tech. 5: 265–268.

Silva, R.C.F.S., D.G. Almeida, J.M. Luna, R.D. Rufino, V.A. Santos et al. 2014. Applications of biosurfactants in the petroleum industry and the remediation of oil spills. Int. J. Mol. Sci. 15: 12523–12542.

Silvanito Alves Barbosa and Roberto Rodrigues De Souza. 2011. Biosurfactant production for development of biodegradable detergent. PI 1102592-1 A2.

Singh, P. and Y. Patil. 2018. Biosurfactant production: Emerging trends and promising strategies. Journal of Applied Microbiology 126(2): 1–13.

Sivapathasekaran, C. and R. Sen. 2013. Performance evaluation of an ANN–GA aided experimental modeling and optimization procedure for enhanced synthesis of marine biosurfactant in a stirred tank reactor. J. Chem. Technol. Biotechnol. 88: 794–799.

Smyth, T.J., A. Perfumo, R. Marchant, I.M. Banat, M. Chen et al. 2010. Directed microbial biosynthesis of deuterated biosurfactants and potential future application to other bioactive molecules. Appl. Microbiol. Biotechnol. (in press).

Soetaert, W., M.S. De, K. Saerens, S. Roelants, B.I. Van et al. 2013. Modified sophorolipid production by yeast strains and uses. EP 2580321 A1.

Solaiman, D.K.Y., R.D. Ashby, T.A. Foglia, A. Nuñez, W.N. Marmer et al. 2003. Fermentation-based processes for the conversion of fats, oil and derivatives into biopolymers and biosurfactants. *In*: Proceedings of 31st United States-Japan Cooperative Program in Natural Resources (UJNR), Protein Resources Panel (Tsukuba: Eastern Regional Research Center, ARS, USDA), V1–V10.

Song, B., W. Zhu, R. Song, F. Yan, Y. Wang et al. 2019. Exopolysaccharide from Bacillus vallismortis WF4 as an emulsifier for antifungal and antipruritic peppermint oil emulsion. Int. J. Biol. Macromol. 125: 436–444.

Sudesh, K., H. Abe and Y. Doi. 2000. Synthesis, structure and properties of polyhydroxyalkanoates: Biological polyesters. Prog. Polymer. Sci. 25: 1503–1555.

Susumu, H. and M. Zenrou. 1998. Bacterial preparation for agricultural use. US 5733355.

Sutyak, K.E., R.E. Wirawan, A.A. Aroutcheva and M.L. Chikindas. 2008. Isolation of the Bacillus subtilis antimicrobial peptide subtilosin from the dairy product derived Bacillus amyliquefaceiens. Journal of Applied Microbiology 104: 1067–1074.

Tanaka, Y., T. Fukui and T. Negi. 1991. Novel *A. calcoaceticus* and novel biosurfactant. JP 03130073.

Tang, J.S., H. Gao, K. Hong, Y. Yu, M.M. Jiang, H.P. Lin et al. 2007. Complete assignments of 1H and 13C NMR spectral data of nine surfactin isomers. Magn. Reson. Chem. 45: 792–796.

Teichmann, B., U. Linne, S. Hewald, M. Marahiel, M. Bolker et al. 2007. A biosynthetic gene cluster for a secreted cellobiose lipid with antifungal activity from *Ustilago maydis*. Molecular Microbiology 66: 525–533.

Thavasi, R., R. Marchant and I.M. Banat. 2011. Biosurfactant applications in agriculture. pp. 313–326. *In*: Kosaric, N. and F.V. Sukan (eds.). Biosurfactants: Production and Utilization—Processes, Technologies, and Economics. New York, NY: CRC Press.

Tokumoto, Y., N. Nomura, H. Uchiyama, T. Imura, T. Morita et al. 2009. Structural characterization and surface-active properties of a succinoyl trehalose lipid produce by Rhodococcus sp. SD-74. J. Oleo Sci. 58: 97–102.

Toren, A., S. Navon-Venezia, E.Z. Ron and E. Rosenberg. 2001. Emulsifying activities of purified alasan proteins from acinetobacter radioresistens KA53. Appl. Environ. Fticrobiol. 67: 1102–1106.
Toren, A., G. Segal, E.Z. Ron and E. Rosenberg. 2002. Structure-function studies of the recombinant protein bioemulsifier AlnA. Environ. Fticrobiol. 4: 257–261.
Tullock, P., A. Hill and J.F.T. Spencer. 1967. A new type of marocyclic lactone from torulopsis apicola. J. Chem. Soc. Chem. Commun., 584–586.
Van Hoogmoed, C.G., ft. van der Kuijl-Booij, H.C. van der Mei and H.J. Buscher. 2000. Inhibition of streptococcus mutans NS adhesion to glass with and without a salivary conditioning film by biosurfactant-releasing streptococcus mitis strains. Appl. Environ. Microbiol. 66: 659–663.
Vanessa, Walter, S. Christoph and H. Rudolf. 2010. Screening concepts for the isolation of biosurfactant producing microorganisms. Article in Advances in Experimental Medicine and Biology 672: 1–13.
Vater, J., B. Kablitz, C. Wilde, P. Franke, N. Mehta, S. Cameotra et al. 2002. Matrix-assisted laser desorption ionization-time of flight mass spectrometry of lipopeptide biosurfactants in whole cells and culture filtrates of Bacillus subtilis C-1 isolated from petroleum sludge. Appl. Environ. Microbiol. 68: 6210–6219.
Wayman, ft., A.D. Jenkins and A.G. Kormady. 1984. Biotechnology for oil and fat industry. J. Am. Oil Chem. Soc. 61: 129–131.
Weber, L., C. Doge, G. Haufe, R. Hommel and H.P. Kleber. 1992. Oxygenation of hexadecane in the biosynthesis of cyclic glycolipids in Torulopsis apicola. Biocata 5: 267–292.
Yakimov, M.M., H.L. Fredrickson and K.N. Timmis. 1996. Effect of heterogeneity of hydrophobic moieties on surface activity of lichenysin A, a lipopeptide biosurfactant from bacillus licheniformis BAS50. Biotechnol. Appl. Biochem. 23: 13–18.
Yakimov, M.M., M.M. Amro, M. Bock, K. Boseker, H.L. Fredrickson et al. 1997. The potential of *Bacillus licheniformis* strains for in situ enhanced oil recovery. J. Petrol. Sci. Eng. 18: 147–160.
Yeh, M.S., Y.H. Wei and J.S. Chang. 2005. Enhanced production of surfactin from Bacillus subtilis by addition of solid carriers. Biotechnol. Progress. 21(4): 1329–1334.
Zinjarde, S.S. and A. Pant. 2002. Emulsifier from a tropical marine yeast Yarrowia lipolytica NCIM 3589. J. Basic. Microbiol. 42: 67–73.
Zouari, R., S. Ellouze-Chaabouni and D. Ghribi-Aydi. 2014. Optimization of Bacillus subtilis SPB1 biosurfactant production under solid-state fermentation using by-products of a traditional olive mill factory. Achievements in the Life Sciences 8: 162–169.
Zouari, R., D. Moalla-Rekik, Z. Sahnoun, T. Rebai, S. Ellouze-Chaabouni, D. Ghribi-Aydi et al. 2016. Evaluation of dermal wound healing and *in vitro* antioxidant efficiency of *Bacillus subtilis* SPB1 biosurfactant. Biomed. Pharmacother. 84: 878–891.
Zukerberg, A., A. Diver, Z. Peeri, D.L. Gutnick, E. Rosenberg et al. 1979. Emulsifier of arthrobacter RAG-1: Chemical and physical properties. Appl. Environ. Fticrobiol. 37(3): 414–420.

7

Biosurfactant as an Antimicrobial and Biodegradable Agent

A Review

Veena Kumari

1. Introduction

Biosurfactants can be defined as the surface-active antimicrobial and biodegradable biomolecules produced by microorganisms having variable applications. Now a days, due to their unique properties like specificity, low toxicity and relative ease of preparation, these surface-active biomolecules have attracted more importance due to their wide range of applications. Having a wide range of unique functional properties, biosurfactants were used in several industries including organic chemicals, petroleum, petrochemicals, mining, metallurgy (mainly bioleaching), agrochemicals, fertilizers, foods, beverages, cosmetics, pharmaceuticals and many others. Surfactants (surface-active materials) are part of the most versatile group of chemicals potentially used in various industries which include detergents, paints, paper products, pharmaceuticals, cosmetics, petroleum, food, and water treatment (Mahamallik and Pal 2017, Varjani and Upasani 2017, Phulphoto 2020). Currently, commercial surfactants are synthesized from petrochemicals, animal fats, plants, and microorganisms. Studies have shown that majority of the production markets rely on petrochemicals (De Almeida et al. 2016). The development of new strategies for replacing petroleum, coal, and natural gas-based products with renewable, biodegradable, and sustainable green energies are the novel challenges for humans and environmental protection agencies (Ajala et al. 2015, Shaban and Abd-Elaal 2017). They can be used as

Department of Botany, Buddha P.G. College Kushinagar, Gorakhpur, India.

emulsifiers as well as demulsifiers, wetting agents, foaming agents, spreading agents, functional food ingredients and detergents. The interfacial surface tension reducing properties of biosurfactants provide them an important role to play in oil recovery and bioremediation of heavy crude oil (Volkering et al. 1998).

Usually, surfactants extracted from organic compounds contain both hydrophilic (water-loving) and hydrophobic (water-hating) or oil-loving compartments (Fracchia et al. 2012). Natural surfactants (saponins) belong to the group of secondary metabolites which are abundantly found in various plants and some marine organisms (Boruah and Gogoi 2013, Cheeke 2010). Generally, they can be found in different parts of plants, including seeds, leaves, roots, flowers, and fruits (Cheeke 2010). The word saponin is a Latin word which means foaming agents from the plants. Saponins are soluble in water with higher molecular weights ranging between 600–2000 Da (Cheeke 2010).

The major functions of biosurfactants are to increase the surface area of hydrophobic substrates (Rosenberg and Ron 1999). Biosurfactants are also used to increase the bioavailability of hydrophobic substrates through solubilization/desorption. They play an eminent role in regulation of the attachment and removal of microorganisms from the surfaces. Biosurfactants possess both hydrophilic and hydrophobic regions which help in aggregation of interfaces between fluids with different polarities such as hydrocarbons and water (Banat 1995, Karanth et al. 1999) hence, they decrease interfacial surface tension (Volkering et al. 1998). They also help in enhancement of the nutrient transport across membranes and also affect the various host-microbe interactions.

Biosurfactants have several advantages including their biodegradability, biocompatibility and digestibility. The biosurfactants can be used in environmental cleanup by biodegradation and detoxification of industrial effluents and in bioremediation of contaminated soil and water. Their specificity and availability as raw materials also make them most valuable surfactants (Olivera et al. 2003). Thus due to their unique and distinct properties the biosurfactants are also suitable for commercial applications. The distinctive properties of microbial surfactants are related to their surface activity, tolerance to pH, temperature and ionic strength, biodegradability, low toxicity, emulsifying and demulsifying ability and antimicrobial activity (Chakrabarti 2012). The aim of the present review is to provide information about biosurfactants and their potential application as eco-friendly products and alternative compounds to the synthetic origin surfactants. The important features of biosurfactants are discussed as follows:

2. Efficient Role of Biosurfactant in Reducing Surface Tension

A Biosurfactant plays an important role in reducing surface tension and the interfacial tension. *B. subtilis* produce an important chemical which reduces surface tension of water from 25 to 30 mN m^{-1} and interfacial tension of water/hexadecane interface to less than 1 mN/m^{-1} (Cooper et al. 1981). *Pseudomonas aeruginosa* produce rhamnolipids which decrease surface tension of water to 25 mN m^{-1} and interfacial tension of water/hexadecane to a value less than 1 mN m^{-1} (Syldatk et al. 1985). In general, biosurfactants are more effective and efficient and their Critical Micelle

Concentration (CMC) is about several times lower than chemical surfactants (Desai and Banat 1997).

3. Biosurfactants and Stressed Environments

The biosurfactant production from extremophiles has gained attention in previous decades for their considered commercial interest. Most of the biosurfactants and their surface activities are resistant towards environmental factors such as temperature and pH. McInerney et al. 1990 reported that lichenysin from *Bacillus licheniformis* was found to be resistant to temperatures up to 50°C, pH between 4.5 and 9.0 and NaCl and Ca concentrations up to 50 and 25 g L^{-1}, respectively. Another biosurfactant produced by *Arthrobacter protophormiae* was found to be both thermostable (30–100°C) and pH (2 to 12) stable (Singh and Cameotra 2004). Since, industrial processes involve exposure to extremes of temperature, pH and pressure, it is necessary to isolate novel microbial products that are able to function under these stressed conditions (Cameotra and Makkar 2004). The pH plays an important role in growth and development of micro-organisms under stressed environmental conditions.

4. Biosurfactant Activity with Low Toxicity

Biosurfactants are generally considered as low or non-toxic products and are appropriate for pharmaceutical, cosmetic and food uses. Poremba et al. (1991) have demonstrated the higher toxicity of the chemically derived surfactant (Corexit) which displayed a LC50 against *Photobacterium phosphoreum* and was found to be 10 times lower than of rhamnolipids. Flasz et al. (1998) compared the toxicity and mutagenicity profile of biosurfactants from *Pseudomonas aeruginosa* and chemically derived surfactants and indicated the biosurfactant as non-toxic and non-mutagenic. The low toxicity profile of biosurfactant, sophorolipids from *Candida bombicola* made them useful in food industries for preservation of food (Cavalero and Cooper 2003).

5. Biosurfactant as an Antiadhesive Agent

Hood and Zottola (1995) proposed that a biofilm can be described as a group of bacteria/other organic matter that have colonized/accumulated on any surface. The first step on biofilm establishment is bacterial adherence over the surface which is affected by various factors including type of microorganism, hydrophobicity and electrical charges of surface, environmental conditions and ability of microorganisms to produce extracellular polymers that help cells to anchor the surfaces Zottola (1994). The biosurfactants can be used in altering the hydrophobicity of the surface which in turn affects the adhesion of microbes over the surface. A surfactant from *Streptococcus thermophilus* slows down the colonization of other thermophilic strains of *Streptococcus* over steel which are responsible for fouling. Similarly, a biosurfactant from *Pseudomonas fluorescens* inhibited attachment of *Listeria monocytogenes* onto a steel surface as reported by Chakrabarti in 2012.

6. Biosurfactant as a Biodegradable Agent

Microbial derived compounds can be easily degraded when compared to synthetic surfactants (Mohan et al. 2006) and, are suitable for environmental applications such as bioremediation/biosorption (Mulligan et al. 2001). The increasing environmental concern forces us to search for alternative products such as biosurfactants (Cameotra and Makkar 2004). Synthetic chemical surfactants impose environmental problems and hence, biodegradable biosurfactants from marine microorganisms caused concerned for the biosorption of poorly soluble polycyclic aromatic hydrocarbon, phenanthrene contaminated in aquatic surfaces (Olivera et al. 2003). Lee et al. (2008) controlled the blooms of marine algae, *Cochlodinium* using the biodegradable biosurfactant sophorolipid with the removal of 90% efficiency in 30 min of treatment.

7. Types and Chemical Composition of Biosurfactants

Biosurfactants are generally categorized on the basis of their **chemical composition** which are discussed as follows:

A. Glycolipids: They are carbohydrates linked to long-chain aliphatic acids or hydroxy aliphatic acids by an ester group. Biosurfactants are majorly glycolipids. Among the glycolipids, the best known are rhamnolipids, trehalolipids and sophorolipids (Jarvis and Johnson 1949). The sources and properties of the different glycolipids are discussed below:

- **Rhamnolipids:** Rhamnolipids are glycolipids, in which, one or two molecules of rhamnose are linked to one or two molecules of hydroxydecanoic acid. It is the widely studied biosurfactants which are the principal glycolipids produced by *P. aeruginosa* (Edwards and Hayashi 1965).
- **Trehalolipids:** These are associated with most species of *Mycobacterium, Nocardia* and *Corynebacterium*. Trehalose lipids from *Rhodococcus erythropolis* and *Arthrobacter* spp. lowered the surface and interfacial tensions in culture broths from 25–40 and 1–5 mNm, respectively (Asselineau and Asselineau, 1978).
- **Sophorolipids:** These are glycolipids which are produced by yeasts and consist of a dimeric carbohydrate sophorose linked to a long-chain hydroxyl **fatty acid** by glycosidic linkage. Sophorolipids, generally a mixture of at least six to nine different hydrophobic sophorolipids (Gautam and Tyagi 2006) and lactone form of the sophorolipid are preferable for many applications (Hu and Ju 2001).

B. Lipopeptides and lipoproteins: These consist of a lipid attached to a polypeptide chain (Rosenberg and Ron 1999). Several biosurfactants have shown antimicrobial action against various bacteria, algae, fungi and viruses by Besson et al. (1976) who reported the antifungal property and Singh and Cameotra (2004) who reported the antibacterial property of the lipopeptide, iturin which was produced by *Bacillus subtilis*. Iturin from *B. subtilis* was found to be active even after autoclaving, in a pH range of 5–11 and with a shelf life of 6 months at –18°C.

Surfactin: The cyclic lipopeptide surfactin is one of the most powerful biosurfactant composed of a seven amino-acid ring structure coupled to a fatty-acid chain via lactone linkages (Arima et al. 1968, Ramirez-vargas et al. 2019, Singh 2019). Previous studies have reported the various physicochemical properties of surfactin from *B. subtilis*. They found that surfactins are able to reduce the surface tension and interfacial tension of water. The inactivation of herpes and retrovirus was also observed with surfactin.

Lichenysin: *Bacillus licheniformis* produces several biosurfacants which exhibit excellent stability under extreme temperature, pH and salt conditions which are similar to surfactin. McInerney et al. (1990) reported that lichenysin from *B. licheniformis* are able to reduce the surface tension and interfacial tension of water to 27 and 0.36 mN m^{-1}, respectively.

C. Fatty acids, phospholipids and neutral lipids: Several bacteria and yeast produce large quantities of **fatty acid**s and phospholipid surfactants during growth on n-alkanes. In *Acinetobacter* spp. 1-N, phosphatidyl ethanolamine-rich vesicles are produced which form optically clear micro-emulsions of alkanes in water. These biosurfactants are essential for medical applications. Gautam and Tyagi (2006) reported that the deficiency of phospholipid protein complex is found to be the major cause for the respiration failure in prematurely born children. They have also suggested that the isolation and cloning of the genes responsible for such surfactants can be employed in their fermentative production.

D. Polymeric biosurfactants: These are the best-studied polymeric biosurfactants including emulsan, liposan, alasan, lipomanan and other polysaccharide-protein complexes. Emulsan is an effective emulsifying agent for hydrocarbons in water, even at a concentration as low as 0.001–0.01% (Hatha et al. 2007). Liposan is an extracellular water-soluble emulsifier synthesized by *Candida lipolytica* and is composed of 83% carbohydrate and 17% protein (Cooper and Paddock 1984, Chakrabarti 2012). The application of such a polymeric biosurfactant, liposan, as emulsifier in food and cosmetic industries was also discussed by Chakrabarti in 2012.

E. Particulate biosurfactants: These form the extracellular membrane vesicles partition to form a microemulsion which plays an important role in alkane uptake by microbial cells. Vesicles of *Acinetobacter* spp. strain HO1-N with a diameter of 20–50 nm and a buoyant density of 1.158 cubic g cm are of proteins, phospholipids and lipo-polysaccharides (Kaeppeli and Finnerty 1979, Chakrabarti 2012).

Table 1. Types and chemical composition of biosurfactants.

Types of Biosurfactant	Chemical Composition
A. Glycolipid ex. Rhamnolipid, Trehalolipid, Sophorolipid	Carbohydrate linked to long chain aliphatic acid
B. Lipopeptide and Lipoprotein ex. Surfactin and Lichenisin	Lipid attached to polypeptide chain
C. Phospholipid	Lipid attached to phosphoric acid
D. Polymeric biosurfactant ex. Liposan	Polysachharide protein complex
E. Particulate biosurfactant	Protein, phospholipid and lipopolysachharide

8. Origin and Sources of Biosurfactants

Many of the biosurfactant producing microorganisms are found to be hydrocarbon degraders (Willumsen and Karlson 1996, Volkering et al. 1998). However, in the past decades, many studies have shown the effects of microbially produced surfactants not only on bioremediation but also on enhanced oil recovery (Volkering et al. 1998, Tabatabaee et al. 2005).

Table 2. Biosurfactants produced from microbial sources.

Microorganisms	Biosurfactants	References
A. BACTERIA		
1. *Serratia marcescens*	Serrawettin	Lai et al. 2009
2. *Rodotorula glutinis, R. graminis*	Polyol lipids	Amaral et al. 2006
3. *Nicardia erythropolis, Corynebacterium* spp, *Mycobacterium* spp	Trehalose lipid	Muthusamy et al. 2008
4. *Pseudomonas fluorescens, Leuconostoc mesenteriodes*	Viscosin	Banat et al. 2010
5. *Pseudomonas* spp *Thiobacillus thiooxidans Agrobacterium* spp	Ornithine	Banat et al. 2010
6. *Pseudomonas aeruginosa, Pseudomonas chlororaphids, Serratia rubidea*		
7. *Pseudomonas fluorescens, Debaryomyces polymorphus*	Rhamnolipids	Jadhav et al. 2011
8. *Pseudomonas aeruginosa*		
8. *Lactobacillus fermentum*		
B. FUNGI	Carbohydrate-lipid	Nerukar et al. 2009
1. *Torulopsis bombicola*		
2. *Candida lipolytica*	Protein PA	Hisatsuka et al. 1971
3. *Candida bombicola*	Diglycosyl diglyceride	Mulligan et al. 2001
4. *Candida batistae*		
5. *Aspergillus ustus*	Sophorose lipid	Kim et al. 1997
6. *Trichosporon ashii*	Protein lipid polyglyceride-complex	Sarubbo et al. 2007
	Sophorolipid	Casas et al. 1997
	Sophorolipid	Konishi et al. 2008
	Glycolipoprotein	Aljandro et al. 2011
	Sophorolipid	Chandran and Das 2010

A. Bacterial biosurfactants: Bacteria make use of a wide range of organic compounds as a source of carbon and energy for their growth. When the carbon source is in an insoluble form like a hydrocarbon, bacteria make their diffusion possible into the cell by producing a variety of substances, i.e., biosurfactants. Some of the bacteria and yeasts excrete ionic surfactants which emulsify the CxHy substance in the growth medium. A few examples of this group of biosurfactants are rhamnolipids that are produced by different *Pseudomonas* spp. (Burger et al. 1963, Guerra-Santos et al. 1986). Sophorolipids that are produced by several *Torulopsis* spp. (Cutler and Light 1979, Cooper and Paddock 1984). Some other microorganisms are able to change the structure of their cell wall which is achieved by producing nonionic or lipopolysaccharides surfactants in their cell wall. Some examples of this group are: *Rhodococcus erythropolis* and various *Mycobacterium* spp. and *Arthrobacter* spp. which produce nonionic trehalose corynomycolates (Ristau and Wanger 1983, Kilburn and Takayama 1981, Kretschmer et al. 1982). There are lipopolysaccharides, such as emulsan, produced by *Acinetobacter* spp. (Kretschmer et al. 1982 and lipoproteins such as surfactin and subtilisin, that are produced by *Bacillus subtilis* (Cooper et al. 1981).

B. Fungal biosurfactants: Some fungi are also known to produce biosurfactants. Among them, *Candida bombicola* (Casas et al. 1997) *Candida lipolytica* (Sarubbo et al. 2007), *Candida ishiwadae* (Thanomsub et al. 2004), *Candida batistae* (Konishi et al. 2008), *Aspergillus ustus* (Alejandro et al. 2011), and *Trichosporon ashii* (Chandran and Das 2010) are the explored ones. Many of these are known to produce biosurfactant on low cost raw materials. The major type of biosurfactants produced by these strains is sophorolipids (glycolipids). *Candida lipolytica* produces cell wall-bound lipopolysaccharides when it is growing on n-alkanes (Rufino et al. 2007).

9. Applications of Biosurfactant in Heavy Metal Ion Removal

Studies have been done on the environmental applications of biosurfactants due to their diverse structure, better physicochemical properties, eco-friendly characteristics, suitability for many purposes which include remediation of hydrophobic organic compounds (HOCs) from soil, and removal of heavy metals from contaminated soil, root colonization. Heavy metals are becoming part of serious environmental problems. Contaminated soils contain lead (Pb), mercury (Hg), arsenic (As), cadmium (Cd), chromium (Cr), zinc (Zn), copper (Cu), and nickel (Ni) as heavy metals which can create many health problems such as inorganic chemical hazards for humans, animals, and plants (Li and Qian 2017, Liu et al. 2017, Tang et al. 2015). In the past, chemical surfactants had been used to treat heavy metal-contaminated soils and solubilizing HOCs. However, the chemical surfactants themselves are known to expose toxic substances and may cause other environmental issues due to their degradability in the soil (Liu et al. 2017). In comparison with chemical surfactants, biosurfactants derived from plants and microorganisms have shown better performance and considered suitable for removing heavy metals from contaminated soils (Luna et al. 2016, Tang et al. 2017). Essentially, there are three main steps involved in the removal of heavy metals from the soil through washing with biosurfactant solution. The heavy metals adsorbed on the surface of soil particles separate through the sorption of biosurfactant molecules at the interfaces between sludge/wet soil and metal in aqueous solution. Then, the metal is absorbed by biosurfactants and trapped within the micelle through electrostatic interactions. Finally, the biosurfactant can be recovered through the method of membrane separation (Ibrahim et al. 2016). The following chemical reaction describes the reaction between heavy metal ions and the functional group of biosurfactants.

(1) Soil–Men++R–(COOH)m+H2O→Soil+R–O–Men+–(COOH)m+2H

where Me^{n+} represents metal ions and R–$(COOH)_m$ are the surfactant molecules. As it can be seen that surfactants enhanced the extraction of heavy metals from the soil because of the existing carboxylic functional group in biosurfactants which acts as organic ligands (Tang et al. 2017). Since soil particles and other organic matter have negative charges on their surfaces, cationic materials can easily be adsorbed to negative charges of the soil surface.

The traditional methods of removing heavy metals from contaminated soil such as washing with water, organic and inorganic acids, metal-chelating agents, soil replacement, thermal desorption, and chemical surfactants had been used. However, these methods showed improper removal of heavy metals from the soil (Shah et al. 2016). Previous studies had reported that the remediation technique using biosurfactants is the best method to eliminate heavy metals from the soil with about 100% efficiency. Hong et al. (2002) studied the efficiency of biosurfactants for removing heavy metals from sludge and soil and achieved removal rates of 90–100%

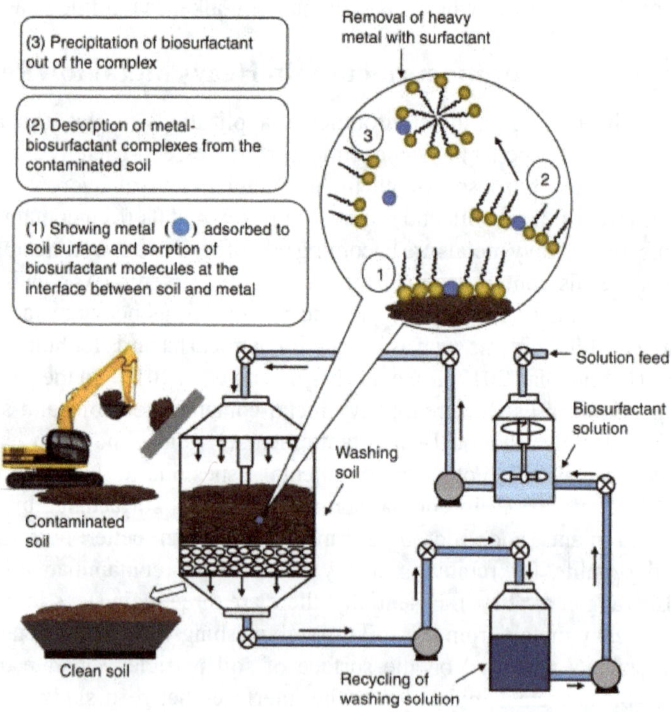

Figure 1. Mechanism of heavy metal removal from contaminated soil using biosurfactant; Contaminated soil—washing of soil—biosurfactant solution mixed—metal adsorbed to soil surface and sorption of biosurfactant molecules at the interface between soil and metal—biosurfactant complex formed from the contaminated soil—desorption of metal.

for Cu, Zn, Cr, and Cd. In addition, natural surfactants are found to be effective in treating contaminated soils with crude oil and diesel (Da Rosa et al. 2015).

Recently it has been observed that biosurfactants have attracted the interest of cosmetic and pharmaceutical industries because of their potential use as detergents, wetting, emulsifying, foaming and solubilizing agents, and many other useful applications Marchant and Banat 2012b. The uses of biosurfactants in these two industries are very wide since they are one of the main essential components in producing products such as shampoo, hair conditioners, soaps, shower gels, toothpastes, creams, moisturizers, cleansers, and many other skin care and healthcare products (Boruah and Gogoi 2013, Chakraborty et al. 2015). It has been reported that the use of chemically based surfactants in cosmetic formulations is one of the most challenging problems due to their potential risk of skin allergy and irritation (Bujak et al. 2015, Nitschke and Costa 2007). However, the excellent characteristics of bio-based surfactants make them an excellent component as a green product for cosmetics (Lee et al. 2017). Recently, it has been found that biosurfactants are very useful for skin moisturizing similar to ceramides, Lee et al. 2017. Ceramides are epidermal lipids important for skin barriers and dryness. The depletion of ceramides in *stratum corneum* (horny layer) layer of the skin can cause chronic skin diseases such as psoriasis, atopic dermatitis, and aged skin due to water loss and barrier dysfunction in the epidermis (Meckfessel and Brandt 2014, Tessema et al. 2017). The epidermis consists of four layers, namely stratum corneum, stratum granulosum, stratum spinosum, and the innermost stratum basale. Stratum corneum is the outermost surface layer of the skin that forms a barrier between the external environment and the internal body (Van Smeden et al. 2014). This layer is responsible for maintaining the skin barrier function and preventing excessive water loss from the skin. It has been found that ceramides are effective treating damaged skin, preventing skin roughness, and dryness. Ceramides are sphingolipids (glycosylceramides) that represent about 50% of intercellular lipids in the stratum corneum. Ferreira et al. 2017, discovered that microorganism-based surfactants besides having a good ability as emulsifiers could be applied as ceramides to enhance skin roughness and get rid of ceramide deficiency in the skin. It is also claimed that the combination of biosurfactants in cosmetics such as skin care lotions and moisturizing creams can improve the quality of the product and helps in improving roughness, Ferreira et al. 2017.

10. Biosurfactants as Emulsifiers

The efficiency of biosurfactants as emulsifiers is well known and widely used in the emulsification process. The applications of biosurfactants in different industries particularly in pharmaceutical and cosmetic formulations were found to have a satisfactory result due to their low toxicity and higher biodegradability. Generally, biosurfactants with higher molecular mass are considered effective emulsifiers to stabilize the emulsions for a long period of time (Nitschke and Costa 2007). In terms of preparing nano-emulsions, higher homogenizing speed would be effective to disperse the phase in nano-sizes (Bhadoriya et al. 2013). There are two types of emulsions: oil-in-water (O/W) and water-in-oil (W/O). In an O/W type, the continuous phase is water referring to as water-based emulsions. However, the oil-based emulsion is a W/O type, whereby the oil acts as the continuous phase. The main function of an emulsifier surfactant is to accumulate between phases and lower the surface and interfacial tensions of the emulsion which finally results in the formation of an emulsion. Moreover, the cosmetic and pharmaceutical formulations deal with both emulsion types (Masmoudi et al. 2005). In cosmetics, plant-based essential oils are considered as an important component for moisturizing and anti-aging purposes (Ferreira et al. 2017). In order to stabilize these types of emulsions, the presence of surfactants as emulsifiers is necessary. Particularly, bio-based surfactants are safe as they can cure in a natural way (Vijayakuma and Saravanan 2015).

Biosurfactants may act as emulsifiers or de-emulsifiers. An emulsion can be described as a heterogeneous system, consisting of one immiscible liquid dispersed in another in the form of droplets, whose diameter in general exceeds 0.1 mm. Emulsions are generally of two types: oil-in-water (o/w) or water-in-oil (w/o) emulsions. They possess a minimal stability which may be stabilized by additives such as biosurfactants and can be maintained as stable emulsions for months to years (Velikonja and Kosaric 1993). Liposan is a water-soluble emulsifier synthesized by *Candida lipolytica* which has been used to emulsify edible oils by coating droplets of oil, thus forming stable emulsions. These liposans were commonly used in cosmetics and food industries for making oil/water emulsions for making stable emulsions (Cirigliano and Carman 1985).

The surfactants can have various other functions in food industries, apart from their obvious role as agents that decrease surface and interfacial tensions, thus facilitating the formation and stabilization of emulsions. For example, to control the aggregation of fat globules, stabilization of aerated systems, improvement of texture and shelf-life of products containing starch, modification of rheological properties of wheat dough and improvement of constancy and texture of fat-based products (Kachholz and Schlingmann 1987, Arumugam and Shereen 2020). In bakery and ice-cream formulations, biosurfactants act by controlling the consistency, slowing staling and solubilizing the flavor oils; they are agents during cooking of fats and oil. Improvement in the stability of dough, volume, texture and conservation of bakery products is obtained by the addition of rhamnolipid surfactants (Van Haesendonck and Vanzeveren 2004). The study also suggested the use of rhamnolipids to improve the properties of butter cream and frozen confectionery products. L-Rhamnose has substantial potential as a forerunner for flavoring.

11. Removal of Oil and Petroleum Contamination

Itoh and Suzuki (1972) were the first to show that hydrocarbon culture media stimulated the growth of a rhamnolipid producing strain of *P. aeruginosa*. Recent research findings confirmed the effects of biosurfactants on hydrocarbon biodegradation by increasing microbial accessibility to insoluble substrates and thus enhancing their biodegradation (Zhang and Miller 1992, Hunt et al. 1994). Various experiments have been conducted on the effects of biosurfactants on hydrocarbons; enhancing their water solubility and increasing the displacement of oily substances from soil particles. Thus, biosurfactants increase the apparent solubility of these organic compounds at concentrations above the Critical Micelle Concentration (CMC) which enhance their availability for microbial uptake (Chang et al. 2008). For these reasons, inclusion of biosurfactants in a bioremediation treatment of a hydrocarbon polluted environment could be really promising, facilitating their assimilation by microorganisms (Calvo et al. 2009).

Many of the biosurfactants known today have been studied to examine their possible technical applications (Nayak et al. 2009). Most of these applications involve their efficiency in bioremediation, dispersion of **oil spill**s and enhanced oil recovery. *Alcanivorax* and *Cycloclasticus* genera are highly specialized hydrocarbon degraders in marine environments. *Alcanivorax borkumensis* utilizes aliphatic hydrocarbons as its main **carbon source** for growth and produces an anionic glucose lipid biosurfactant and thus potentials of *Alcanivorax* strains during bioremediation of hydrocarbon pollution in marine habitats have been studied (Olivera et al. 2009) thus, this property needs to be studied extensively in soil to ensure its effectiveness. Several species of *P. aeruginosa* and *B. subtilis* produce rhamnolipid, a commonly isolated glycolipid biosurfactant and surfactin, a lipoprotein type biosurfactant, respectively; these two biosurfactants have been shown by Whang et al. (2008) to increase solubility and bioavailability of a petrochemical mixture and also stimulate indigenous microorganisms for enhanced biodegradation of diesel contaminated soil. *Gordonia* species BS29 grows on aliphatic hydrocarbons as the sole **carbon source** and has been found to produce Bioemulsan which effectively degrades crude oil, Polycyclic Aromatic Hydrocarbons (PAH) and other recalcitrant branched hydrocarbons from contaminated soils. The rate of biodegradation is dependent on the physicochemical properties of the biosurfactants and not by the effects on microbial metabolism (Franzetti et al. 2008).

12. Biosurfactant in Removal of Toxic Pollutants

Removal of toxic pollutants occurs through the process of bioremediation which involves the acceleration of natural biodegradative processes in contaminated environments by improving the availability of materials (e.g., nutrients and oxygen), conditions (e.g., pH and **moisture content**) and prevailing microorganisms. Thus, bioremediation usually consists of the application of nitrogenous and phosphorous fertilizers, adjusting the pH and water content, if necessary, supplying air and often adding bacteria. The addition of emulsifiers is advantageous when bacterial growth is slow (e.g., at cold temperatures or in the presence of high concentrations of pollutants) or when the pollutants consist of compounds that are difficult to degrade, such as PAHs. Bioemulsifiers can be applied as an additive to stimulate the bioremediation process, however with advanced genetic technologies it is expected that the increase in bioemulsifier concentration during bioremediation would be achieved by the addition of bacteria that overproduce bioemulsifiers. This approach has been recently used successfully in the cleaning of oil pipes. Cultures of *A. radioresistens* (Navon-Venezia et al. 1995) which produce the bioemulsifier alasan but are unable to use hydrocarbons as a **carbon source**, were added to a mixture of oil-degrading bacteria to enhance oil bioremediation.

Table 3. Application of Biosurfactants for human welfare.

	Field	Uses as an Agent
A.	Pharmaceuticals	1. An antimicrobial
		2. An antifungal agent
		3. An anticancer
		4. An antiviral
		5. An antiadhesive
B.	In food industry	1. An emulsifier
		2. Food preservative
		3. Anti adhesive
C.	In petroleum	1. As emulsifier
		2. As de-emulsifier
		3. Oil recovery enhancer
D.	In waste water treatment	1. Biocomposite
		2. Heavy metal remover
		3. Bio-adsorbant lubricant
E.	In textile industry	1. As a bleaching assistant
F.	In agriculture	1. In soil remediation
		2. As herbicide
		3. As a fungicide
		4. As a bactericide
		5. As a pesticide
		6. In root colonization
		7. In heavy metal removal

Persistent organic pollutants found in oil containing wastewater and sediments, such as PAHs (phenanthrene, crysene) are also hydrophobic in nature and thus water solubility of PAHs normally decreases with the increasing number of rings in molecular structures. This property induces the low bioavailability of these organic compounds that is a crucial factor in the biodegradation of PAHs. The water solubility of some PAHs can be improved by addition of biosurfactants owing to their amphipathic structure by several folds (Yin et al. 2009). In addition, most hydrocarbons exist in strongly adsorbed forms when they are introduced into soils. Thus, their removal efficiency can be limited in low mass transfer phases. However, additions of solubilization agents, such as biosurfactants to the system enhances the **bioavailability** of low solubility and highly sorptive compounds (Shin et al. 2004).

13. Role of Biosurfactant in Mechanism of Bioremediation

There are at least two ways in which biosurfactants are involved in bioremediation: increasing the surface area of hydrophobic water-insoluble substrates and increasing the bioavailability of hydrophobic compounds.

A. Increasing the surface area of hydrophobic water insoluble substrates: For bacteria growing on hydrocarbons, the growth rate can be limited by the interfacial surface area between water and oil (Sekelsky and Shreve 1999). When the surface area becomes limiting, biomass increases arithmetically rather than exponentially. The evidence that emulsification is a natural process brought about by extracellular agents is indirect and there are certain conceptual difficulties in understanding how emulsification can provide an (evolutionary) advantage for the microorganism producing the emulsifier. Stated briefly, emulsification is a cell-density-dependent phenomenon: that is, the greater the number of cells, the higher the concentration of extracellular product. The concentration of cells in an open system, such as an oil-polluted body of water, never reaches a high enough value to effectively emulsify oil. Furthermore, any emulsified oil would disperse in the water and not be more available to the emulsifier-producing strain than to competing microorganisms.

One way to reconcile the existing data with these theoretical considerations is to suggest that the emulsifying agents do play a natural role in oil degradation but not in producing macroscopic emulsions in the bulk liquid. If the emulsion occurs at, or very close to, the cell surface and no mixing occurs at the microscopic level, then each cluster of cells creates its own microenvironment and no overall cell-density dependence would be expected.

B. Increasing the bioavailability of hydrophobic water-insoluble substrates: The low water solubility of many hydrocarbons, especially the Polycyclic Aromatic Hydrocarbons (PAHs), is believed to limit their availability to microorganisms which is a potential problem for bioremediation of contaminated sites. It has been assumed that surfactants would enhance the bioavailability of hydrophobic compounds. Several non-biological surfactants have been studied and both negative and positive effects of the surfactants on biodegradation were observed. For example, the addition of the surfactant Tergitol NP-10 increased the dissolution rate of solid-phase phenanthrene and resulted in an overall increase in the growth of a strain of *Pseudomonas stutzeri*, Grimberg et al. 1996. A similar effect was obtained by the addition of Tween 80 to two *Sphingomonas* strains, the rate of fluoranthene mineralization was almost doubled. By contrast, the same surfactant inhibited the rate of fluoranthene mineralization by two strains of *Mycobacterium* (Willumsen et al. 2001) and no stimulation was observed in other studies using several surfactants (Bruheim and Eimhjellen 1998).

14. Conclusion

On the basis of the above discussion we can conclude that biosurfactants act as tools for human welfare in modern life. Biosurfactants show several properties which could be useful in many fields of the food industry; recently, their antiadhesive activity has attracted attention as a new tool to inhibit and disrupt the biofilms formed in food contact surfaces. The combination of particular characteristics such as emulsifying, antiadhesive and antimicrobial activities presented by biosurfactants suggests potential application as multipurpose ingredients or additives. Scant information regarding toxicity, combined with high production costs seems to be the major cause for the limited uses of biosurfactants in the food area. However, the use of agroindustrial wastes can reduce the biosurfactants production costs as well as the waste treatment expends and also renders a new alternative for food and food-related industries not only for valorizing their wastes but also for becoming microbial surfactant producers. Biosurfactants obtained from Generally Regarded As Safe (GRAS) microorganisms like lactobacilli and yeasts are of great promise for food and medicine applications though, much more research is already required in this field. The prospect of new types of surface-active compounds from microorganisms can contribute to the detection of different molecules in terms of structure and properties but the toxicological aspects of new and current biosurfactants should be emphasized in order to certify the safety of these compounds for food utilization.

A promising approach seems to be the application of inoculants of biosurfactant producing bacteria in phytoremediation of hydrocarbon polluted soil to improve the efficiency of this technology. Application of the biosurfactants in phytoremediation on a large scale requires studies to identify their potential toxic effect on plants. Although the biosurfactants are thought to be ecofriendly, some experiments indicated that under certain circumstances they can be toxic to the environment. Nevertheless, careful and controlled use of these interesting surface active molecules will surely help in the enhanced cleanup of the toxic environmental pollutants and provide us with a clean environment.

References

Ajala, O.E., F. Aberuagba, T.E. Odetoye and A.M. Ajala. 2015. Biodiesel: Sustainable energy replacement to petroleum-based diesel fuel—A review. ChemBioEng. Reviews 2: 145–156.

Alejandbro, C.S., H.S. Humberto and J.F. Maria. 2011. Production of glycolipids with antimicrobial activity by *Ustilago maydis* FBD12 in submerged culture. Afr. J. Microbiol. Res. 5: 2512–2523.

Amaral, P.F.F., J.M. da Silva, M. Lehocky, A.M.V. Barros-Timmons, M.A.Z. Coelho, I.M. Marrucho and J.A.P. Coutinho. 2006. Production and characterization of a bioemulsifier from *Yarrowia lipolytica.* Process Biochem. 41: 1894–1898.

Arima, K., A. Kakinuma and G. Tamura. 1968. Surfactin, a crystalline peptidelipid surfactant produced by *Bacillus subtilis*: Isolation, characterization and its inhibition of fibrin clot formation. Biochem. Biophys. Res. Commun. 3: 488–494.

Arumugam, A. and M.F. Shereen. 2020. Potential food application of a biosurfactant produced by frotiers. J. of Microbiology (Accepted).

Asselineau, C. and J. Asselineau. 1978. Trehalose-containing glycolipids. Prog. Chem. Fats Lipids 16: 59–99.

Banat, I.M. 1995. Biosurfactants production and possible uses in microbial enhanced oil recovery and oil pollution remediation: A review. Bioresour. Technol. 51: 1–12.

Banat, I.M. 1995. Characterization of biosurfactants and their use in pollution removal—State of the art (Review). Acta Biotechnologica 15: 251–267.

Banat, I.M., R.S. Makkar and S.S. Cameotra. 2000. Potential commercial applications of microbial surfactants. Applied Microbiology and Biotechnology 53: 495–508.

Banat, I.M., A. Frbanzetti, I. Gandolfi, G. Bestetti, M.G. Martinotti et al. 2010. Microbial biosurfactants production, applications and future potential. Applied Microbiol. Biotechnol. 87: 427–444.

Besson, F., F. Peypoux, G. Michel and L. Delcambe. 1976. Characterization of iturin A in antibiotics from various strains of *Bacillus subtilis*. J. Antibiot. 29: 1043–1049.

Bhadoriya, S.S., N. Madoriya, K. Shukla and M.S. Parihar. 2013. Biosurfactants: A new pharmaceutical additive for solubility enhancement and pharmaceutical development. Biochemical Pharmacology 2: 113.

Boruah, B. and M. Gogoi. 2013. Plant based natural surfactants. Asian Journal of Home Science 8: 759–762.

Bruheim, P. and K. Eimhjellen. 1998. Chemically emulsified crude oil as substrate for bacterial oxidation: Differences in species response. Can. J. Microb. 442: 195–199.

Bujak, T., T. Wasilewski and Z. Nizioł-Łukaszewska. 2015. Role of macromolecules in the safety of use of body wash cosmetics. Colloids and Surfaces B: Biointerfaces 135: 497–503.

Burger, M.M., L. Glaser and R.M. Burton. 1963. The enzymatic synthesis of a rhamnose-containing glycolipid by extracts of *Pseudomonas aeruginosa*. J. Biol. Chem. 238: 2595–2602.

Calvo, C., M. Manzanera, G.A. Silva-Castro, I. Uad, J. Gonzalez-Lopez et al. 2009. Application of bioemulsifiers in soil oil bioremediation processes. Future prospects. Sci. Total Environ. 407: 3634–3640.

Cameotra, S.S. and R.S. Makkar. 2004. Recent applications of biosurfactants as biological and immunological molecules. Curr. Opin. Microbiol. 7: 262–266.

Casas, J.A., S.G. de Lara and F. Garcia-Ochoa. 1997. Optimization of a synthetic medium for *Candida bombicola* growth using factorial design of experiments. Enzyme Microb. Technol. 21: 221–229.

Cavalero, D.A. and D.G. Cooper. 2003. The effect of medium composition on the structure and physical state of sophorolipids produced by *Candida bombicola* ATCC 22214. J. Biotechnol. 103: 31–41.

Chakrabarti, S. 2012. Bacterial biosurfactant: Characterization, antimicrobial and metal remediation properties. Ph.D. Thesis, National Institute of Technology.

Chakraborty, S., M. Ghosh, S. Chakraborti, S. Jana, K.K. Sen et al. 2015. Biosurfactant produced from *Actinomycetes nocardiopsis* A17: Characterization and its biological evaluation. International Journal of Biological Macromolecules 79: 405–412.

Chandran, P. and N. Das. 2010. Biosurfactant production and diesel oil degradation by yeast species *Trichosporon asahii* isolated from petroleum hydrocarbon contaminated soil. Int. J. Eng. Sci. Technol. 2: 6942–6953.

Chang, M.W., T.P. Holoman and H. Yi. 2008. Molecular characterization of surfactant-driven microbial community changes in anaerobic phenanthrene-degrading cultures under methanogenic conditions. Biotechnol. Lett. 30: 1595–1601.

Cheeke, 2010. Actual P.R., and potential applications of *Yucca schidigera* and *Quillaja saponaria* saponins in human and animal nutrition. Journal of Animal Science 13: 115–126.

Cirigliano, M.C. and G.M. Carman. 1985. Purification and characterization of liposan, a bioemulsifier from *Candida lipolytica*. Applied Environ. Microbiol. 50: 846–850.

Cooper, D.G., C.R. Macdonald, S.J.B. Duff and N. Kosaric. 1981. Enhanced production of surfactin from *Bacillus subtilis* by continuous product removal and metal cation additions. Applied Environ. Microbiol. 42: 408-412.

Cooper, D.G. and D.A. Paddock. 1984. Production of a biosurfactant from *Torulopsis bombicola*. Applied Environ. Microbiol. 47: 173–176.

Cutler, A.J. and R.J. Light. 1979. Regulation of hydroxydocosanoic acid sophoroside production in *Candida bogoriensis* by the levels of glucose and yeast extract in the growth medium. J. Biol. Chem. 254: 1944–1950.

Da Rosa, C.F.C., D.M.G. Freire and H.C. Ferraz. 2015. Biosurfactant microfoam: Application in the removal of pollutants from soil. Journal of Environmental Chemical Engineering 3: 89–94.

De Almeida, D.G., R. Soares, DaSilva, C.F. de, J.M. Luna, R.D. Rufino et al. 2016. Biosurfactants: Promising molecules for petroleum biotechnology advances. Frontiers in Microbiology 7: 1–14.

Desai, J.D. and I.M. Banat. 1997. Microbial production of surfactants and their commercial potential. Microbiol. Mol. Biol. Rev. 61: 47–64.

Edwards, J.R. and J.A. Hayashi. 1965. Structure of a rhamnolipid from *Pseudomonas aeruginosa*. Arch. Biochem. Biophys. 111: 415–421.

Ferreira, A., X. Vecino, D. Ferreira, J.M. Cruz, A.B. Moldes et al. 2017. Novel cosmetic formulations containing a biosurfactant from *Lactobacillus paracasei*. Colloids and Surfaces B: Biointerfaces 155: 522–529.

Flasz, A., C.A. Rocha, B. Mosquera and C. Sajo. 1998. A comparative study of the toxicity of a synthetic surfactant and one produced by *Pseudomonas aeruginosa* ATCC 55925. Med. Sci. Res. 26: 181–185.

Fracchia, L., M. Cavallo and M.M.I. Giovanna. 2012. Biosurfactants and bioemulsifiers biomedical and related applications—Present status and future potentials. Biomedical Science Engineering and Technology, 325–370.

Franzetti, A., G. Bestetti, P. Caredda, P. La Colla, E. Tamburini et al. 2008. Surface-active compounds and their role in the access to hydrocarbons in *Gordonia* strains. FEMS Microbiol. Ecol. 63: 238–248.

Gautam, K.K. and V.K. Tyagi. 2006. Microbial surfactants: A review. J. Oleo Sci. 55: 155–166.

Grimberg, S.J., W.T. Stringfellow and M.D. Aitken. 1996. Quantifying the biodegradation of phenanthrene by *Pseudomonas stutzeri* P16 in the presence of a nonionic surfactant. Applied Environ. Microbiol. 62: 2387–2392.

Guerra-Santos, L., O. Kappeli and A. Fiechter. 1986. Dependence of *Pseudomonas aeruginosa* continuous culture biosurfactant production on nutritional and environmental factor. Applied Microbiol. Biotechnol. 24: 443–448.

Hatha, A.A.M., G. Edward and K.S.M.P. Rahman. 2007. Microbial biosurfactants-review. J. Mar. Atmos. Res. 3: 1–17.

Hisatsuka, K.I., T. Nakahara, N. Sano and K. Yamada. 1971. Formation of rhamnolipid by *Pseudomonas aeruginosa* and its function in hydrocarbon fermentation. Agric. Biol. Chem. 35: 686–692.

Hong, K.J., S. Tokunaga and T. Kajiuchi. 2002. Evaluation of remediation process with plant-derived biosurfactant for recovery of heavy metals from contaminated soils. Chemosphere 49: 379–387.

Hood, S.K. and E.A. Zottola. 1995. Biofilms in food processing. Food Control 6: 9–18.

Hu, Y. and L.K. Ju. 2001. Purification of lactonic sophorolipids by crystallization. J. Biotechnol. 87: 263–272.

Hunt, W.P., K.G. Robinson and M.M. Ghosh. 1994. The role of biosurfactants in biotic degradation of hydrophobic organic compounds. pp: 318–322. *In*: Hinchee, R.E., R.E. Hoeppel and R.N. Miller (eds.). Hydrocarbon Bioremediation. CRC Press, Boca Raton, FL.

Ibrahim, W.M., A.F. Hassan and Y.A. Azab. 2016. Biosorption of toxic heavy metals from aqueous solution by *Ulva lactuca* activated carbon. Egyptian Journal of Basic and Applied Sciences 3: 241–249.

Itoh, S. and T. Suzuki. 1972. Effect of rhamnolipids on growth of *Pseudomonas aeruginosa* mutant deficient in n-paraffin-utilizing ability. Agric. Biol. Chem. 36: 2233–2235.

Jadhav, M., S. Kalme, D. Tamboli and S. Govindwar. 2011. Rhamnolipid from *Pseudomonas desmolyticum* NCIM-2112 and its role in the degradation of Brown 3REL. J. Basic Microbiol. 51: 385–396.

Jarvis, F.G. and M.J. Johnson. 1949. A glycolipid produced by *Pseudomonas aeruginosa*. J. Am. Chem. Soc. 71: 4124–4126.

Kachholz, T. and M. Schlingmann. 1987. Possible food and agricultural applications of microbial surfactants: An assessment. pp: 183–210. *In*: Kosaric, N., W.L. Carns and N.C.C. Gray (eds.). Biosurfactants and Biotechnology. Marcel Dekker, New York, USA.

Kaeppeli, O. and W.R. Finnerty. 1979. Partition of alkane by an extracellular vesicle derived from hexadecane-grown Acinetobacter. J. Bacteriol. 140: 707–712.

Karanth, N.G.K., P.G. Deo and N.K. Veenanadig. 1999. Microbial production of biosurfactants and their importance. Curr. Sci. 77: 116–126.

Kilburn, J.O. and K. Takayama. 1981. Effects of ethanebutol on accumulation and secretion of trehalose mycolates and free mycolic acid in *Mycobacterium smegmatis*. Antimicrob. Agents Chemother. 20: 401–404.

Konishi, M., T. Fukuoka, T. Morita, T. Imura, D. Kitamoto et al. 2008. Production of new types of sophorolipids by *Candida batistae*. J. Oleo Sci. 57: 359–369.

Kretschmer, A., H. Bock and F. Wagner. 1982. Chemical and physical characterization of interfacial-active lipids from *Rhodococcus erythropolis* grown on n-alkanes. Applied Environ. Microbiol. 44: 864–870.

Lai, C.C., Y.C. Huang, Y.H. Wei and J.S. Chang. 2009. Biosurfactant-enhanced removal of total petroleum hydrocarbons from contaminated soil. J. Hazard. Mater. 167: 609–614.

Lee, S.M., J.Y. Lee, H.P. Yu and J.C. Lim. 2017. Synthesis of environment friendly biosurfactants and characterization of interfacial properties for cosmetic and household products formulations. Colloids and Surfaces A: Physicochemical and Engineering Aspects 536: 224–233.

Lee, Y.J., J.K. Choi, E.K. Kim, S.H. Youn, E.J. Yang et al. 2008. Field experiments on mitigation of harmful algal blooms using a Sophorolipid: Yellow clay mixture and effects on marine plankton. Harmful Algae 7: 154–162.

Li, X. and P. Qian. 2017. Identification of an exposure risk to heavy metals from pharmaceutical-grade rubber stoppers. Journal of Food and Drug Analysis 25: 723–730.

Liu, Z., Z. Li, H. Zhong, G. Zeng, Y. Liang et al. 2017. Recent advances in the environmental applications of biosurfactant saponins: A review. Journal of Environmental Chemical Engineering 5: 6030–6038.

Luna, J.M., R.D. Rufino and L.A. Sarubbo. 2016. Biosurfactant from *Candida sphaerica* UCP0995 exhibiting heavy metal remediation properties. Process Safety and Environmental Protection 102: 558–566.

Mahamallik P. and A. Pal. 2017. Degradation of textile wastewater by modified photo-Fenton process: Application of Co (II) adsorbed surfactant-modified alumina as heterogeneous catalyst. Journal of Environmental Chemical Engineering 5: 2886–2893.

Marchant, R. and I.M. Banat. 2012. Microbial biosurfactants: Challenges and opportunities for future exploitation. Trends in Biotechnology 30: 558–565.

Masmoudi, H., Y. Le, P. Piccerelle and J. Kister. 2005. The evaluation of cosmetic and pharmaceutical emulsions aging process using classical techniques and a new method: FTIR. International Journal of Pharmaceutics 289: 117–131.

McInerney, M.J., M. Javaheri and D.P. Nagle Jr. 1990. Properties of the biosurfactant produced by *Bacillus licheniformis* strain JF-2. J. Ind. Microbiol. Biotechnol. 5: 95–101.

Meckfessel, M.H. and S. Brandt. 2014. The structure, function, and importance of ceramides in skin and their use as therapeutic agents in skin-care products. Journal of the American Academy of Dermatology 71: 177–184.

Mohan, P.K., G. Nakhla and E.K. Yanful. 2006. Biokinetics of biodegradation of surfactants under aerobic, anoxic and anaerobic conditions. Water Res. 40: 533–540.

Mulligan, C.N., R.N. Yong and B.F. Gibbs. 2001. Remediation technologies for metal-contaminated soils and groundwater: An evaluation. Eng. Geol. 60: 193–207.

Muthusamy, K., S. Gopalakrishnan, T.K. Ravi and P. Sivachidambaram. 2008. Biosurfactants: Properties, commercial production and application. Curr. Sci. 94: 736–747.

Navon-Venezia, S., Z. Zosim, A. Gottlieb, R. Legmann, S. Carmeli et al. 1995. Alasan, a new bioemulsifier from *Acinetobacter radioresistens*. Applied Environ. Microbiol. 61: 3240–3244.

Nayak, A.S., M.H. Vijaykumar and T.B. Karegoudar. 2009. Characterization of biosurfactant produced by *Pseudoxanthomonas* sp. PNK-04 and its application in bioremediation. Int. Biodeter. Biodegrad. 63: 73–79.

Nerurkar, A.S., K.S. Hingurao and H.G. Suthar. 2009. Bioemulsfiers from marine microorganisms. J. Sci. Ind. Res. 68: 273–277.

Nitschke, M. and S.G.V.A.O. Costa. 2007. Biosurfactants in food industry. Trends in Food Science and Technology 18: 252–259.

Olivera, N.L., M.G. Commendatore, O. Delgado and J.L. Esteves. 2003. Microbial characterization and hydrocarbon biodegradation potential of natural bilge waste microflora. J. Ind. Microbiol. Biotechnol. 30: 542–548.

Olivera, N.L., M.L. Nievbas, M. Lozada, G. del Prado, H.M. Dionisi et al. 2009. Isolation and characterization of biosurfactant-producing *Alcanivorax* strains: Hydrocarbon accession strategies and alkane hydroxylase gene analysis. Res. Microbiol. 160: 19–26.

Phulphoto, I.A. 2020. Production and characterization surfactin like biosurfactant. Colloids Surf. B. 175: 56–63.

Poremba, K., W. Gunkel, S. Lang and F. Wagner. 1991. Toxicity testing of synthetic and biogenic surfactants on marine microorganisms. Environ. Toxicol. Water Qual. 6: 157–1.

Ramirez-Vargas et al. 2019. Biosurfactants and synthetic surfactants. Frontiers in Microbiology (Accepted).

Ristau, E. and F. Wanger. 1983. Formation of novel trehalose lipids from *Rhodococcus erythropolis* under growth RR limiting conditions. Biotech. Lett. 5: 95–100.

Rosenberg, E. and E.Z. Ron. 1999. High- and low-molecular-mass microbial surfactants. Applied Microbiol. Biotechnol. 52: 154–162.

Rufino, R.D., L.A. Sarubbo and G.M. Campos-Takaki. 2007. Enhancement of stability of biosurfactant produced by *Candida lipolytica* using industrial residue as substrate. World J. Microbiol. Biotechnol. 23: 729–734.

Sarubbo, L.A., C.B.B. Farias and G.M. Campos-Takaki. 2007. Co-utilization of canola oil and glucose on the production of a surfactant by *Candida lipolytica*. Curr. Microbiol. 54: 68–73.

Sekelsky, A.M. and G.S. Shreve. 1999. Kinetic model of biosurfactant-enhanced hexadecane biodegradation by *Pseudomonas aeruginosa*. Biotechnol. Bioeng. 63: 401–409.

Shaban, S.M. and A.A. Abd-Elaal. 2017. Studying the silver nanoparticles influence on thermodynamic behavior and antimicrobial activities of novel amide Gemini cationic surfactants. Materials Science and Engineering C 76: 871–885.

Shah, A., S. Shahzad, A. Munir, M.N. Nadagouda, G.S. Khan et al. 2016. Micelles as soil and water decontamination agents. Chemical Reviews 116: 6042–6074.

Shin, K.H., K.W. Kim and E.A. Seagren. 2004. Combined effects of pH and biosurfactant addition on solubilization and biodegradation of phenanthrene. Applied Microbiol. Biotechnol. 65: 336–343.

Singh, P. and S.S. Cameotra. 2004. Potential applications of microbial surfactants in biomedical sciences. Trends Biotechnol. 22: 142–146.

Singh, P. 2019. Biosurfactant production: Emerging trends and promising strategies. (Abs): 41.

Syldatk, C., S. Lang, F. Wagner, V. Wray and L. Witte. 1985. Chemical and physical characterization of four interfacial-active rhamnolipids from *Pseudomonas* spec. DSM 2874 grown on n-alkanes. Zeitschrift Naturforschung C 40: 51–60.

Tabatabaee, A., M.A. Mazaheri, A.A. Noohi and V.A. Sajadian. 2005. Isolation of biosurfactant producing bacteria from oil reservoirs. Iran. J. Environ. Health Sci. Eng. 2: 6–12.

Tang, J., J. He, T. Liu and X. Xin. 2017. Removal of heavy metals with sequential sludge washing techniques using saponin: Optimization conditions, kinetics, removal effectiveness, binding intensity, mobility and mechanism. RSC Advances 7: 33385–33401.

Tang, Z., L. Zhang, Q. Huang, Y. Yang, Z. Nie et al. 2015. Contamination and risk of heavy metals in soils and sediments from a typical plastic waste recycling area in North China. Ecotoxicology and Environmental Safety 122: 343–351.

Tessema, E.N., T. GebreMariam, S. Lange, B. Dobner, R.H.H. Neubert et al. 2017. Potential application of oat-derived ceramides in improving skin barrier function: Part 1. Isolation and structural characterization. Journal of Chromatography B 1065-1066: 87–95.

Thanomsub, B., T. Watcharachaipong, K. Chotelersak, P. Arunrattiyakorn, T. Nitoda et al. 2004. Monoacylglycerols: Glycolipid biosurfactants produced by a thermotolerant yeast, *Candida ishiwadae*. J. Applied Microbiol. 96: 588–592.

Van Haesendonck, I.P.H. and E.C.A. Vanzeveren. 2004. Rhamnolipids in bakery products. W.O. 2004/040984, International Application Patent (PCT), Washington, DC., USA.

Van Smeden, J., M. Janssens, G.S. Gooris and J.A. Bouwstra. 2014. The important role of stratum corneum lipids for the cutaneous barrier function. Biochimica et Biophysica Acta—Molecular and Cell Biology of Lipids 1841: 295–313.

Varjani, S.J. and V.N. Upasan. 2017. Critical review on biosurfactant analysis, purification and characterization using rhamnolipid as a model biosurfactant. Bioresource Technology 232: 389–397.

Vecino, X., J.M. Cruz and A.B. Moldes. 2015. Wastewater treatment enhancement by applying a lipopeptide biosurfactant to a lignocellulosic biocomposite. Carbohydrate Polymers 131: 186–196.

Velikonja, J. and N. Kosaric. 1993. Biosurfactants in food applications. pp. 419–448. *In*: Kosaric, N. and F.V. Sukan (eds.). Biosurfactants: Production: Properties: Applications, Chapter 16, CRC Press, New York, ISBN-13.

Vijayakuma, S. and V. Saravanan. 2015. Biosurfactants—Types sources and applications. Research Journal of Microbiology 10: 181–192.

Volkering, F., A.M. Breure and W.H. Rulkens. 1998. Microbiological aspects of surfactant use for biological soil remediation. Biodegradation 8: 401–417.

Whang, L.M., P.W.G. Liu, C.C. Ma and S.S. Cheng. 2008. Application of biosurfactants, rhamnolipid and surfactin, for enhanced biodegradation of diesel-contaminated water and soil. J. Hazard. Mater. 151: 155–163.

Willumsen, P.A. and U. Karlson. 1996. Screening of bacteria, isolated from PAH-contaminated soils, for production of biosurfactants and bioemulsifiers. Biodegradation 7: 415–423.

Willumsen, P., J. Nielsen and U. Karlson. 2001. Degradation of phenanthrene-analogue azaarenes by *Mycobacterium gilvum* strain LB307T under aerobic conditions. Applied Microbiol. Biotechnol. 56: 539–544.

Yin, H., J. Qiang, Y. Jia, J. Ye, H. Peng et al. 2009. Characteristics of biosurfactant produced by *Pseudomonas aeruginosa* S6 isolated from oil-containing wastewater. Process Biochem. 44: 302–308.

Zhang, Y. and R.M. Miller. 1992. Enhanced octadecane dispersion and biodegradation by a *Pseudomonas rhamnolipid* surfactant (biosurfactant). Applied Environ. Microbiol. 58: 3276–3282.

Zottola, E.A. 1994. Microbial attachment and biofilm formation: A new problem for the food industry? Food Technol. 48: 107–114.

8

Biosurfactant Mediated Synthesis and Stabilization of Nanoparticles

Muhammad Bilal Sadiq,[1,*] *Muhammad Rehan Khan*[2] and *R Z Sayyed*[3]

1. Introduction

Biosurfactants are natural surface-active agents and are mainly produced by various microorganisms. Due to a relatively simple structure, stable nature and natural origin biosurfactants are preferred over the synthetic chemical surfactants and are employed in the food, pharmaceutical and agriculture sectors (Rawat et al. 2020). Biosurfactants have amphiphilic nature, low toxicity, biodegradable nature and low critical micelle concentration. These characteristics enable biosurfactants as an alternative to synthetic surface-active agents (Elakkiya et al. 2020).

Due to nontoxic nature and antimicrobial potential against a wide range of microbes, metallic nanoparticles (NPs), such as silver and zinc oxide NPs have been used in various fields. NPs have been synthesized by various methods however the application of microbial surfactants in the green synthesis of NPs is gaining interest due to their ability to act as reducing agents, stability, and associated lower toxicity (Eswari et al. 2018). Rane et al. (2017) reported that microbial biosurfactants obtained from *Bacillus subtilis* ANR 88 efficiently synthesized gold and silver NPs without any conventional chemical surfactants or reducing agents. Moreover, the synthesized NPs were of uniform size and shape with better stability.

[1] School of Life Sciences, Forman Christian College (A Chartered University), Lahore, 54600, Pakistan.

[2] Department of Agricultural Science, University of Naples Federico II, Via Università 133, 80055, Portici (NA), Italy.

[3] Department of Microbiology, PSGVP Mandal's, Arts, Science, and Commerce College, Shahada, Maharashtra 425409, India.

* Corresponding author: bilalsadiq@fccollege.edu.pk; m.bilalsadiq@hotmail.com

NPs have a tendency to form aggregates due to their large surface area and biosurfactants act as agents that adsorb onto the surface of NPs and prevent their aggregation by stearic hinderance (Sangeetha et al. 2013). Biosurfactants have emerging and diverse applications in nanotechnology. NPs are associated with high surface energy and tend to produce toxic byproducts, if synthetic chemicals are used in the formulation or stabilization (Santos et al. 2016). Biosurfactants can stabilize the NPs without any toxicity. Reddy et al. (2009) reported that silver NPs were stable for 60 days by using surfactin as stabilizer.

Joanna et al. (2018) reported that biosurfactants produced by *Bacillus subtilis* improved the stability of biogenic silver NPs and also led to synergistic antimicrobial effects against various bacterial and fungal pathogens. The classes of natural compounds involved in the stabilization of biogenic NPs are summarized in Figure 1. Rhamnolipids obtained from *Pseudomonas aeruginosa* were reported to produce stable zinc oxide NPs. Rhamnolipids have a simple structure, low molecular weight and high affinity for metallic NPs, which enables them to form a micelle like aggregate on the surface of NPs and hence promote NPs stabilization (Singh et al., 2014). This chapter summarizes the various types of biosurfactants used in the nanotechnology and influence of various types of biosurfactants in the formulation and stabilization of NPs.

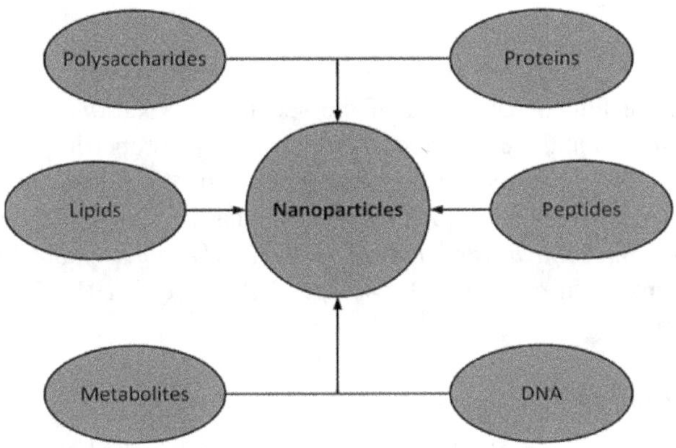

Figure 1. Classes of natural compounds involved in the stabilization of biogenic NPs (adopted and modified from Piacenza et al. (2018).

2. Biosurfactants and their Types

Microbes including yeast, fungi, and bacteria are the potential producers of biosurfactants. A majority of biosurfactants are produced by bacteria. Biosurfactants are surface active biochemical compounds possessing both hydrophobic and hydrophilic properties, produced on microbial cell surfaces, containing distinctive non-polar and polar moieties which allow them to shape micelles that concentrate at the interface between fluids of different polar characteristics (i.e., oil and water) (Roy 2017). Biosurfactants are generally classified on the basis of biochemical compositions and microbial origin in comparison to chemically synthesized surfactants (which are classified on the basis of polarity index) (Mahanti et al. 2017). Biosurfactants on the basis of their molecular mass can be classified into two major groups. The higher molecular mass surfactants include lipopolysaccharides and lipoproteins, while low molecular mass surfactants include glycolipids and lipopeptides.

2.1 *Glycolipids*

Glycolipids are the most frequently occurring biosurfactants which are carbohydrate in nature and linked to long-chain hydroxyaliphatic and aliphatic acids by either an ester or ether group. Among the glycolipids, the best known are sophorolipids, trehalolipids, and rhamnolipids (Mnif and Ghribi 2016).

2.2 *Sophorolipids*

These types of glycolipids are produced by yeasts (i.e., *Torulopsis apicola*, *T. bombicola*, and *T. petrophlium*) and consist of sophorose (a dimeric carbohydrate) linked by glycosidic linkage to a long-chain hydroxyl fatty acid. Generally, sophorolipids are a combination of at least six to nine varied hydrophobic sophorolipids. Furthermore, it has been observed that lactone form of the sophorolipid is essential for various useful applications (Chakrabarti 2012, Elshikh et al. 2017).

2.3 *Trehalolipids*

Trehalolipids are linked with most of the species of *Nocardia*, *Corynebacterium*, and *Mycobacterium* and are reported to possess notable properties with a wide range of applications as emulsifiers in bioremediation, microbial enhanced oil recovery, hydrocarbon solubilizer, and in food industries. It has been reported that trehalolipids obtained from *Anthrobacter* and *Rhodococcus erythropolis* reduced the interfacial and surface tension in culture broths (Bages-Estopa et al. 2018, Vijayakumar and Saravanan 2015).

2.4 *Rhamnolipids*

Rhamnolipids are vastly studied glycolipids which are principal glycolipids produced by *Pseudomonas aeruginosa*. In these glycolipids, one or two molecules of rhamnose are connected to one or two molecules of hydroxydecanoic acid through glycosidic linkages (Chong and Li 2017).

2.5 *Polymeric Biosurfactants*

These types of biosurfactants include alasan, liposan, lipomanan, emulsan and other polysaccharide-protein complexes. *Candida lipolytica* has have been reported to synthesize an extracellular hydrophilic emulsifier (liposan) which is composed of 17% protein and 83% carbohydrate and is widely used in food and cosmetic industries for producing stable oil/water emulsions (Santos et al. 2016).

2.6 Lipoproteins and Lipopeptides

These biosurfactants comprise of a lipid connected to a polypeptide chain (Chakrabarti 2012). Some of the best examples of these biosurfactants are iturin, surfactin, and arthrofacin (Sun et al. 2019). These biosurfactants are high in demand for environmental as well as for pharmacological research and the two most important biosurfactants in this category are surfactin and lichenysin.

2.6.1 Surfactin

Bacillus subtilis have been reported to produce this important cyclic lipopeptide biosurfactant which consist of seven amino acid ring structures attached to a fatty acid chain through lactone linkages and can reduce surface and interfacial tension of water (Wu et al. 2019).

2.6.2 Lichenysin

This biosurfactant is produced by *B. licheniformis* and exhibit excellent stability under extreme pH, temperature and saline conditions (Zhu et al. 2017).

2.7 Particulate Biosurfactants

These biosurfactants play an essential role in the hydrocarbon uptake by the cells by forming extracellular membrane vesicles (Singh et al. 2020).

Many yeast and bacterial species during growth on n-alkanes produce huge amounts of phospholipid and un-saturated fatty acid surfactants and are used for therapeutic applications (Mahanti et al. 2017).

3. Sources of Biosurfactants

3.1 Bacterial Biosurfactants

As microbes utilize organic compounds as a carbon source for their growth, these organic compounds become carriers for diffusion into the microbial cells by generating a number of substances known as bacterial biosurfactants (Chakrabarti 2012). It has been reported that some of the microbes can alter the structure of their cell walls by utilizing non-ionic lipopolysaccharide bacterial biosurfactants, i.e., trechlose corbynomycolates produced by some *Arthrobacter* species, *Mycobacterium* species and *Rhodococcus erythropolic* (Mahanti et al. 2017). Most of the research on biosurfactants is focused on microbial isolates from soil including *Bacillus* and *Pseudomonas* species, however, biosurfactants from marine microbes (i.e., *Halomonas*, *Alteromonas*, *Pseudoalteromonas*, and *Colwellia*) are relatively unexplored areas as they are difficult to culture in the lab (Mapelli et al. 2017). Furthermore, biosurfactants from marine microbes have also been explored due to their non-pathogenicity and their possible applications in the synthesis of metallic nanoparticles (NPs) (Tripathi et al. 2018).

3.2 Fungal Biosurfactants

Researchers are screening low cost biosurfactants since bacterial strains are not cheaper, for this reason biosurfactants from fungal species are a better option. Among fungi,*Candida bombicola*, *Candida ishiwade*, *Candida lipolytica*, *Candida batistae*, *Trichosporon ashii*, *Aspergillus ustus*, and *Yarrowia* are the major source of fungi based biosurfactants and sophorolipids are a major type of biosurfactants produced by these fungal species (Luft et al. 2020).

4. Biosurfactant Mediated Synthesis and Stabilization of NPs

Although, synthesis of NPs through biological approaches (i.e., microbes and extracts) is superior to the approaches which utilize reducing agents, biological approaches still generate NPs at a moderate rate (i.e., reaction time between 24–120 hours) in comparison to reduction methods. Due to rising interest in the synthesis of biologically active NPs, researchers have tried to develop environmentally friendly methods, such as the use of biosurfactants as capping agents for the synthesis of metallic NPs as an alternative to bacterial mass or extracts (due to complex downstream processes and longer reduction periods) (Satpute et al. 2018). Since biosurfactants play an essential role in the aggregation and stabilization process, they can be used as an alternative green option for enhancing both NPs stabilization and synthesis (Bezza et al. 2020, Reddy et al. 2009). Spherical NPs synthesis in water-in-oil emulsion systems is one of the most prominent biosurfactant mediated process. In a microemulsion system, the droplet functions as a micro-reactor and contains water soluble molecules inside. Principally, a decrease in particle size could be observed due to a decrease in droplet size, due to the increase in the concentration of the biosurfactant. One of the major mechanisms of the action of biosurfactants is to adsorb onto metallic NPs, preventing subsequent aggregation and surface stabilizing of the NPs (Joanna et al. 2018). It has been reported that the thickness of the adsorbed layer and the type of surfactant (i.e., polymeric, ionic and non-ionic etc.) influences the mechanism of surfactant adsorption (Dinets and Maksin 2019). In a microemulsion-mediated reduction method, the interactive forces (elastic, repulsive, osmotic, and Van der Waals) among reverse micelles result in the transfer of reactants due to collision of micelles. As a result, the monomeric metal nuclei begin to form in the micelle and grow in size which is based on the microemulsion's water core (Liu 2012). Many laboratory-based samples as well as commercially available biosurfactants have been evaluated as modifier and stabilizer in the synthesis of NPs (Table 1). The biological applications of metallic gold NPs are restricted due their unstable nature and permeability concerns. However, lipopeptide biosurfactant obtained from *Acinetobacter junii* B6 was reported to form a protective layer around the gold NPs to ensure their stability and permeability (Ohadi et al. 2020).

Marine derived biosurfactants can be used in the synthesis and stabilization of NPs. Marine microbes have developed the ability to tolerate high metal concentration in the surrounding environment by adopting various mechanisms such as, impermeable cell membranes, toxin ions efflux, oxidation and reduction of ions and production of exopolysaccharides. Marine bacteria have the capability to bind heavy metals due to their ability to produce biosurfactants. Therefore, marine microbes based biosurfactants can be used for the synthesis and stabilization of safe and environment friendly metallic NPs (Tripathi et al. 2018).

Table 1. Summary of NPs synthesized from biosurfactants.

Biosurfactant Type	Source	Nanoparticle Synthesized/stabilized	References
Rhamnolipid	*Pseudomonas aeruginosa*	ZnO	Singh et al. (2014)
Rhamnolipid	*Pseudomonas aeruginosa* strain BS-161R	Ag	Kumar et al. (2010)
Rhamnolipid	*Pseudomonas aeruginosa* TEN01	Ag	Elakkiya et al. (2020)
Rhamnolipid	*Pseudomonas aeruginosa* PTCC 13401	Au	Bayeea et al. (2020)
Lipopeptide	*Acinetobacter Junii* B6	Au	Ohadi et al. (2020)
Lipopeptide	–	Iron oxide	Sungsuwan (2017)
Surfactin (lipopeptide)	*Bacillus subtilis*	Ag	Eswari et al. (2018)
Surfactin	*Bacillus subtilis* ANR 88	Ag and Au	Rane et al. (2017)
Yeast biosurfactant	*Yarrowia lipolytica* MTCC 9520	Ag	Radha et al. (2020)
Glycolipid	*Williopsis saturnus* NCIM 3298	Ag and Au	Mohite et al. (2017)
Mannosylerythritol lipids	–	Ag	Ga'al et al. (2020)

4.1 NPs Synthesized from Glycolipid Biosurfactants

Khalid et al. (2019) prepared silver NPs by using rhamnolipids. After mixing equal volumes of rhamnolipids and silver nitrate solution, the solution was stirred rigorously to allow the biosurfactant to make its complex with silver and subsequently form silver NPs. Khalid et al. (2019) also prepared iron oxide NPs through co-precipitation. The biosurfactant coated NPs were characterized by using Zetasizer and field emission scanning electron microscopy. A negative zeta potential was observed for formulated silver (–56.3 mV) and iron oxide (–31.2 mV) NPs due to the presence of negatively charged carboxylate group of the rhamnolipid present on the NP surface (Sangeeta et al. 2013).

Singh et al. (2014) synthesized and stabilized ZnO NPs by using *Pseudomonas aerugniosa* rhamnolipids through the sol-gel method. Equal concentrations of rhamnolipid and $ZnNO_3$ were mixed and incubated to obtain a biosurfactant-NP complex. A few drops of NaOH were finally added into the solution to obtain colloidal NPs (Figure 2). Biosurfactant stabilized NPs were characterized by using x-ray diffraction, scanning electron microscopy and Fourier transform infrared spectroscopy (FTIR). Generally, rhamnolipid biosurfactants are dispersed in aqueous solution in the form of core-shell spherical micellar structure, in which the shell represents –OH groups while the core indicates the hydrophobic (C-C) chains. The lipophilic alkyl chains of rhamnolipid molecule in the presence of oxygen gets attached to the primary ZnO crystallite surface. The pre-formed ZnO crystallite starts to form rhamnolipid stabilized ZnO NPs due to high surface energy. ZnO NPs continue to be synthesized inside the core of rhamnolipid micelles by means of nucleation and growth of NPs. After the addition of NaOH in the aqueous solution

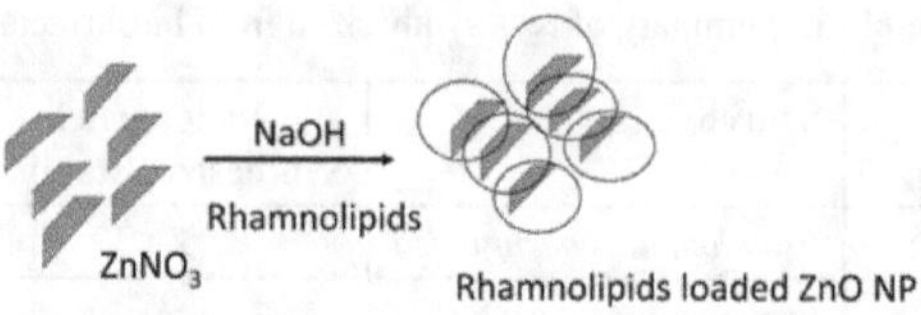

Figure 2. Stabilization of ZnO nanoparticles by rhamnolipids.

of $ZnNO_3$, ZnOH is formed, which was dehydrated to form crystalline ZnO (Singh et al. 2014, Sangeetha et al. 2013). It has been reported that ZnO NPs formulated with rhamnolipid biosurfactants displayed far better stability at room temperature as compared to bare NPs due to the fact that rhamnolipids possess a double bond which forms a disordered shell and expedites the dissolution, furthermore, longer tails of rhamnolipids stabilize NPs better due to formation of aggregates (micelle like) on the NP surface as compared to carboxymethyl cellulose (Singh et al. 2014).

Shikha et al. (2020) synthesized sophorolipid mediated gold NPs by adding chloroauric and sodium borohydride into the sophorolipid solution. Characterization of the biosurfactant capped NPs was done by using UV-visible spectrophotometry, Zetasizers, transmission electron microscopy and FTIR.

4.2 NPs Synthesized from Lipopeptides

Rangarajan et al. (2018) synthesized silver NPs by *Bacillus* lipopeptides as biosurfactants via two methods using $NaBH_4$ and reported that at low lipopeptide concentration (25 mg/L) the silver NPs were spherical in shape. Furthermore, when concentration was increased upto 62.5 mg/L, NPs were observed to be surrounded by lipopeptide micelles, which offered stability to these NPs. However, when the concentration of lipopeptide increased beyond that limit, a few number of small NPs were observed to be attached with lipopeptide vesicles due to the fact that at high lipopeptide concentration, more than an anticipated number of silver ions were concentrated at micellar surface via electrostatic interactions with COO^- of the amino acids. This resulted in NP growth quicker than in the surrounding medium (Ewais 2014). It can be concluded from this study that lipopeptides stabilize silver NPs by more than one mechanism. Bezza et al. (2020) similarly prepared stabilized silver NPs by using the chemical reduction method in lipopeptide biosurfactant reverse micelles. It was reported that adsorbed lipopeptides provide colloidal stability to the NPs during synthesis and preservation through steric or electrostatic stabilization (Tran and Le 2013). The adsorption of nonionic biosurfactants on the surface of NPs results in steric repulsion. The adsorption of biosurfactants on the surface of NPs restricts their free movements in the suspending medium, and prevents the interactions between NPs and their aggregation (Figure 3).

Nonionic biosurfactants are characterized by two segments; anchor segment which interacts with the NPs and a tail segment which has interactions with the dispersion medium and stabilize NPs by inhibiting their interactions. Whereas ionic biosurfactants exhibit a combined mechanism of steric and electrostatic repulsions in the stabilization of NPs (Figure 4) (Piacenza et al. 2018).

Zhao et al. (2019) synthesized silver NPs by using lipopeptides obtained from *Bacillus subtilis* and observed that iturin showed the best efficacy among other

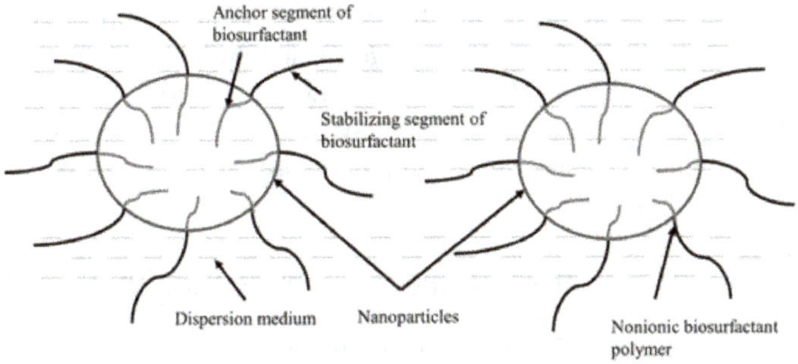

Figure 3. Adsorption of nonionic biosurfactant on the surface of nanoparticles and stabilization by steric repulsion (adopted from Piacenza et al. 2018).

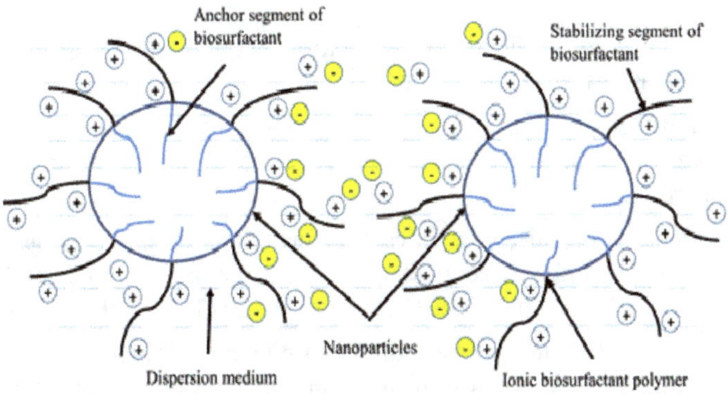

Figure 4. Adsorption of ionic biosurfactant on the surface of nanoparticles and stabilization of nanoparticles by steric and electrostatic repulsions (adopted from Piacenza et al. 2018).

lipopeptides (fengycin and surfactin) to synthesize NPs. It was observed that iturin along with its polypeptide chain and tyrosine were the major functional groups which were involved in the synthesis of NPs. Furthermore, under alkaline conditions a reduction in Ag+ was observed due to functional (phenolic) group of tyrosine (Zhao et al. 2019) which led to the silver NP formulation.

Ohadi et al. (2020) synthesized gold NPs by using lipopeptide biosurfactant derived from *Acinetobacter junii* B6. The UV visible absorption spectrum of NPs at 550 nm confirmed the formulation of gold NPs. FTIR studies revealed lipopeptides are able to attract NPs by free amine groups through electrostatic attraction of negatively charged COO^- groups. These negatively charged lipopeptides could change the surface properties of NPs and stabilize them. Furthermore, the lipopeptides contain functional groups (of amine, imide, and micellar structure) which are responsible for the stability of gold NPs (Singh et al. 2011).

4.3 *NPs Synthesized from Plant Based Biosurfactants*

Recently plant extracts have also been utilized as a source of biosurfactant molecules for the synthesis and stabilization of NPs. Sur et al. (2018) synthesized stable silver NPs using extracts of Shikakai and Reetha. Shivaji et al. (2019) utilized waste tea leaves as a source of biosurfactants to formulate cadmium sulfide quantum dots

(2.5–4 nm). FTIR confirmed phytochemicals (vitamins, amino acids, minerals, caffeine, and polyphenols) in the production and stabilization of quantum dots. Even though the correct mechanism for synthesis and stabilization of NPs is not known, however, by employing Gas Chromatography Mass Spectrometry, High Resolution Mass Spectroscopy, FTIR and Near Magnetic Resonance studies we can predict the mechanism of synthesis.

5. Advantages of Biosurfactants Over other NP Stabilizers

Biosurfactants offer many advantages over chemical surfactants, including:

a. Raw material availability-Cheap raw materials can be used to produce biosurfactants as compared to their chemical counterparts (Merchant and Banat 2012).
b. Biosurfactants show effectiveness at extreme conditions (i.e., high salinity, pH and temperature) (Plaza et al. 2006).
c. Biodegradability-Biosurfactants due to their simple chemical structure and low toxicity, are ecofriendly and are easily degraded as compared to chemical surfactants (Roy 2017).
d. Biosurfactant molecules contain certain functional groups, thus can be used for specific food and pharmaceutical applications, detoxification of pollutants, and de-emulsification of emulsions (Campos et al. 2013).
e. Because biosurfactants are digestible and biocompatible as compared to other NP stabilizers, they are widely used in food, cosmetic and pharmaceutical industries (Campos et al. 2013).

6. Conclusion

The use of biosurfactants in the field of nanotechnology is gaining interest due to associated benefits such as stabilization ability, biodegradable nature, low toxicity and synergistic effect in biological applications. There are diverse sources of biosurfactants including extremophiles, mesophiles and various other microbial and plant sources. The metabolites from various groups of microorganisms should be explored for their possible application as biosurfactants and as an additive or synergistic agent in nano-formulations for food, feed, pharmaceutical and agriculture applications.

References

Bages-Estopa, S., D.A. White, J.B. Winterburn, C. Webb, P.J. Martin et al. 2018. Production and separation of a trehalolipid biosurfactant. Biochemical Engineering Journal 139: 85–94.

Bayeea, P., H. Amani, G.D. Najafpoura and H. Kariminezhadb. 2020. Experimental investigations on behavior of rhamnolipid biosurfactant as a green stabilizer for the biological synthesis of gold nanoparticles. International Journal of Engineering 33: 1054–1060.

Bezza, F.A., S.M. Tichapondwa and E.M. Chirwa. 2020. Synthesis of biosurfactant stabilized silver nanoparticles, characterization and their potential application for bactericidal purposes. Journal of Hazardous Materials 393: 122319.

Campos, J.M., T.L. Montenegro Stamford, L.A. Sarubbo, J.M. de Luna, R.D. Rufino et al. 2013. Microbial biosurfactants as additives for food industries. Biotechnology Progress 29(5): 1097–1108.

Chakrabarti, S. 2012. Bacterial biosurfactant: Characterization, antimicrobial and metal remediation properties (Doctoral dissertation).

Chong, H. and Q. Li. 2017. Microbial production of rhamnolipids: Opportunities, challenges and strategies. Microbial Cell Factories 16(1): 137.

Dinets, O. and V. Maksin. 2019. Adsorption of ionic and non-ionic surfactants mixtures from aqueous solutions at the surface of carbon sorbents. Publishing House "Baltija Publishing".

Elakkiya, V.T., P. Suresh Kumar, N.S. Alharbi, S. Kadaikunnan, J.M. Khaled et al. 2020. Swift production of rhamnolipid biosurfactant, biopolymer and synthesis of biosurfactant-wrapped silver nanoparticles and its enhanced oil recovery. Saudi Journal of Biological Sciences 27(7): 1892–1899.

Elshikh, M., I. Moya-Ramírez, H. Moens, S.L.K.W. Roelants, W. Soetaert et al. 2017. Rhamnolipids and lactonic sophorolipids: natural antimicrobial surfactants for oral hygiene. Journal of Applied Microbiology 123(5): 1111–1123.

Eswari, J.S., S. Dhagat and P. Mishra. 2018. Biosurfactant assisted silver nanoparticle synthesis: A critical analysis of its drug design aspects. Advances in Natural Sciences: Nanoscience and Nanotechnology 9(4): 045007.

Ewais, H.A. 2014. Kinetics and mechanism of the formation of silver nanoparticles by reduction of silver (I) with maltose in the presence of some active surfactants in aqueous medium. Transition Metal Chemistry 39(5): 487–493.

Joanna, C., L. Marcin, K. Ewa and P. Grażyna. 2018. A nonspecific synergistic effect of biogenic silver nanoparticles and biosurfactant towards environmental bacteria and fungi. Ecotoxicology 27(3): 352–359.

Khalid, H.F., B. Tehseen, Y. Sarwar, S.Z. Hussain, W.S. Khan et al. 2019. Biosurfactant coated silver and iron oxide nanoparticles with enhanced anti-biofilm and anti-adhesive properties. Journal of Hazardous Materials 364: 441–448.

Kumar, C.G., S.K. Mamidyala, B. Das, B. Sridhar, G.S. Devi et al. 2010. Synthesis of biosurfactant-based silver nanoparticles with purified rhamnolipids isolated from *Pseudomonas aeruginosa* BS-161R. Journal of Microbiology Biotechnology 20(7): 1061–1068.

Liu, Q. 2012. Functional Nanostructure Synthesis and Properties (Doctoral dissertation).

Luft, L., T.C. Confortin, I. Todero, G.L. Zabot, M.A. Mazutti et al. 2020. An overview of fungal biopolymers: bioemulsifiers and biosurfactants compounds production. Critical Reviews in Biotechnology, 1–22.

Mahanti, P., S. Kumar and J.K. Patra. 2017. Biosurfactants: an agent to keep environment clean. In Microbial Biotechnology. Springer, Singapore, pp. 413–428.

Mapelli, F., A. Scoma, G. Michoud, F. Aulenta, N. Boon et al. 2017. Biotechnologies for marine oil spill cleanup: indissoluble ties with microorganisms. Trends in Biotechnology 35(9): 860–870.

Marchant, R. and I.M. Banat. 2012. Biosurfactants: A sustainable replacement for chemical surfactants? Biotechnology Letters 34(9): 1597–1605.

Mnif, I. and D. Ghribi. 2016. Glycolipid biosurfactants: main properties and potential applications in agriculture and food industry. Journal of the Science of Food and Agriculture 96(13): 4310–4320.

Mohite, P., A.R. Kumar and S. Zinjarde. 2017. Relationship between salt tolerance and nanoparticle synthesis by Williopsis saturnus NCIM 3298. World Journal of Microbiology and Biotechnology 33(9): 163.

Ohadi, M., H. Forootanfar, G. Dehghannoudeh, T. Eslaminejad, A. Ameri et al. 2020. Biosynthesis of gold nanoparticles assisted by lipopeptide biosurfactant derived from Acinetobacter junii B6 and evaluation of its antibacterial and cytotoxic activities. BioNanoScience, 1–10.

Piacenza, E., A. Presentato and R.J. Turner. 2018. Stability of biogenic metal (loid) nanomaterials related to the colloidal stabilization theory of chemical nanostructures. Critical Reviews in Biotechnology 38(8): 1137–1156.

Płaza, G.A., I. Zjawiony and I.M. Banat. 2006. Use of different methods for detection of thermophilic biosurfactant-producing bacteria from hydrocarbon-contaminated and bioremediated soils. Journal of Petroleum Science and Engineering 50(1): 71–77.

Radha, P., P. Suhazsini, K. Prabhu, A. Jayakumar, R. Kandasamy et al. 2020. Chicken tallow, a renewable source for the production of biosurfactant by *Yarrowia lipolytica* MTCC9520, and its application in silver nanoparticle synthesis. Journal of Surfactants and Detergents 23(1): 119–135.

Rane, A.N., V.V. Baikar, V. Ravi Kumar and R.L. Deopurkar. 2017. Agro-industrial wastes for production of biosurfactant by *Bacillus subtilis* ANR 88 and its application in synthesis of silver and gold nanoparticles. Frontiers in Microbiology 8: 492.

Rangarajan, V., G. Dhanarajan, P. Dey, D. Chattopadhya, R. Sen et al. 2018. Bacillus lipopeptides: powerful capping and dispersing agents of silver nanoparticles. Applied Nanoscience 8(7): 1809–1821.

Rawat, G., A. Dhasmana and V. Kumar. 2020. Biosurfactants: the next generation biomolecules for diverse applications. Environmental Sustainability, 1–17.

Reddy, A.S., C.Y. Chen, C.C. Chen, J.S. Jean, C.W. Fan et al. 2009. Synthesis of gold nanoparticles via an environmentally benign route using a biosurfactant. Journal of Nanoscience and Nanotechnology 9(11): 6693–6699.

Roy, A. 2017. Review on the biosurfactants: properties, types and its applications. Journal of Fundamentals of Renewable Energy and Applications 8: 1–14.

Sangeetha, J., S. Thomas, J. Arutchelvi, M. Doble, J. Philip et al. 2013. Functionalization of iron oxide nanoparticles with biosurfactants and biocompatibility studies. Journal of Biomedical Nanotechnology 9(5): 751–764.

Santos, D.K.F., R.D. Rufino, J.M. Luna, V.A. Santos, L.A. Sarubbo et al. 2016. Biosurfactants: multifunctional biomolecules of the 21st century. International Journal of Molecular Sciences 17(3): 401.

Satpute, S.K., S.S. Zinjarde and I.M. Banat. 2018. Recent updates on biosurfactant/s in Food industry. In Microbial Cell Factories. Taylor & Francis, pp. 1–20.

Shikha, S., S.R. Chaudhuri and M.S. Bhattacharyya. 2020. Facile one pot greener synthesis of sophorolipid capped gold nanoparticles and its antimicrobial activity having special efficacy against gram negative Vibrio cholerae. Scientific Reports 10(1): 1–13.

Shivaji, K., M.G. Balasubramanian, A. Devadoss, V. Asokan, C.S. De Castro et al. 2019. Utilization of waste tea leaves as bio-surfactant in CdS quantum dots synthesis and their cytotoxicity effect in breast cancer cells. Applied Surface Science 487: 159–170.

Singh, B.N., A.K.S. Rawat, W. Khan, A.H. Naqvi, B.R. Singh et al. 2014. Biosynthesis of stable antioxidant ZnO nanoparticles by *Pseudomonas aeruginosa* rhamnolipids. PLoS One 9(9): e106937.

Singh, B.R., S. Dwivedi, A.A. Al-Khedhairy and J. Musarrat. 2011. Synthesis of stable cadmium sulfide nanoparticles using surfactin produced by *Bacillus amyloliquifaciens* strain KSU-109. Colloids and Surfaces B: Biointerfaces 85: 207–213. https://doi.org/10.1016/j.colsurfb.2011.02.030.

Singh, S., V. Kumar, S. Singh, D.S. Dhanjal, S. Datta et al. 2020. Biosurfactant-based bioremediation. In Bioremediation of Pollutants. Elsevier, pp. 333–358.

Sun, D., J. Liao, L. Sun, Y. Wang, Y. Liu et al. 2019. Effect of media and fermentation conditions on surfactin and iturin homologues produced by Bacillus natto NT-6: LC–MS analysis. AMB Express 9(1): 120.

Sungsuwan, S. 2017. Lipopeptide-Coated Iron Oxide Nanoparticles and Engineered Qβ Virus Like Particles as Potential Glycoconjugate-Based Synthetic Anticancer Vaccines. Michigan State University.

Sur, U.K., B. Ankamwar, S. Karmakar, A. Halder, P. Das et al. 2018. Green synthesis of silver nanoparticles using the plant extract of Shikakai and Reetha. Materials Today: Proceedings 5(1): 2321–2329.

Tran, Q.H. and A.T. Le. 2013. Silver nanoparticles: synthesis, properties, toxicology, applications and perspectives. Advances in Natural Sciences: Nanoscience and Nanotechnology 4(3): 033001.

Tripathi, L., V.U. Irorere, R. Marchant and I.M. Banat. 2018. Marine derived biosurfactants: a vast potential future resource. Biotechnology Letters 40(11-12): 1441–1457.

Vijayakumar, S. and V. Saravanan. 2015. Biosurfactants-types, sources and applications. Research Journal of Microbiology 10(5): 181.

Wu, Q., Y. Zhi and Y. Xu. 2019. Systematically engineering the biosynthesis of a green biosurfactant surfactin by Bacillus subtilis 168. Metabolic Engineering 52: 87–97.

Zhao, X., L. Yan, X. Xu, H. Zhao, Y. Lu et al. 2019. Synthesis of silver nanoparticles and its contribution to the capability of *Bacillus subtilis* to deal with polluted waters. Applied Microbiology and Biotechnology 103(15): 6319–6332.

Zhu, C., F. Xiao, Y. Qiu, Q. Wang, Z. He et al. 2017. Lichenysin production is improved in codY null Bacillus licheniformis by addition of precursor amino acids. Applied Microbiology and Biotechnology 101(16): 6375–6383.

9

Biosurfactants A Sustainable Approach Towards Environmental Clean-up

Shagufta Siddiqui,[1,*] *Rajendrasinh Jadeja*[1] and *Zeba Usmani*[2]

1. Introduction

In the present times of a heavily industrialized world, mankind as a whole is hugely dependent on fossil fuels and their derivatives for every aspect of their lifestyle. Petroleum and its products have become an indispensable part of modern economy too; as they have entered every sphere of necessity ranging from fuels, pesticides and cosmetics to pharmaceuticals. The advancements in the field of science & technology have further extended the scope of exploration & extraction of these natural resources even in the remotest areas of the globe which were once inaccessible! Unfortunately, with the advent of these new techniques on one side even though the production of petroleum has increased, on the other side the instances of accidents during transportation & drilling as well as generation of a humongous amount of petrochemical waste has become a regular menace!

It is now a well understood fact that petroleum and its derivatives, especially polyaromatic hydrocarbon compounds (PAH) are known pollutants and quite toxic to the environment. Their toxicity to aquatic organisms is majorly affected by the metabolism of the organisms and photo-oxidation of these compounds. Also, generally they tend to become more toxic in the presence of ultraviolet light. Various studies have indicated that PAHs have moderate to acute toxicity to aquatic life and birds. Some adverse effects that can manifest on these organisms include tumours

[1] Department of Environmental Studies, Faculty of Science, The Maharaja Sayajirao University of Baroda, Vadodara, Gujarat, India-390002.

[2] Department of Chemistry and Biotechnology, Tallin University of Technology, Estonia.

* Corresponding author: shagufta.siddiqui-env@msubaroda.ac.in

as well as impaired reproduction, development, and immunity. Mammals too can absorb these carcinogenic & teratogenic compounds through various routes, e.g., inhalation, dermal contact, and ingestion leading to serious irreversible ailments (Beyer et al. 2010, Veltman et al. 2012).

Petroleum hydrocarbons are moderately persistent in the environment, recalcitrant in nature and have been reported to be biomagnified as well as bio-accumulated in the food chain. Thus, the concentrations of these hydrocarbons are expected to be much higher in fish and molluscs compared to the actual environment they were taken from. According to data from the National Oceanic and Atmospheric Administration (NOAA), there were 137 oil spills in 2018 in the U.S. alone; making it an average of 11 oil spills per month. From 1970 to 2019, almost fifty percent of the large spills that took place; were in the open seas, and mostly occurred due to accidents when a ship hit another vessel or got damaged by hitting some solid barrier under water. The 2010 Deepwater Horizon spill, killed 11 people and spewed an estimated 205 million gallons of oil into the Gulf of Mexico, and is one of the worst oil spills ever that killed a lot of faunal biodiversity (Peter Mwai, Aug. 2020, BBC report).

The hydrocarbons present in the oil spills pose a great threat to the ecological balance of the environment and their dangers can't be overlooked. PAHs, endocrine disrupting chemicals and certain heavy metals like arsenic and manganese (by products of unconventional oil and natural gas operations) have been reported to cause significant neurological and developmental health problems in infants, children and young adults (Ellen et al. 2018). On the other hand, plants can absorb PAHs from their surroundings through their roots and translocation to other plant parts. But their uptake rates are generally governed by variables like concentration, water solubility, and their physicochemical state as well as soil type. However, in plants the PAH-induced phytotoxicity effects are much rare. Certain plants contain substances that can provide protection against PAH effects (root exudates can covert complex PAHs into simpler and less harmful counterparts). Other plants can synthesize PAHs that act as growth hormones (Beyer et al. 2010). However, this accumulation in plants can enter food chains and disrupt the ecological balance.

Today, biological treatment is considered as an interesting and wholesome approach towards removal of petroleum contamination. Bioremediation is a technique in which biological systems such as microorganisms are applied to destroy or transform (degrade) harmful chemicals. Since a decade or so, employing hydrocarbon degrading bacteria to clean a petroleum contaminated soil has become very prevalent and is also economical as well as an efficient method that leads to non toxic end products. Bioremediation can be broadly divided into *in situ* and *ex situ* types, which refers to treatment at the site of contamination itself and treatment by taking the contaminated medium away from the site, respectively. As shown in Figure 1 during *in situ* treatment the natural course of intrinsic biodegradation where only monitoring is undertaken, is called natural attenuation. It is a fairly cheaper way to undertake bioremediation without much input cost but is much slower compared to others. However, this process can be enhanced through engineered biodegradation by introducing certain amendments like biostimulation (altering the environment for stimulating the activity of an indigenous microbial population), bioaugmentation (bioengineered microbes are introduced to the site of contamination), biosurfactant (use of surface active compounds having hydrophobic and hydrophilic moiety), bioventing (providing a regulated stream of air/nutrient for enhanced biodegradation in the unsaturated zone), biosparging (regulated stream of air/nutrients for enhanced biodegradation in the saturated zone) and bioslurping (combines elements of bioventing and vacuum-enhanced pumping to recover by-products).

For any bioremediation taking place, there is interaction between environmental factors, organisms and contaminants as shown in Figure 2. These interactions can be physiological factors, biodegradability potential or bioavailability of the substrate.

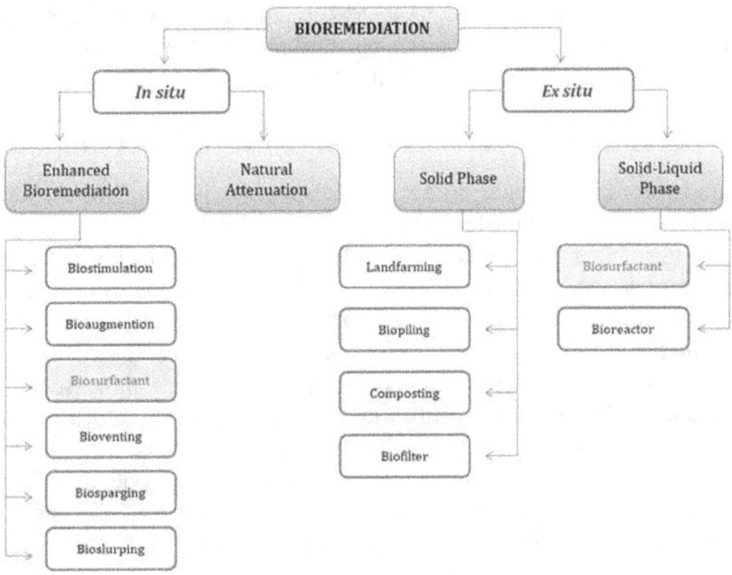

Figure 1. Types of bioremediations and various enhancements.

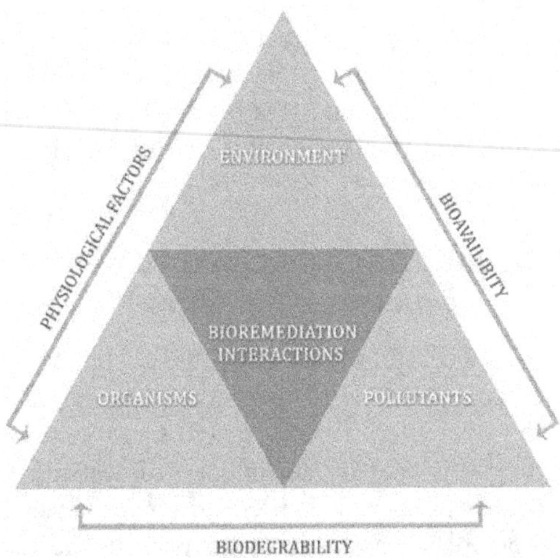

Figure 2. Interactions between physical and biological components during bioremediation.

Biosurfactants aid in degradation of pollutants as it possesses property of decreasing the surface tension and/or interfacial tension between two miscible or immiscible liquids. It further blocks hydrogen bond formation and increases hydrophilic/hydrophobic interactions owing to its cell surface hydrophobicity (Anyanwu et al. 2011, Darvishi et al. 2011). These properties make the biosurfactants immensely useful in industrial applications like chemistry, cosmetics, horticulture, agriculture, food production, food processing, and pharmaceuticals. Due to their multifunctional approach, biosurfactants are gaining popularity for biodegradation purposes too. Their potential uses in hydrocarbon bioremediation, microbial enhanced oil recovery (MEOR), nanotechnology (biosynthesis of metallic nanoparticles), medicines, commercial laundry detergents, food, textiles, petrochemicals, paper and paint industries, and pollution control can be attributed to their substrate specificity, diverse functions, and rapid controlled inactivation.

2. Bioremediation of Petroleum Hydrocarbon Compounds

Biodegradation of petroleum products consisting primarily of hydrocarbons is a complex procedure. The biodegradability of hydrocarbon compounds decreases with an increasing number of carbons and branching chains further making it complicated. The extent of biodegradability of polyaromatic hydrocarbons can be summarized in decreasing order as; linear alkanes > branched alkanes > low-molecular-weight alkyl aromatics > monoaromatics > cyclic alkanes > polyaromatics > asphaltenes; asphaltenes being the toughest compounds to be degraded biologically (Varjani and Upasini 2017). In environments with extremely loaded hydrocarbon pollution (e.g., oil spills) the total petroleum hydrocarbon concentrations often prove lethal to microbial activity thereby hampering the potential for their biodegradation (Admon et al. 2001). In such polluted places, additional amendments like application of biosurfactants can be very helpful (Figure 5). Even though synthetic surfactants are widely used, their detrimental effect on the environment cannot be overlooked. The application of biosurfactants over chemical surfactants has recently attracted lot of attention majorly due to favourable reasons like higher as well as stable foaming capacity, higher selectivity and specific activity even in extreme environmental conditions with fluctuating temperature, pH and salinity, unique structural properties for application in environment clean up, relative non toxicity, higher biodegradability and ability to be synthesized from renewable feedstocks (Banat et al. 2010, Zhao et al. 2016).

3. Role of Biosurfactants in Enhancement of Petroleum Hydrocarbon Biodegradation

There have been many successful remediation studies that have employed synthetic surfactants for bioremediation. It is rampantly used in extraction of petroleum oil from the reservoir, too. However, with the recent surge in sustainable technologies employing environmentally friendly techniques of remediation of hydrocarbon contaminated sites; the focus is now shifted on biosurfactants! Biosurfactants are basically obtained from plants and microorganisms in the form of saponins and glycolipids respectively (Aparna et al. 2011). These biosurfactant compounds have surfactant properties similar to those of synthetic surfactants but are themselves biodegradable and harmless to the environment. There are numerous advantages of employing biosurfactants in comparison to synthetic ones; like low toxicity, biodegradability, compatibility with microflora as well as surface activity efficiency even in extreme environmental conditions having wide temperature ranges, pH and salinity fluctuations. The major advantage of employing biosurfactants over the synthetic surfactants is their environmental compatibility and feasibility.

Basically, certain metabolites with surface activity consisting of fatty acids, glycolipids, lipopeptides, lipopolysaccharides and lipoproteins which are produced by microorganisms in aqueous solutions containing hydrophobic compounds are termed as biosurfactants. The structures and chemical compositions of the above mentioned surface active compounds varies greatly and is majorly dependent on the culture of microbes used, the nutrient medium and the physical growth conditions (Mukherjee et al. 2006). Some common types of biosurfactants and their applications along with the producing microorganisms have been elaborated in Table 1.

Table 1. Types of biosurfactants, their application and microbes producing it.

Biosurfactants	Application	Producing Organism
Rhamnolipid	Environmental application in bioremediation, treating hydrocarbons & oil spill cleanups, detergent and pharmaceutical industry	*Pseudomonas aeruginosa* S2, *Pseudomonas aeruginosa* BS20, *Pseudomonas aeruginosa* UFPEDA 614, *Pseudoxanthomonas* sp. PNK-04, *Pseudomonas alcaligenes* and *Burkholderia plantari* DSM 9509
Lipopeptides	Environmental and biomedical applications, potential in pharmaceuticals, environmental protection, cosmetics, oil recovery, bioremediation of marine oil pollution, biocontrol agents, biomedicine, MEOR, industrial lipopeptide production	*Pseudomonas libanensis* M9-3, *Bacillus subtilis* strain ZW-3, *Rhodococcus* sp. TW53, *Bacillus subtilis* BS5, *Azotobacter chroococcum, Bacillus subtilis* HOB2, *Nocardiopsis alba* MSA10, *Pseudomonas koreensis, Pseudomonas fluorescens* BD5, *Brevibacterium aureum* MSA13 and *Bacillus velezensis* H3
Glycolipids	Bioremediation application in marine oil pollution	*Pseudozyma hubeiensis, R. wratislaviensis* BN38 and *Nocardiopsis lucentensis* MSA04
Glucolipids and trehalose lipids	Oil spill cleanup operations	*Rhodococcus erythropolis* 3C-9
Mannosylerythritol lipid	Promising yeast biosurfactants, washing detergents, emulsifiers and/or washing detergents, bioremediation processes in the marine environment	*Pseudozyma siamensis* CBS 9960, *Pseudozyma graminicola* CBS 10092 and *Calyptogena soyoae*
Mannosylmannitol lipid,	Emulsifiers and/or washing detergents	*Pseudozyma parantarctica*
Sophorolipids	Environmental applications	*Candida bombicola*
Trehalose tetraester	Bioremediation of oil-contaminated environments	*Micrococcus luteus* BN56

Source: Makkar et al. (2011); Souza et al. (2014).

The initial step of interaction between the oily phase of the petroleum pollutants and the microorganism involves a direct contact physical between them (Uad et al. 2010, Souza et al. 2014). But this simple to appear interaction is strictly dependent upon the cell wall structure, i.e., its surface hydrophobicity. During the phase of this direct contact, the hydrocarbons molecules penetrate into the cells of the microbes as submicroscopic droplets (Varjani 2014a). The probable limitation of diffusion of the substrate during transport to the cell is overcome by the Surfactant activity and hydrophobicity by facilitating the interaction between microbial cells and the insoluble hydrocarbon substrate (Kavitha et al. 2014, Varjani and Upasani 2016). This is further explained diagrammatically in Figure 5. Microbes possessing oil degradation ability along with surface active agent production are favoured for *in situ* bioremediation of petroleum hydrocarbons. These hydrocarbon degrading microbes produce a range of surface active biomolecules that either remain attached to cell surface or can be released as extracellular molecules (Van Hamme et al. 2003, Sajna et al. 2015, Varjani and Upasani 2016). The amendments using biosurfactants very effectively reduce the interfacial tensions of oil and water *in situ*, as well as the viscosity of the oily components. It has been reported that the production of biosurfactants is an autecological characteristic which increases the hydrocarbon bioavailability, contact between the microbial cell and substrate as well as mass transport in the bioremediating microbes (Varjani et al. 2014b, Waigi et al. 2015).

4. Mechanism of Action of Biosurfactants

Mainly two mechanisms are responsible for the ultimate biodegradation of oil based hydrocarbons by biosurfactants. The initial phase requires enhanced bioavailability of the hydrophobic substrate to the microorganisms, resulting in surface tension reduction of the medium around the microbial cell along with reduction in interfacial tension between the microbial cell wall and hydrocarbon molecules. The second phase of this mechanism involves interaction between the biosurfactant and cell surface which promotes modifications in the membrane, facilitating hydrocarbon adherence (hydrophobicity increase) and reducing the lipopolysaccharide index of the cell wall without causing any damage to the membrane. Thus, biosurfactants facilitate hydrophobic & hydrophilic interactions by blocking the formation of hydrogen bridges as well as cause molecular rearrangement and reduction in the surface tension of the liquid, increase in its surface area and thereby promoting bioavailability and consequent biodegradability of the pollutant (Aparna et al. 2011).

The structure of a typical surfactant monomer consists of a hydrophilic head and hydrophobic tail (Figure 3a). The hydrophobic (lipophilic) end of this monomer molecule is composed of proteins as well as peptides with hydrophobic parts, or carbonated chains (10–18 carbons), whereas esters, hydroxyls, phosphates, carboxyl or carbohydrates characterize the hydrophilic part. Many such monomers attach together in certain conditions and lead to the formation of micelles, a polymer (Figure 3c). A lot many factors are responsible for the formation of these micelles and are mainly governed by the prevailing physical conditions of the medium. Surfactants when at low concentrations do bring about changes in the interfacial tension of the fluid phases but when their concentration keeps increasing a critical concentration is reached, and beyond this critical concentration no further change in the interfacial properties is observed. These micelles are similar to the monomers and they are amphipathic molecules too, which have well arranged hydrophobic and hydrophilic portions. The presence of these heads and tails in surfactants help in acting between fluids of different polarities (oil/water and water/oil) thereby making the hydrophobic substrates accessible (hydrocarbon pollutant). This interaction causes remarkable physical changes by reducing the surface tension of the substrate, increasing the area of contact of insoluble compounds (heavy chain hydrocarbons), their mobility, bioavailability to the microorganisms and thereby biodegradation (Rabiei et al. 2013, Souza et al. 2014, Zhao et al. 2016). As shown in Figure 4, this concentration is called critical micelle concentration (CMC) because beyond this concentration, the surfactant molecules start forming micelles which enables the surfactant to reduce surface and interfacial tensions resulting further to increased solubility and bioavailability of hydrophobic hydrocarbon compounds for biodegradation (Desai

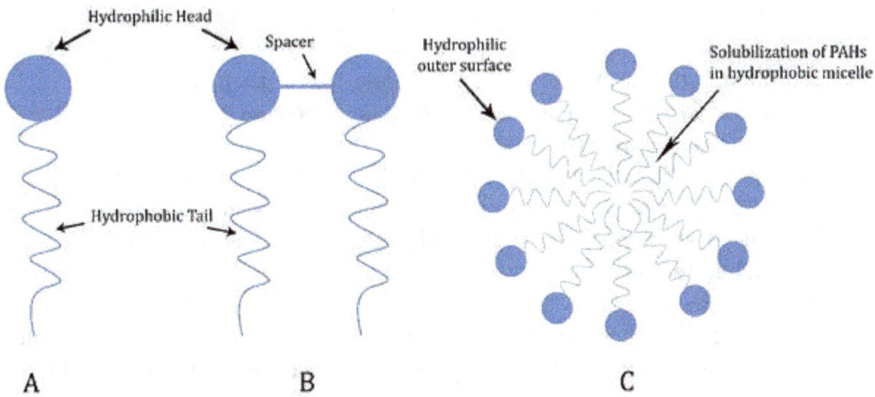

Figure 3. Structures of Biosurfactant monomers, Gemini surfactants and Micelles.

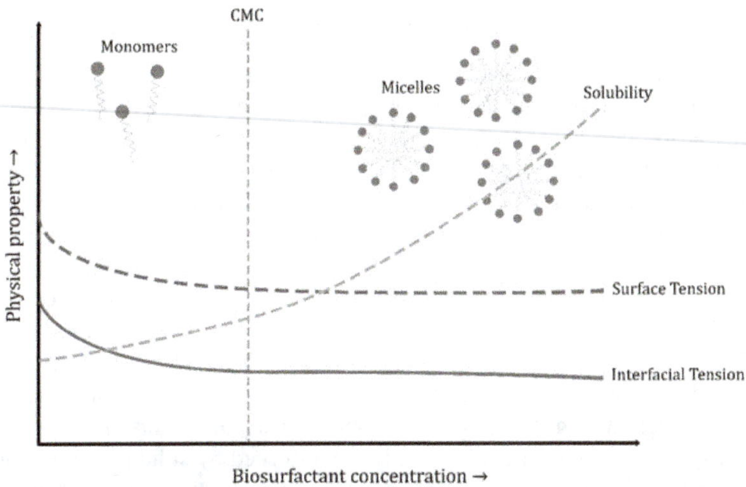

Figure 4. Micelles formation and Critical Micelle concentration.

and Banat 1997, Rabiei et al. 2013). When biosurfactants reduce the surface tension and interfacial tension repulsive forces act between two non-similar phases and help the phases to mix and interact seamlessly (Desai and Banat 1997, Mulligan 2009).

Lately, a group of novel surface-active compounds called 'gemini surfactants' having more than one hydrophobic and hydrophilic group, have gained popularity (Figure 3b). Gemini surfactants are formed when two conventional single chain monomers are stringed together by a spacer chain. These spacers are mainly formed by alkyl carbon chains, rigid phenyl groups, polystyrene chain and polar polyethers. The hydrophilic portions can be made up of anionic sulphates, carboxylates, phosphates, cationic quaternary ammoniums, non-ionic polyethers, polysaccharides, and complicated hydrophilic oligomers, whereas the hydrophobic portions are usually made up of long chains of hydrocarbons. Gemini surfactants possess much lower CMC values compared to their corresponding monomeric surfactant units.

The production of biosurfactants generally takes place during the exponential growth phase or stationary phase of microbial growth, when the cell density is at its maximum (Suwansukho et al. 2008). Thus, the production of biosurfactants can be indirectly evaluated by surface tension (related to the forces of attraction and repulsion between the molecules of a fluid). In order to keep the molecules interconnected on the liquid surface, a strong intermolecular force of attraction is exerted by the micelles on the substrate molecules that are closer to the surface (called surface tension). Hence, the fluids tend to decrease their surface area, creating spherical drops that present lower surface/volume ratios (emulsion) (Aparna et al. 2012, Pirollo et al. 2008). Physically speaking, the surface tension is the necessary energy required to dilate the surface of a liquid. The higher the attraction between liquid molecules, the higher is the work required to be done in order to increase the distance between them. Also, it is an established fact that the less water-soluble hydrophobic molecules, adsorb more strongly to soil components compared to the hydrophilic molecules (Jain et al. 2013). Ultimately it can be said that higher the force of cohesion between the molecules is; the higher the surface tension will be!

5. Factors Influencing the Optimum Activity of Biosurfactants

There are numerous scientific and common factors to be seriously considered for biosurfactant enhanced remediation to be optimally implemented. These factors can be either adsorption behaviour of the biosurfactants onto the soil/sludge/substrate, their solubilising and desolubilising ability on the contaminants, or even their toxicity and biodegradability. Other common factors like the cost of surfactants and scale of the contaminated area and feasibility of the bioremediation should also be simultaneously considered.

A desirable biosurfactant must possess a lower CMC value along with a strong ability to desorb contaminants. It should help in reducing the cost of the overall remediation process by functioning even in smaller dosages. Generally, the factors discussed in Table 2, directly influence the activity of biosurfactants during remediation of hydrocarbon pollutants. When surfactants are added into the aqueous system, some of them inevitably get adsorbed by solid particles. Hence, if the adsorption of biosurfactants is greater, lesser will be their contribution towards

Table 2. Factors influencing Biosurfactant activity during Bioremediation of hydrocarbons.

Factors	Influence on Biodegradability	Reference
Biosurfactant concentration	Increased concentration of biosurfactant enhances hydrocarbon solubilising potential up to a certain level.	Husein and Ismail (2013)
Physical temperature	Solubility is enhanced proportional to the temperature but not much change in solubility takes place between 10-40°C.	Peng et al. (2015)
pH of the medium	Increase in pH from acidic to alkaline has a positive effect on the surface tension property of biosurfactants. CMC formation is hampered below a pH of 4.	Iglesias et al. (2014)
Salinity	Increase in salt significantly enhances interaction of the liphophilic group of biosurfactants with aqueous phase, improving solubility of hydrocarbon compounds.	Iglesias et al. (2014)
Hydrophilic and hydrophobic balance (HLB) value	Biosurfactants with a lower HLB number have a better micelle volume and hence a better PAh solubility than those with a higher HLB number.	Chun et al. (2002)
Dissolved organic matter	Hydrocarbon compounds bind with dissolved organic matter causing reduction in the freely dissolved hydrocarbon compounds and increase their solubility.	Tejeda-Agredano et al. (2014)
Mechanism of solubilisation	Structure of Biosurfactant, nature of the hydrocarbon compound, bioavailability, surface active property as well as CMC formation are responsible.	Lamichhane et al. (2017)
Co-solute formations	Synergistic co-solubilisation of more hydrophobic PAH compounds enhances their degradation.	Liang et al. (2016)
Octanol-water partitioning co-efficient of PAH	Improved solubility and inhibited solubility of PAH is observed with higher and lower octanol-water partition PAH coefficients respectively.	Liang et al. (2014)

pollutant solubilization. Additionally, the hydrophobicity of the substrate is enhanced when higher amounts of biosurfactants gets adsorbed. These further result into the already removed and solubilized organic compounds again getting re-adsorbed on the solid (soil/dirt) surface (Paria 2008).

Thus, selection of the appropriate biosurfactant during a remediation process greatly depends upon the adsorption behaviour of the biosurfactant on the soil particles. This adsorption behaviour of the biosurfactant is in turn governed by its molecular structure, and is the most influencing factor. According to Zhang (2013), the type of biosurfactants, and their molecular structure like the number of hydroxyl groups, length of hydrophobic chains, and presence of different substituent groups can greatly influence the molar solubilization ratio (MSR) of the pollutants. Molar solubilisation ratio is the moles of contaminant solubilised per mole of biosurfactant. It directly co-relates with the solubility of the pollutant in an aqueous medium with respect to biosurfactant concentration and can be calculated by the equation $MSR = (C_{mic} - C_{CMC})/(C_{surf} - CMC)$ (where Cmic is the total solubility of the organic pollutant (in moles per liter) in the micelle solution at a biosurfactant concentration more than the CMC, CCMC is the solubility of the organic pollutant (in moles per liter) at the CMC which can be approximately rounded off to the aqueous solubility of the organic pollutant, and Csurf is the biosurfactant concentration at which Cmic is evaluated in moles per liter) (Lee et al. 2013, Zhu et al. 2012, Chakraborty et al. 2009, Oliveira et al. 2009, Gao et al. 2013, Laha et al. 2009).

One of the key factors for biodegradation through a biosurfactant is the bioavailability of petroleum hydrocarbon molecules to the surfactant producing microbes (Souza et al. 2014). These hydrocarbon pollutants transport to the microbial cells through (a) interaction of microbial cells to hydrocarbon pollutants dissolved in aqueous phase, (b) direct contact of cells with hydrocarbons, and (c) interaction of the cells with hydrocarbon droplets much smaller than the cells (Kavitha et al. 2014, Yalcin et al. 2011). The bioavailability of these pollutants also depends on their chemical properties like extent of hydrophobicity, carbon chain, solubility & volatility; environmental conditions like temperature, pH, salinity and biological activity of the microbes (Varjani et al. 2014b).

6. Biosurfactants—As a Sustainable Approach Towards Environmental Clean-up

Unlike conventional methods that employ chemically active substances; bioremediation takes into usage natural biological processes to cleanse or remove pollutants. Bioremediation helps in removing environmental pollutants through mineralization or biochemical solubilisation thus making it a very ecologically sound option. There are various means by which biodegradation of petroleum pollutants takes place; but natural microbial population characterize one of the significant mechanisms by which hydrophobic contaminants can be removed or disintegrated from ecological components (Korjus 2014, Mani and Kumar 2014, Mnif et al. 2015, Yin et al. 2016, Zhao et al. 2018). Due to the innate characteristics of petroleum hydrocarbons like low water solubility, strong adhesion to soil particles, and low biological availability their mass transfer proportion is hampered leading to hampered biodegradation. Thus, Biosurfactants are a feasible amendment of bioremediation due to its universal action on land as well as in sea. This also makes them convenient to use and employed in a wide range of environmental clean ups which are eco accommodating, financially productive, and support various enhancements (Cameotra and Makkar 2010, Banat et al. 2010, Nitschke et al. 2005, Singh et al. 2007, Van Hamme et al. 2006).

Biosurfactants can be easily obtained and produced on a large scale from waste materials & by-products of certain industrial operations. A lot of these waste products are generally released into the environment leading to organic pollution & microbial overload in water bodies & land. Thus, for the large-scale commercial production of biosurfactant compounds, industrial waste products can be used as cheap raw materials or substrates, reducing the pressure on natural resources. Industrial wastes such as corn steep liquor, sugarcane molasses, soap stock, animal fat, starch waste and others are low-cost materials generated that are employed in the production of biosurfactants. Other few promising, inexpensive and easily accessible carbon and energy substrates for biosurfactant synthesis are the wastes generated from oil processing industries like sunflower, coconut oil, olive oil, and canola oil among others. There have also been instances of using soybean oil waste, cassava flour, molasses, and whey for producing rhamnolipids by *Pseudomonas aeruginosa* LBI strains (Nitschke et al. 2010).

The biosurfactants are environmentally compatible as they help in biodegradation of contaminants while they are getting self destructed during the process, leaving behind no harmful residues. Along with better solubilization and desorption of organic contaminants, they also lead to stimulation of microbial diversity which is positive for the *in situ* bioremediation of organic contaminants. When compared to synthetic chemical surfactants, biosurfactants mostly possess a larger molecular structure and ligand groups, endowing them with extraordinary surface active properties!

7. Summary

With increasing pollution of natural environmental components like land, water & air, optimum abatement is the necessity of the hour. Petroleum extraction, transportation, refining, usage & ultimate disposal has put further pressure on these components. Throughout the life cycle of crude oil, there is contamination of environmental components at each stage. It is almost impossible to completely do away with the usage of petroleum and its products; we can certainly try towards reducing the undesirable outcomes of it. Petroleum oil spills due to accidents & failures lead to unimaginable loss to biodiversity and contamination of land as well as water bodies. Figure 5 has shown a schematic diagram of oil spill biodegradation by bacteria in the presence of biosurfactants. Till a decade ago, conventional methods were the most preferred choice of degrading these pollutants or treating oil spills; but bioremediation is now much simpler with the latest technological advancements in the fields of bioremediation and environmental awareness. Various enhancements to bioremediation have been tried and tested with a great success rate, throughout the globe. In this chapter biosurfactant enhanced remediation of petroleum hydrocarbons was discussed in detail. We briefly tried to cover the structure, types, formation,

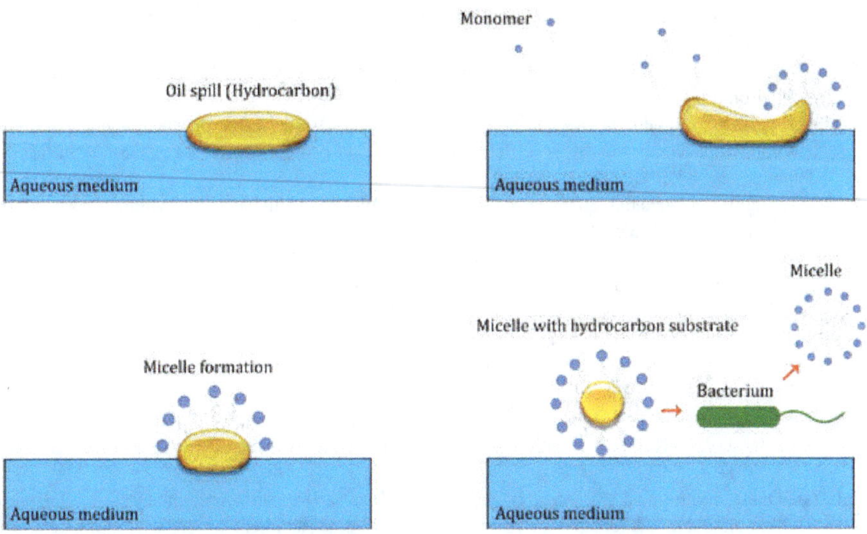

Figure 5. Mechanism of oil spill cleanup using Biosurfactants.

production and remediation mechanism of biosurfactant induced degradation of hydrocarbons. Biosurfactants can be a good option due to many favourable reasons like easy availability, production from agro waste and their eco friendly approach. With all the discussion earlier in this chapter, it can be concluded that biosurfactants are a much better substitute for synthetic surfactants and are a sustainable approach towards wholesome environmental cleanup strategies!

Acknowledgements

The authors are thankful to the University Grants Commission of India (UGC) and the Department of Environmental Studies, The Maharaja Sayajirao University of Baroda, Gujarat, India, for their support and provided facilities.

References

Admon, S., M. Green and Y. Avnimelech. 2001. Biodegradation kinetics of hydrocarbon in soil during land treatment of oily sludge. Bioremediation J. 5(3): 193–209.

Anyanwu, C., S. Obi and B. Okolo. 2011. Lipopeptide biosurfactant production by *Serratia marcescens* NSK-1 strain isolated from petroleum-contaminated soil. J. Appl. Sci. Res. 7: 79–87.

Aparna, A., G. Srinikethan and S. Hedge. 2011. Effect of addition of biosurfactant produced by *Pseudomonas* sp. on biodegradation of crude oil. *In*: 2nd International Proceedings of Chemical, Biological & Environmental Engineering, Vol. 6 IACSIT Press, Singapore, pp. 71–75.

Aparna, A., G. Srinikethan and S. Hegde. 2012. Isolation, screening and production of biosurfactant by Bacillus clausii5B. Res. Biotechnol. 3(2).

Banat, I., A. Franzetti, I. Gandolfi, G. Bestetti, M. Martinotti et al. 2010. Microbial biosurfactants production, applications and future potential. Appl. Microbiol. and Biotechnol. 87: 427–444.

Beyer, J., G. Jonsson, C. Porte, M.M. Krahn, F. Ariese et al. 2010. Analytical methods for determining metabolites of polycyclic aromatic hydrocarbon (PAH) pollutants in fish bile: A review. Environ. Toxicol. Pharmacol. 30: 224–244.

Cameotra, S.S. and R.S. Makkar. 2010. Biosurfactant-enhanced bioremediation of hydrophobic pollutants. Pure Appl. Chem. 82: 97–116.

Chakraborty, S., D. Shukla, A. Jain, B. Mishra, S. Singh et al. 2009. Assessment of solubilization characteristics of different surfactants for carvedilol phosphate as a function of pH. J. Colloid Interface Sci. 335: 242–249.

Chun, C.L., J. Lee and J. Park. 2002. Solubilization of PAH mixtures by three different anionic surfactants. J. Environ. Pollut. 118: 307–313.

Darvishi, P., S. Ayatollahi, D. Mowla and A. Niazi. 2011. Biosurfactant production under extreme environmental conditions by an efficient microbial consortium, ERCPPI-2. Colloids Surfaces B Biointerfaces 84: 292–300.

Desai, J.D. and I.M. Banat. 1997. Microbial production of surfactants and their commercial potential. Microbiology and Molecular Biology Reviews 61(1): 47–64.

Ellen Webb, Julie Moon, Larysa Dyrszka, Brian Rodriguez, Caroline Cox et al. 2018. Neurodevelopmental and neurological effects of chemicals associated with unconventional oil and natural gas operations and their potential effects on infants and children. Rev. Environ. Health. 33(1): 3–29.

Gao, H.P., L. Xu, Y.M. Cao, J. Ma, L.Y. Jia et al. 2013. Effects of hydroxypropyl-cyclodextrin and -cyclodextrin on the distribution and biodegradation of phenanthrene in NAPL-water system. Int. Biodeterior. Biodegrad. 83: 105–111.

Hussein, Taghreed A. and Z. Ismail, Zainab. 2013. Desorption of selected PAHs as individuals and as a ternary PAH mixture within a water-soil-nonionic surfactant system. Environ. Technol. 34(3): 351–361.

Iglesias, Olalla, Sanrom_an, M. Angeles and Pazos, Marta. 2014. Surfactant-enhanced solubilization and simultaneous degradation of phenanthrene in marine sediment by electro-fenton treatment. Industrial Eng. Chem. Res. 53(8): 2917–2923.

Jain, R.M., K. Mody, N. Joshi, A. Mishra, B. Jha et al. 2013. Effect of unconventional carbon sources on biosurfactant production and its application in bioremediation. International Journal of Biological Macromolecules 62: 52–58.

Kavitha, V., A.B. Mandal and A. Gnanamani. 2014. Microbial biosurfactant mediated removal and/or solubilization of crude oil contamination from soil and aqueous phase: an approach with Bacillus licheniformis MTCC 5514. Int. Biodeterior. Biodegrad. 94: 24–30.

Korjus, H. 2014. Polluted soils restoration, climate change and restoration of degraded land. Colegio de Ingenieros de Montes, pp. 411–480.

Laha, S., B. Tansel and A. Ussawarujikulchai. 2009. Surfactant–soil interactions during surfactant-amended remediation of contaminated soils by hydrophobic organic compounds: A review. J. Environ. Manage. 90: 95–100.

Lamichhane, Shanti, K.C. Bal Krishna and Sarukkalige, Ranjan. 2017. Surfactant-enhanced remediation of polycyclic aromatic hydrocarbons: A review. Journal of Environmental Management 199: 46–61.

Lee, Y.C., M.H. Choi, J.I. Han, Y.L. Lim, M. Lee et al. 2013. A low-foaming and biodegradable surfactant as a soil-flushing agent for diesel-contaminated soil. Sep. Sci. Technol. 48: 1872–1880.

Liang, Xujun, Zhang, Menglu, Guo, Chuling, Abel, St_ephane, Yi, Xiaoyun et al. 2014. Competitive solubilization of low-molecular weight polycyclic aromatic hydrocarbons mixtures in single and binary surfactant micelles. Chem. Eng. J. 244: 522–530.

Liang, Xujun, Guo, Chuling, Wei, Yanfu, Lin, Weijia, Yi, Xiaoyun et al. 2016. Cosolubilization synergism occurrence in codesorption of PAH mixtures during surfactant-enhanced remediation of contaminated soil. Chemosphere 144: 583–590.

Makkar, R.S., S.S. Cameotra and I.M. Banat. 2011. Advances in utilization of renewable substrates for biosurfactant production. AMB Exp. 1: 1–19.

Mani, D. and C. Kumar. 2014. Biotechnological advances in bioremediation of heavy metals contaminated ecosystems: An overview with special reference to phytoremediation. Int. J. Environ. Sci. Technol. 11: 843–872.

Mnif, I., S. Mnif, R. Sahnoun, S. Maktouf, Y. Ayedi et al. 2015. Biodegradation of diesel oil by a novel microbial consortium: Comparison between co-inoculation with biosurfactant-producing strain and exogenously added biosurfactants. Environ. Sci. Pollut. Res. Int. 22: 14852–14861.

Mukherjee, S., P. Das and R. Sen. 2006. Towards commercial production of microbial surfactants. Trends Biotechnol. 24: 509–515.

Mulligan, C.N. 2009. Recent advances in the environmental applications of biosurfactants. Curr. Opin. Colloid Interface Sci. 14(5): 372–8.

Nitschke, M., S.G. Costa and J. Contiero. 2005. Rhamnolipid surfactants: An update on the general aspects of these remarkable biomolecules. Biotechnol. Prog. 21: 1593–1600.

Nitschke, M., S. Costa and J. Contiero. 2010. Structure and applications of a rhamnolipid surfactant produced in soybean oil waste. Appl. Biochem. Biotechnol. 160: 2066–2074.

Oliveira, F.J.S., L. Vazquez, N.P. de Campos and F.P. de Franca. 2009. Production of rhamnolipids by a Pseudomonas alcaligenes strain. Process Biochem. 44: 383–389.

Paria, S. 2008. Surfactant-enhanced remediation of organic contaminated soil and water. Adv. Colloid Interface Sci. 138: 24–58.

Peng, Xin, Yuan, Xing-zhong, Liu, Huan, Zeng, Guang-ming, Chen, Xiao-hong et al. 2015. Degradation of polycyclic aromatic hydrocarbons (PAHs) by laccase in rhamnolipid reversed micellar system. Appl. Biochem. Biotechnol. 176(1): 45–55.

Peter Mwai, BBC Report, August 2020. https://www.bbc.com/news/world-53757747.

Pirollo, M.P.S., A.P. Mariano, R.B. Lovaglio, S.G.V.A.O. Costa, V. Walter et al. 2008. Biosurfactant synthesis by *Pseudomonas aeruginosa* LB1 isolated from a hydrocarbon-contaminated site. J. Appl. Microbiol. 105: 1484–90.

Rabiei, A., M. Sharifinik, A. Niazi, A. Hashemi, S. Ayatollahi et al. 2013. Core flooding tests to investigate the effects of IFT reduction and wettability alteration on oil recovery during MEOR process in an iranian oil reservoir. Applied Microbiology and Biotechnology 97(13): 5979–5991.

Sajna, K.V., R.K. Sukumaran, L.D. Gottumukkala and A. Pandey. 2015. Crude oil biodegradation aided by biosurfactants from Pseudozyma sp. NII 08165 or its culture broth. Bioresour. Technol. 191: 133–139.

Singh, A., J.D. Van Hamme and O.P. Ward. 2007. Surfactants in microbiology and biotechnology: Part 2. Application aspects. Biotechnol. Adv. 25: 99–121.

Souza, E.C., T.C. Vessoni-Penna and R.P. de Souza Oliveira. 2014. Biosurfactant enhanced hydrocarbon bioremediation: An overview. Int. Biodeterior. Biodegrad. 89: 88–94.

Suwansukho, P., V. Rukachisirikul, F. Kawai and A. H-Kittikun. 2008. Production and applications of biosurfactant from Bacillus subtilis MUV4. Songklanakarin J. Sci. Technol. 30: 87–93.

Tejeda-Agredano, M.C., P. Mayer and J.J. Ortega-Calvo. 2014. The effect of humic acids on biodegradation of polycyclic aromatic hydrocarbons depends on the exposure regime. Environ. Pollut. 184: 435–442.

Uad, I., G.A. Silva-Castro, C. Pozo, J. Gonzalez-Lopez, C. Calvo et al. 2010. Biodegradative potential and characterization of bioemulsifiers of marine bacteria isolated from samples of seawater, sediment and fuel extracted at 4000 m of depth Prestige wreck. Int. Biodeterior. Biodegrad. 64: 511–518.

Van Hamme, J.D., A. Singh and O.P. Ward. 2003. Recent advances in petroleum microbiology. Microbiol. Mol. Biol. Rev. 67: 503–549.

Van Hamme, J.D., A. Singh and O.P. Ward. 2006. Physiological aspects: Part 1 in a series of papers devoted to surfactants in microbiology and biotechnology. Biotechnol. Adv. 24: 604–620.

Varjani, S.J. 2014a. Hydrocarbon Degrading and Biosurfactants (Bioemulsifiers) Producing Bacteria from Petroleum Oil Wells (Ph.D. Thesis, Kadi Sarva Vishwavidyalaya).

Varjani, S.J., D.P. Rana, S. Bateja, M.C. Sharma, V.N. Upasani et al. 2014b. Screening and identification of biosurfactant (bioemulsifier) producing bacteria from crude oil contaminated sites of Gujarat, India. Int. J. Innovative Res. Sci. Eng. Technol. 3(2): 9205–9213.

Varjani, S.J. and V.N. Upasani. 2016. Core flood study for enhanced oil recovery through *ex-situ* bioaugmentation with thermo- and halo-tolerant rhamnolipid produced by *Pseudomonas aeruginosa* NCIM 5514. Bioresour. Technol. 220: 175–182.

Varjani, S.J. and V.N. Upasani. 2017. Critical review on biosurfactant analysis, purification and characterization using rhamnolipid as a model biosurfactant. Bioresource Technology 232: 389–397.

Veltman, K., M.A.J. Huijbregts, H. Rye and E.G. Hertwich. 2012. Including impacts of particulate emissions on marine ecosystems in life cycle assessment: The case of offshore oil and gas production. Integr. Environ. Assess. Manage. 7: 678–686.

Waigi, M.G., K. Fuxing, G. Carspar, L. Wanting and G. Yanzheng. 2015. Phenanthrene biodegradation by sphingomonads and its application in the contaminated soils and sediments: A review. Int. Biodeterior. Biodegrad. 104: 333–349.

Yalcin, E., K. Cavusoglu and E. Ozen. 2011. Hydrocarbon degradation by a new Pseudomonas sp., strain RW-II, with polycationic surfactant to modify the cell hydrophobicity. Environ. Technol. 32(15): 1743–1747.

Yin, K., M. Lv, Q. Wang, Y. Wu, C. Liao et al. 2016. Simultaneous bioremediation and biodetection of mercury ion through surface display of carboxylesterase E2 from *Pseudomonas aeruginosa* PA1. Water Res. 103: 383–390.

Zhang, C., H. Yan, F. Li, X. Hu, Q. Zhou et al. 2013. Sorption of short-and long-chain perfluoroalkyl surfactants on sewage sludges. J. Hazard. Mater. 260: 689–699.

Zhao, F., J.D. Zhou, F. Ma, R.J. Shi, S.Q. Han et al. 2016. Simultaneous inhibition of sulfate-reducing bacteria, removal of H2S and production of rhamnolipid by recombinant Pseudomonas stutzeri Rhl: applications for microbial enhanced oil recovery. Bioresour. Technol. 207: 24–30.

Zhao, G., Y. Sheng, C. Wang, J. Yang, Q. Wang et al. 2018. *In situ* microbial remediation of crude oil-soaked marine sediments using zeolite carrier with a polymer coating. Mar. Pollut. Bull. 129: 172–178.

Zhao, Y.-H., L.-Y. Chen, Z.-J. Tian, Y. Sun, J.-B. Liu et al. 2016. Characterization and application of a novel bioemulsifier in crude oil degradation by acinetobacter Beijerinckii ZRS. Journal of Basic Microbiology 56(2): 184–195.

Zhu, J.X., W. Shen, Y.H. Ma, L.Y. Ma, Q. Zhou et al. 2012. The influence of alkyl chain length on surfactant distribution within organo-montmorillonites and their thermal stability. J. Therm. Anal. Calorim. 109: 301–309.

10

Cyanobacterial Biosurfactants in the Bioremediation of Oil Industries Effluents

Krishnakant Das,[1] *Sushrirekha Das*,[1,*]
Manas Kumar Sinha,[2] *Manorama Das*,[1]
Mahidhar Bolem[1] and *Nityasundar Pal*[1]

1. Introduction

In recent times, the increase in the rate of pollution becomes a key focus of concern. Eco-friendly processes need to be developed to clean up the environment without creating toxic products. Industrial effluents are the major sources of environmental pollution. They contain certain toxic chemicals which are hazardous to human health. Their direct exit to rivers and other water sources causes pollution not only to water sources but also to ecological and economical sources. Different industries like rubber, mining, petrochemical and also some food industries discharge toxic substances directly to water bodies. The discharges of industries could be treated with algae (Oswald 1992). Industries generate massive by-products, solid wastes, high amounts of wastewater rich in organic wastes with different pollutant loads of and emissions into the air. Petrochemical industries generate aqueous effluent containing various conventional pollutants as well as specific petrochemicals and intermediates. The present study focuses on the bioremediation of oil refinery effluents by using algae. The industrial effluents contain several types of chemicals such as dispersants, leveling agents, acids, alkalis, carriers and various dyes, phenols, carbonates, alcohols, cyanides, and heavy metals among others (Cooper 1995). The use of silicon by different industries results in the increase in the toxicity in their industrial effluents. This causes a serious threat to the environment and human life. Even in low concentrations, these elements have a toxic effect upon daphnia, thus disrupting

[1] Marsco Aqua Clinics Pvt. Ltd., (AOC), Balasore, Odisha, India.
[2] National Freshwater Fish Brood Bank, Kausalyagang, Bhubaneswar, Odisha, India.
* Corresponding author: sushrirekha3@gmail.com

the food chain for fish life and possibly inhibiting photosynthesis of various aquatic plants (Bosnic et al. 2000). Release of these effluents into aquatic ecosystems alters the pH, increases the BOD and COD and gives the water intense colorations (Ajayi and Osibanjo 1980). Treatment of effluents includes physiochemical methods such as filtration, specific coagulation, use of activated carbon and chemical flocculation (Olukanni et al. 2006). Conventional treatment of effluents for the purpose of detoxification requires application of physical and chemical methods which involve chrome precipitation and sulphide treatment. Due to associated problems in these treatments such as cost and intense experimental set-ups (Do et al. 2002), biological treatment methods using various bacteria and fungi have been widely studied. However, it is now becoming apparent that cyanobacteria also play a major role in degrading organic materials from the ecosystem. Cyanobacteria also called blue-green algae got their name from the word 'cyan' meaning a 'turquoise blue' color. The blue-green color of the algae is due to the presence of a blue-green color pigment C-phycocyanin (C-PC), the pigment used for photosynthesis. Blue green algae (cyanobacteria) are considered as the most primitive photosynthetic prokaryotes which are supposed to have appeared on this planet during the Precambrian period (Ash and Jenkins 2006). They are the oldest known fossils of more than 3.5 billion years old. They are playing an important role in the field of evolution from lower to higher forms. They have the capacity for nitrogen fixation and the occurrence of gas vesicles which are especially important to the success of blue-green species which are a nuisance. The Blue-green variety are not true algae as they have no nucleus (the structure that encloses the DNA) and no chloroplast (the structure that encloses the photosynthetic membranes), the structures which are evidence of being true photosynthetic algae. Cyanobacteria are very susceptible to sudden physical and chemical alterations of light, salinity, temperature and nutrient composition (Boomiathan 2005, Semyalo 2009). The cyanobacteria have also shown immense potential in wastewater and industrial effluent treatment, bioremediation of aquatic and terrestrial habitats and chemical industries. They are also used as biofertilizers, biofuels and become an attractive source of innovative classes of pharmacologically active compounds due to the presence of bioactive compounds and pigments showing interesting and exciting biological activities ranging from antibiotic, immunosuppressant and anticancer, antiviral, and anti-inflammatory to proteinase-inhibiting agents for which they can be used in pharma, food and feed industries (Becker 1992), feed in aquacultures (Lora-Vilchis et al. 2004). They are also used for the production of pigments (Johnson and An 1991) and in agriculture (Metting 1992). Bioremediation, the use of microorganisms or microbial processes, is one among the new technologies. The use of cyanobacteria is considered for low-cost, low maintenance remediation of pollutants in surface water. The bioremediation has proven to be successful in numerous applications especially in treating petroleum contaminated soils (Gazyna et al. 2005). Bioremediation strategies for typical hazardous wastes are illustrated by Satinder et al. in 2006. The previous studies about the removal of nutrients like phosphorous and nitrogenous compounds from waste water by the use of algal cultures was successfully demonstrated (Chan et al. 1979, Neos and Varma 1966, Oswald et al. 1978, Saxena et al. 1974). Therefore, they have been used tremendously for the treatment of sewage water to remove pollutants.

2. Different Species of Cyanobacterial Remediation in Different Fields

Several cyanobacteria have the potential for surviving in different environments. Many species, particularly *Oscillatoria* (Vijayakumar 2005, Manoharan and Subramanian 1992, Boominathan 2000, Fogg and Thake 1987, Hashimato and Furukawa 1989, Manoharan and Subramanian 1993), Phormidium (Blier et al. 1995, Noue and Basseres 1989, Pouliot et al. 1989), *Aphanocapsa* (Boominathan 2000) and *Westiellopsis* (Vijayakumar et al. 2005) have been used in the sewage treatment of various industries. In laboratories, only a few cyanobacterial strains were used by various researchers in waste water treatment (Noue and Prouk 1990, Rai 1992). A few cyanobacteria have the potential to survive after treatment with effluents such as dairy (Boominathan 2000), dye (Vijayakumar 2005), paper mill, sago industries (Kasthuri 2008), sewage (Manoharan 1992), primary settled swine (Canizares et al. 1991) and phenolic compounds (Klekner and Kosaric 1992).

3. Potential Role of Cyanobacteria in Bioremediation

The physiology and biochemistry of the cyanobacterial system change as they are treated with different effluents and their survivability increases, i.e., increase in biomass is a proof of it as per the researchers (Boominathan 2005, Canizares et al. 1991, Manoharan and Subramanian 1992, Vijayakumar 2005). Cyanobacteria have the capacity to degrade aromatic hydrocarbons (Cerniglia and Gibson 1979, Cerniglia et al. 1980, Ellis 1977, Narro et al. 1992) and Xenobiotics (Megharaj et al. 1987). Their higher multiplication rate and re-absorption capacity of heavy metals make them suitable for use in the treatment of industrial effluents. These characters have enhanced the use of this algal biomass in the field of industrial effluent detoxification (Darnall et al. 1986) and have an edge over conventional waste water treatment facilities (Modak and Natarajan 1995). As the cyanobacteria are photosynthetic in nature, their pH is always two units higher than the pH of the surrounding liquids due to which they have the ability to remove heavy metals (Kuenen et al. 1986) and hence are resistant to the transfer of products out of the biofilm (Liehr et al. 1994). The rate of uptake of heavy metals is higher in immobilized cells compared to free living cells. This is ascribed to enhanced photosynthetic energy production in immobilized cells (Affolter and Hall 1986). The adsorption capacity of biomass changes according to the concentration of metal ions and the removal efficiency is more at low concentrations (Pons and Fuste 1993). It is well established that the larger is the surface of algal biomass, more is the capacity for attracting biophilic pesticide molecules, thus, helping in reducing the toxic impact of pollution in aquatic systems (Cerniglia et al. 1980). Algae, particularly cyanobacteria are used for the removal of toxicants from waste water and in bioassays; cyanobacteria are used for testing the toxicity of chemicals.

The effect could be inhibitory, selective or even stimulatory depending on the type, biological property, concentration of pesticides and the algal strains (Alberto and Paniagua 2014). It has been observed that cyanobacteria are also used as insecticides and in biofertilizers which are capable of tolerating high levels of

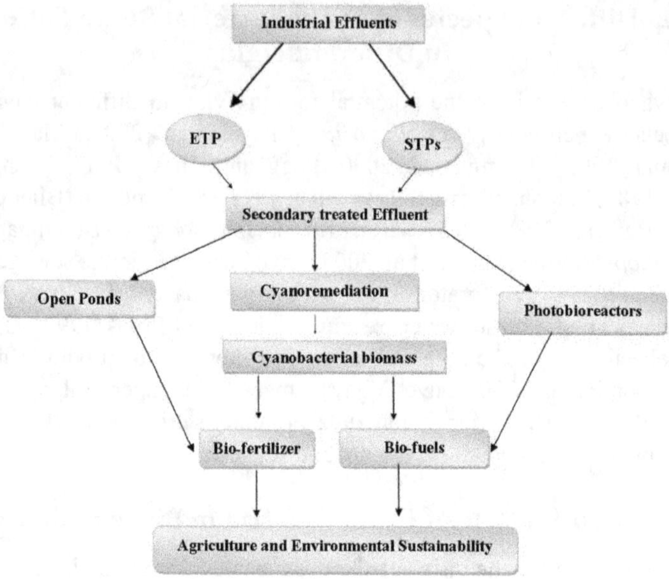

Figure 1. Cyanobacteria: A sustainable and commercial bio-resource in the production of bio-fertilizers and bio-fuels from industrial effluents.

pesticides recommended for agricultural field applications. The removal of metals by microbes is an active process with two phases:

(i) Binding of cations to the negatively charged groups of the cell walls.

(ii) The intracellular uptake by the metabolic process (Gipps and Coller 1980, Khummongkol et al. 1982, Norris and Kelly 1977).

4. Cyanobacterial Exopolysaccharide as a Biosurfactant

Cyanobacteria have the capacity of producing exopolysaccharides with potentially useful properties and these exopolysaccharides are considered as mucilaginous material (Martin and Wyatt 1974, Drews and Weckesser 1982, Painter 1983, Bertocchi et al. 1990). Many cyanobacterial strains possess additional surface structures outside their outer membrane mainly of a polysaccharidic nature. They consist of a thick and slimy layer associated with the cell surface with sharp outlines and structurally coherent to exclude particles. The slime refers to the mucilaginous material surrounding the organism without interfering with the shape of the cells. The synthesis of exocellular polysaccharide by bacterial cells is directly related to the environmental conditions of the producing microorganism (Dudman 1977). The main purpose of the polysaccharide is to serve as a boundary between the bacterial cell and its immediate environment.

More specifically, they have a protective role against desiccation, antibacterial agents (e.g., antibiotics, antibodies, bacteriocins, phages, phagocytic cells, surfactants) or predation by protozoans. The exocellular polysaccharides help the cyanobacteria with the capability to form biofilms on solid surfaces (Costerton et al. 1981, Costerton et al. 1987, Whitfield 1988). Various researches revealed that

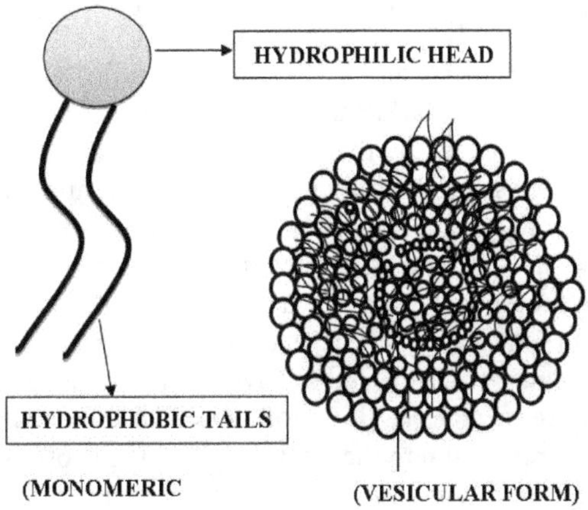

Figure 2. Structure of biosurfactants in monomeric and vesicular form.

the capability of some polysaccharide producing cyanobacteria helped to overcome stress due to desiccation or due to low water activity in desert or saline environments such as desiccation-tolerant *Nostoccommune* strain. Hill et al. 1994 reported that the secreted glycan provides a repository for water, acts as a buffer between the cells and the atmosphere and shows the main component of the mechanism used by this cyanobacterium to tolerate desiccation. Mazor et al. 1996 reported that the release of polysaccharides by cyanobacteria isolated from sand dunes in the Negev Desert (Israel), plays a vital role in maintaining the moisture in desert microbial crusts where for many months, the only source is water. The polysaccharide released by a *Chroococcidilopsis* strain play the role of a buffer compound for the accumulation and the slow release of water (Grilli et al. 1996). *Phormidium* J-1 synthesizes emulcyan, a sulfated heteropolysaccharide, which contains fatty acids and proteins that shows hydrophobicity on the macromolecule (Bar and Shilo 1987). The attachment of *Phormidium* J-1 and several other benthic cyanobacteria is also stimulated by the co-flocculation of polysaccharide-producing cells with sedimentary clay particles (Bar and Shilo 1988, Bar et al. 1989). In case of *Nostoc* sp., the slimy shrouds surrounding trichomes facilitate the homogeneous dispersion of trichomes into the liquid medium which improves utilization of light and nutrient uptake (Martin and Wyatt 1974) (Table 1).

Table 1. A partial list of biosurfactant producing microorganisms.

Microorganism	Biosurfactants	References
Candida bombicola	Sophorolipids	Daverey and Pakshiraja (2009)
Pseudomonas aeruginosa	Rhamnolipids	Kumar et al. (2008)
Candida tropicalis	Lipomannan	Muthuswamy et al. (2008)
Pseudomonas fluorescens	Rhamnolipids	Mahmound et al. (2008)
Bacillus subtilis	Surfactin	Muthuswamy et al. (2008)
Aeromonas sp.	Glycolipid	Ilori et al. (2005)
Bacillus sp.	Glycolipid	Kumar et al. (2008)

5. Composition of Cyanobacterial Biosurfactants

Chemically synthesized surfactants are usually classified according to the nature of their polar groups but biosurfactants are generally categorized mainly by their chemical composition dictated by the different molecules forming the hydrophobic and hydrophilic moieties and their microbial origin. Ten different monosaccharides have been found in the cyanobacterial exopolysaccharides such as hexoses like glucose, galactose and mannose, pentoses like ribose, xylose and arabinose, deoxyhexoses like fucose and rhamnose and the acidic hexoses like glucuronic and galacturonic acid. In some cases, the presence of additional types of Monosaccharides (i.e., methyl sugars and/or amino sugars) have been reported by Bender et al. 1994, Fischer et al. 1997, Tease et al. 1991, Panoff et al. 1988, Gloaguen et al. 1995, Filali et al. 1993. In many released polysaccharides (RPSs), glucose is the most abundant monosaccharide but there are also some other sugars like arabinose, galactose or fucose, which are present at considerable concentrations; ribose polymers have been found in lower numbers. In species like *Mastigocladuslaminosus* and *Phormidium* sp., rhamnose, glucose, xylose, mannose, galactose and fucose are the major components and arabinose and uronic acids like glucuronic and galacturonic acids are present in very low concentrations (Gloaguen et al. 1996). Many cyanobacterial polysaccharides are characterized by their anionic nature, many of them containing

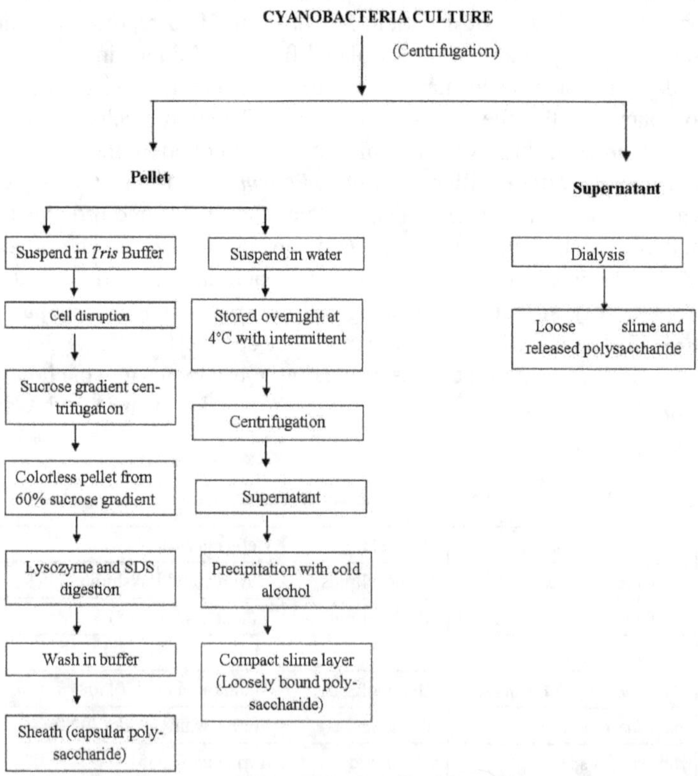

Figure 3. Schematic representation of extraction of EPS from cyanobacteria and green algae.

glucuronic and galacturonic acids, a feature rarely found in the polymers released by strains belonging to other microbial groups. Cyanobacterial RPSs show the presence of one or two pentoses and sugars that are generally absent in other polysaccharides of prokaryotic origin (Sutherland 1994). Cyanobacteria are capable of utilizing renewable and cheap substrates being photoautotrophs and many of them are nitrogen fixers; they utilize the CO_2 emitted by industrial plants as carbon sources. The economy of the process could be increased by extracting more than one useful compound with a multiproduct strategy as already proposed for microalgae (Thepenier et al. 1988). The presence of ester-linked acetyl groups and deoxy sugars like fructose and rhamnose make a contribution to the emulsifying properties of the polysaccharides. Some lipophilic character is shown by small molecules in the macromolecules that otherwise would be highly hydrophilic (Shepherd et al. 1995, Neu et al. 1992).

6. Role of Cyanobacterial Biosurfactants on the Effluents of Oil Industries

Our daily life needs petroleum based products as a major source of energy. Various industries are running by using petroleum and other petrochemical products (Cooney 1984). Accidental spilling occurs regularly during the exploration, production, refining, transport and storage of petroleum and petroleum products (Walker et al. 1975). The hydrocarbons present in the petroleum released by human activities or by accident are a major source of water and soil pollution (Holliger et al. 1997). The direct discharge of petroleum hydrocarbons in marine areas like diesel, crude oil and some distillates is a major global concern because of their toxic effects (Mohammad et al. 2010). The hydrocarbon contamination of soil may get accumulated in the tissues of plants and animals which may lead to death and mutations (Alvarez and Vogel 1991). Various articles and researches have been studied that show the rate of oil biodegradation (Leahy and Colwell 1990, Zobell 1946, Atlas 1981, Bartha 1992, Foght and Westlake 1987). The presence of microbes with their appropriate metabolic capabilities ensures optimal rates of hydrocarbon biodegradation. Hydrocarbons differ in their susceptibility to microbial attacks and generally degrade in the following order of decreasing susceptibility: n-alkanes > branched alkanes > low-molecular weight aromatics > cyclic alkanes (Leahy and Colwell 1990).

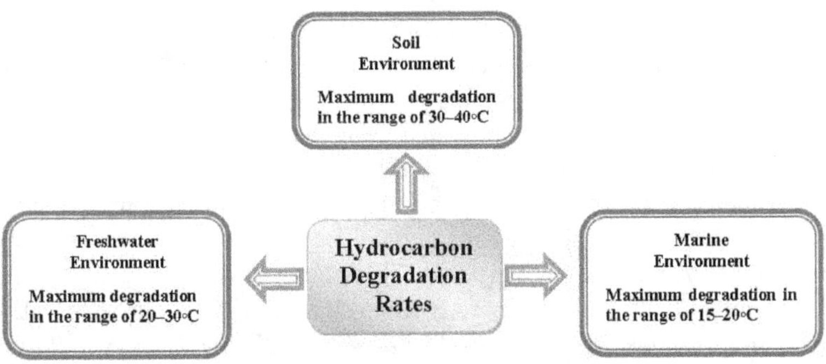

Figure 4. Hydrocarbon degradation rates in soil, fresh water, and marine environments.

In microbial communities, cyanobacteria and other algae are the important members which can grow in both terrestrial and aquatic ecosystems and have been reported to have the capacity to degrade hydrocarbons. Cyanobacteria play a direct or indirect role in the degradation of hydrocarbons (Atlas and Bartha 1992). They are photoautotrophic in nature so, the degradation of the crude oil is a means of environmental detoxification. Some cyanobacteria like *Agmenellum quadruplicatum*, *P. faveolarum*, *Anabaena cylindrica*, *Oscillatoria* sp. strain JCM can degrade different aromatic compounds (Cerniglia et al. 1980). The capacity of algal cells for biodegradation of organic pollutants enhances their proliferation in the presence of the pollutant. Chaillan et al. 2006 reported that in tropical petroleum polluted environments cyanobacterial mats are found. They form a high biodiversity microbial consortium that contains efficient hydrocarbon degraders. Raghukumar et al. 1991 researched that the three cyanobacterial species (*Oscillatoria salina*, *Plectonema terebrans* and *Aphanocapsa* sp.) mixed culture removed over 40% of the crude oil. In a mixed culture, there is potential for use in mitigating oil pollution on seashores either individually or in combination. Raghukumar et al. (1991) reported that cyanobacteria such as *Aphanocapsa* sp and *P. terebrans* were grown as epilithic and endolithic forms on the surface of and inside calcareous shells. So, they can be immobilized on that substratum used in the degradation of oil spills on seashores. Zhang et al. 1999 reported that *Spirulina* sp. has been shown to utilize organic carbon substrates for heterotrophic and mixotrophic, i.e., photo-heterotrophic growth. The photosynthetic production of biosurfactants in seawater is also a great advantage for sensitive environments and for conservation issues of biodiversity in rich marine environments exposed to pollution of petroleum hydrocarbons as well (Morales et al. 2013). Cyanobacterial polysaccharides play a vital role in the emulsification of oil, breaking it into small droplets which are subsequently utilized by the heterotrophs (Cohen 2002). The presence of exopolysaccharides emulsifying/ biosurfactant properties in *Phormidium* sp. is achieved by petroleum hydrocarbons, viz., hexadecane and diesel oil. This shows *Phormidium* sp. holds a significant practical potential for petroleum hydrocarbons bioremediation in coastal-marine environments. *Aphanocapsa* sp. showed better growth in stationary cultures in the presence of crude oil. Philippis and Vincenzini (1998) reported that the gelatinous matrix surrounding the cells protect them from the toxic effect of the crude oil and help in its degradation by acting as a biosurfactant. Sudo et al. (1995) reported that *Aphanocapsahalophytica* produces sulphated exopolysaccharide. Several aliphatic compounds are present in petroleum hydrocarbons such as decane, dodecane, tridecane, tetradecane, pentadecane, hexadecane, heptadecane, octadecane, nonadecane, docosane, tricosine, tetracosine, pentacosane, hexacosane, octacosane, nonacosane, and many more. Cyanobacteria like *Spirulina platensis* and *Nostoc punctiforme* have the capacity to degrade these aliphatic compounds into two units, i.e., acetate and malonate and these compounds have the capability to form aromatic rings through different sequential steps (Bosnic et al. 2000). These cyanobacteria have enzyme complexes which convert acetyl-CoA and malonyl-CoA into the final product. These enzyme complexes have polyketide synthase and poliketide cyclase activities (Dewick 2002). Hasan et al. (1994) reported that cyanobacteria such as *Microcoleus chthonoplastes* and *Phormidium corium* have the capacity to degrade oil in the Arabian Gulf coasts. *Oscillatoria* have the capacity to oxidize biphenyl reported by Cerniglia et al. (1980b).

They degrade branched as well as linear hydrocarbons reported to be effective in the removal of the Bombay High crude oil (Raghukumar 2001). Radwan and Hasan 2001 reported that gravel particles coated with biofilm contain 100 mg of

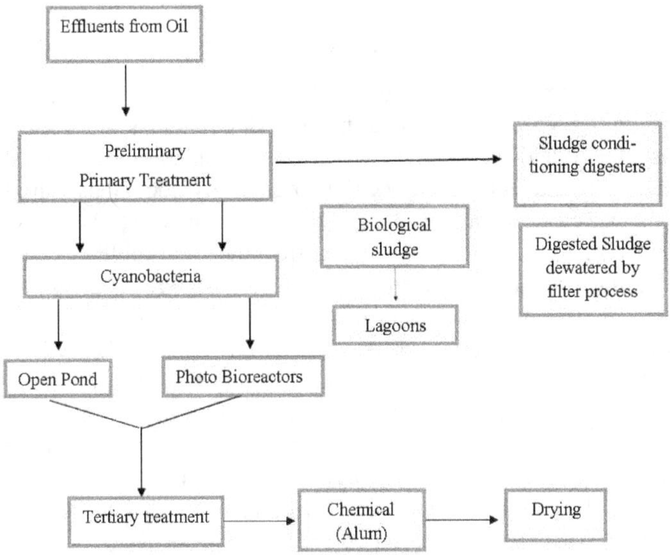

Figure 5. Integrated microbial bio-fuels production using effluent from oil industry treatment plants for algal biomass.

Table 2. Chemical composition of exopolysaccharide sheath of cyanobacteria.

Phormidium foveolarum	Gal, Glc (+), Man, Ara±, Fuc, Rha, Xyl, Rib±
Phormidium ectocarpi	Gal, Glc (+), Man, Rha, Xyl, Fuc
Phormidium minutum	Gal, Glc (+), Man, Rha, Xyl, Ara, Fuc
Oscillatoria amphibian	Gal, Glc (+), Man, Rha, Xyl, Fuc
Oscillatoria corallinae	Gal, Glc (+), Man, Rha, Xyl, Fuc
Lyngba confervoides	Gal, Glc (+), Man, Rha, Xyl, Ara, Fuc
Cyanothece	Gal, Glc (+), Man, Xyl, Fuc, Rha
Phormidium tenue	Ara (+), Rha, Fuc, Xyl, Man, Gal, Glc
Oscillaria sp.	Rib, Xyl, Glc (+)
Arthrospira platensis	Fruc, Fucose, Gal, Glc (+), Man, Rha, Rib, Xyl
Chlamydomonas mexicana	Ara, Rha, Fuc, Rib, Xyl, Man, Gal, Glc (+)
Chlamydomonas sajao	Ara, Rha, Fuc, Rib, Xyl, Man, Gal (+), Glc
Botryococcus braunii	Ara, Rha, Fuc, Gal (+), Fuc, Rha
Desmococcus olivaceus	Ara, Rha, Xyl, Man, Gal (+), Glc
Chlamydomonas reinhardtii	Ara, Rha, Rib, Xyl, Gal, Glc
Johannesbaptistia pellucida	Ara, Rha, Fuc, Xyl, Glc, Gal, Man

cyanobacterial biomass used in the treatment of oil wastes (Bosnic et al. 2000). Raghukumar et al. (1997) shows that oil wastes could be degraded by growing the cyanobacterial culture in plain natural seawater which is an advantage of bioremediation. The using of cyanobacteria for the mitigation of crude oil pollution on the seashore has an advantage because these can be used without addition of any nutrients.

7. Conclusion

Algal bioremediation is considered to be an efficient and environmentally safe technology for inexpensive decontamination of polluted systems. Cyanobacteria are photosynthetic and are believed to utilize carbon as a source of energy from atmospheric CO_2 and have been seen to degrade hydrocarbons. They utilize the carbon in the petroleum hydrocarbons. It seems that cyanobacteria are a cheap source of pollutants removal, used for waste water treatment and could be mass cultured to degrade organic wastes. The treatment with cyanobacteria is a feasible technique for bioremediation of industrial effluents. This cyanobacterial interaction with industrial effluents effectively decreases the organic content like hydrocarbons, and phosphorous, nitrite, sulphur and ammonia content. The requirement of nitrogenous fertilizers with minimal investment can be fulfilled by the use of cyanobacteria as compared to the conventional means. They are widely used for heavy metal removal from waste water. In future, more research on bio-stimulation and bio-augmentation should be carried out in order to establish the best and efficient ways to degrade toxic wastes by optimizing the degradation potential of the isolates.

References

Affolter, D. and D.O. Hall. 1986. Long-term stability of photosynthetic electron transport in polyvinyl foam immobilised bacteria. Photobiochem. Photobiophys. 11: 193–210.

Ajayi, S.O. and O. Osibanjo. 1980. The state of environment in Nig. Pollution studies of textile industries in Nigeria. Monogra 1: 76–86.

Alberto, R.M. and M.J. Paniagua. 2014. Bioremediation of hexadecane and diesel oil is enhanced by photosynthetically produced marine biosurfactants. Department of Marine Biotechnology, Center of Scientific Research and Higher Education from Ensenada, Mexico.

Alvarez, P.J.J. and T.M. Vogel. 1991. Substrate interactions of benzene, toluene, and para-xylene during microbial degradation by pure cultures and mixed culture aquifer slurries. Appl. Environmen. Microbiol. 57(10): 2981–2985.

Ash, N. and M. Jenkins. 2006. Biodiversity and poverty reduction: The importance of biodiversity for ecosystem services. Final report prepared by the United Nations Environment Programme World Conservation Monitoring Centre (UNEP-WCMC) (http://www.unepwcmc.org) for the Department for International Devèlopment (DFID).

Atlas, R.M. 1981. Microbial degradation of petroleum hydrocarbons: An environmental perspective. Microbiol. Rev. 45(1): 180–209.

Atlas, R.M. and R. Bartha. 1992. Hydrocabon biodegradation and oil spill bioremediation. Adv. in Microb. Ecol. 12: 287–338.

Bar-Or, Y. and M. Shilo. 1987. Characterization of macromolecular flocculants produced by Phormidium sp. strain J-1 and by Anabaenopsis circularis PCC 6720. Appl. Environ. Microbiol. 53: 2226–2230.

Bar-Or, Y. and M. Shilo. 1988. The role of cell-bound flocculants in coflocculation of benthic cyanobacteria with clay particles. FEMS Microbiol. Ecol. 53: 169–174.

Bar-Or, Y., M. Kessel and M. Shilo. 1989. Mechanisms for release of the benthic cyanobacterium Phormidium strain J-1 to water column. pp. 214–218. *In*: Cohen, Y. and E. Rosenberg (eds.).

Microbial Mats Physiological Ecology of Benthic Microbial Communities. American Society for Microbiology, Washington, DC.

Bartha, R. and I. Bossert. 1992. The treatment and disposal of petroleum wastes. pp. 553–578. *In*: Atlas, R.M. (ed.). Petroleum Microbiology. Macmillan, New York, NY, USA.

Becker, E.W. 1992. Micro-algae for human and animal consumption. pp. 222–256. *In*: Borowitzka, M.A. and L.J. Borowitzka (eds.). Micro-algal Biotechnology. Cambridge University Press, Cambridge.

Bender, J., E.S. Rodriguez, U.M. Ekanemesang and P. Phillips. 1994. Characterization of metal-binding bioflocculants produced by the cyanobacterial component of mixed microbial mats. Appl. Environ. Microbiol. 60: 2311–2315.

Bertocchi, C., L. Navarini, A. Cesaéro and M. Anastasio. 1990. Polysaccharides from cyanobacteria. Carbohydr. Polym. 12: 127–153.

Blier, R.G. and J. Noue. 1995. Tertiary treatment of cheese factory anaerobic effluent with Phormidiuimbohneri and Micractiniumpusilllum. Bioresour. Technol. 52: 151–155.

Boominathan, M. 2000. Interaction of Spirulina platensis with starchy effluent. Bharathidasan University, Tiruchiraplli, India.

Boominathan, M. 2005. Bioremediation studies on dairy effluent using cyanobacteria. Bharathidasan University, Tiruchirapalli, Tamil Nadu, India.

Bosnic, M., J. Buljan and R.P. Daniels. 2000. Pollutants in tannery effluents. United Nations Industrial Development Organization, 1–26.

Canizares, R.O., L.R. Montes and A.R. Dorminguez. 1991. Aerated swine waste water treatment with K-carrageenan-immobilized Spirulina maxima. Bioresour. Technol. 47: 89–91.

Cerniglia, C.E. and D.T. Gibson. 1979. Algal oxidation of aromatic hydrocarbons: formation of 1-naphthol from naphthalene by *Agmenellum quadruplicatum*, strain PR-6. Biochem. Biophys. Res. Commun. 88: 50–58.

Cerniglia, C.E., C. van Baalen and D.T. Gibson. 1980. Metabolism of naphthalene by the cyanobacterium *Oscillatoria* sp., strain JCM. J. Gen. Microbiol. 116: 485–494.

Chaillan, F., M. Gugger, A. Saliot, A. Coute, J. Oudot et al. 2006. Role of cyanobacteria in the biodegradation of crude oil by a tropical cyanobacterial mat. Chemosphere 62: 1574–1582.

Chan, K.Y., K.H. Wong and P.K. Wong. 1979. Nitrogen and phosphorus removal from sewage effluent with high salinity by Chlorella salina. Environ. Poll. 18: 139–146.

Cohen, Y. 2002. Bioremediation of oil by marine microbial mats. Int. Microbiol. 5: 189–193.

Cooney, J.J. 1984. The fate of petroleum pollutants in fresh water ecosystems. pp. 399–434. *In*: Atlas, R.M. (ed.). Petroleum Microbiology. Macmillan, New York, NY, USA.

Cooper, P. 1995. Colour in Dyehouse Effluent. Society of Dyers and Colourists, The Alden Press, Oxford.

Costerton, J.W., R.T. Irvin and K.J Cheng. 1981. The bacterial glycocalyx in nature and disease. Annu. Rev. Microbiol. 35: 299–324.

Costerton, J.W., K.J. Cheng, G.G. Geesey, J.C. Ladd, M. Dasgupta et al. 1987. Bacterial biofilms in nature and disease. Annu. Rev. Microbiol. 41: 435–464.

Darnall, D.W., B. Greene, M.T. Henzl, J.M. Hosea, R.A. McPherson et al. 1986. Selective recovery of gold and other metal ions from an algal biomass. Environ. Sci. Technol. 20: 206–208.

Daverey, A. and K. Pakshirajan. 2009. Production of sophorolipids by the yeast Candida bombicola using simple and low cost fermentative media. F. Res. Int. 42(4): 499–504.

Dewick, P.M. 2002. Medicinal Natural Products, A Biosynthetic Approach, John Wiley & Sons, Ltd., Chichester.

Do, T., J. Shen, G. Cawood and R. Jeckins. 2002. Biotreatment of textile effluent using *Pseudomonas* spp. Immobilized on polymer supports. *In*: Hardin, I.R., D.E. Akin and J.S. Wilson (eds.). Advances in Biotreatment for Textile Processing. University of Georgia Press.

Drews, G. and J. Weckesser. 1982. Function, structure and composition of cell walls and external layers. pp. 333–357. *In*: Carr, N.G. and B.A. Whitton (eds.). The Biology of Cyanobacteria. Blackwell Scientific, Oxford.

Dudman, W.F. 1977. The role of surface polysaccharides in natural environments. pp. 357–414. *In*: Sutherland, I.W. (ed.). Surface Carbohydrates of the Prokaryotic Cell. Academic Press, New York.

Ellis, B.E. 1977. Degradation of phenolic compounds by fresh-water algae. Plant Sci. Lett. 8: 213–216.

Filali, M., R. Cornet, J.F. Fontaine, T. Fournet, G. Dubertret et al. 1993. Production, isolation and preliminary characterization of the exopolysaccharide of the cyanobacterium Spirulina platensis. Biotechnol. Lett. 15: 567–572.

Fischer, D., U.G. Schloësser and P. Pohl. 1997. Exopolysaccharide production by cyanobacteria grown in closed photobioreactors and immobilized using white cotton towelling. J. Appl. Phycol. 9: 205–213.

Fogg, G.E. and B. Thake. 1987. Algal Cultures and Phytoplankton Ecology. The University Wisconsin Press.

Foght, J.M. and D.W.S. Westlake. 1987. Biodegradation of hydrocarbons in freshwater. pp. 217–230. *In*: Vandermeulen, J.H. and S.R. Hrudey (eds.). Oil in Freshwater: Chemistry, Biology, Countermeasure Technology. Pergamon Press, New York, NY, USA.

Gazyna, P., N.J. Grzegorz, U. Krzysztof and L. Brigmon. 2005. Assessment of genotoxic activity of petroleum hydrocarbon-bioremediated soil. Ecotoxicol. and Environ Safety 62: 415–420.

Gipps, J.F. and B.A.W. Coller. 1980. Effect of physical and culture conditions on uptake of cadmium by Chlorella pyrenoidosa. Aus. J. of Mari. and Freshwater Res. 31: 747–755.

Gloaguen, V., H. Morvan and L. Ho¡mann. 1995. Released and capsular polysaccharides of Oscillatoriaceae (Cyanophyceae, Cyanobacteria). Alg. Studies 78: 53–69.

Gloaguen, V., H. Morvan and L. Ho¡mann. 1996. Metal accumulation by immobilized cyanobacterial mats from a thermal spring. J. Environ. Sci. Health 31: 2437–2451.

Grilli, C., M.D. Billi and E.I. Friedmann. 1996. Effect of desiccation on envelopes of the cyanobacterium Chroococcidiopsis sp. (Chroococcales). Eur. J. Phycol. 31: 97–105.

Hasan, R.H., N.A. Sorkhoh, D. Bader and S.S. Radwan.1994. Utilization of hydrocarbons by cyanobacteria from microbial mats on oily coasts of the Gulf. Appl. of Microbiol. Biotechnol. 41: 615–619.

Hashimato, S. and K. Furukawa. 1989. Nutrient removal from secondary effluent by filamentous algae. J. Ferment. Bioeng. 67: 62–69.

Hill, D.R., A. Peat and M. Potts. 1994. Biochemistry and structure of the glycan secreted by desiccation-tolerant *Nostoc commune* (Cyanobacteria). Protoplasma 182: 126–148.

Holliger, C., S. Gaspard and G. Glod. 1997. Contaminated environments in the subsurface and bioremediation: Organic contaminants. FEMS Microbiol. Rev. 20(4): 517–523.

Ilori, M.O., C.J. Amobi and A.C. Odocha. 2005. Factors affecting biosurfactant production by oil degrading *Aeromonas* spp. isolated from a tropical environment. Chem. 61(7): 985–992.

Johnson, E.A. and G.H. An. 1991. Astraxanthin from microbial sources. Crit. Rev. Biotechnol. 11: 297–326.

Kasthuri, J. 2008. Interaction between Coirpigh effluent and the Cyanobacterium Oscillatoria acuminate. Bharathidasan University, Tiruchirappalli, India.

Khummongkol, D.G., S. Canterford and C. Fryer. 1982. Accumulation of heavy metals in unicellular algae. Biotechnol. Bioeng. 24: 2643–2660.

Klekner, V. and N. Kosaric. 1992. Degradation of phenols by algae. Environ. Tech. 13: 493–501.

Kuenen, J.G., B.B. Jongensen and N.P. Revsbech. 1986. Oxygen microprofiles of trickling filter biofilms. Water Res. 20: 1589–1598.

Kumar, M., V. León, M.A. De Sisto, O.A. Ilzins and L. Luis. 2008. Biosurfactant production and hydrocarbon degradation by halotolerant and thermotolerant *Pseudomonas* sp. W. J. Microbiol. Biotechnol. 24(7): 1047–1057.

Leahy, J.G. and R.R. Colwell. 1990. Microbial degradation of hydrocarbons in the environment. Microbiol. Rev. 54(3): 305–315.

Liehr, S.K., H.J. Chen and S.H. Lin. 1994. Metals removal by algal biofilms. Water Sci. Technol. 30: 59–68.

Lora-Vilchis, M.C., M. Robles-Mungaray and N. Doctor. 2004. Food value of four micro algae for juveniles of Lion's paw scallop Lyropectensubnodosus (Sowerby, 1833). J. World Aquaculture Soc. 35: 297–303.

Mahmound, A., Y. Aziza, A. Abdeltif and M. Rachida. 2008. Biosurfactant production by Bacillus strain injected in the petroleum reservoirs. J. of Ind. Microbiol. Biotechnol. 35: 1303–1306.

Manoharan, C. and G. Subramanian. 1992. Interaction between paper mill effluent and the Cyanobacterium Oscillatoria pseudogeminata. Var. unigranulata. Poll. Res. 11: 73–84.

Manoharan, C. and G. Subramanian. 1992. Sewage-cyanobacterial interaction—A case study. IJEP 12: 254–258.

Manoharan, C. and G. Subramanian. 1993. Feasibility studies on using cyanobacteria in Ossein effluent treatment. Indian J. Environ. Health 35: 88–96.

Martin, T.J. and J.T. Wyatt. 1974. Extracellular investments in blue-green algae with particular emphasis on genus Nostoc. J. Phycol. 10: 204–210.

Mazor, G., G.J. Kidron, A. Vonshak and A. Abeliovich. 1996. The role of cyanobacterial exopolysaccharides in structuring desert microbial crusts. FEMS Microbiol. Ecol. 21: 121–130.

Megharaj, M., K. Venkateswarlu and A.S. Rao. 1987. Metabolism of monocrotophos and quinalphos by algae isolated from soil. Bull. Environ. Contam. Toxicol. 39: 251–256.

Metting, B. 1992. Micro-algae in agriculture. pp. 288–304. *In*: Borowitzka, M.A. and L.J. Borowitzka (eds.). Micro-algal Biotechnology. (Cambridge University Press, Cambridge).

Modak, J.M. and K.A. Natarajan. 1995. Biosorption of metals using nonliving biomass—A review. Min. and Metallurg. Proc. 12: 189–196.

Mohammad, A.Z., A.A. Hamidi, H.I. Mohamed and L. Mohajeri. 2010. Effect of initial oil concentration and dispersant on crude oil biodegradation in contaminated seawater. Bull Environ. Contam. Toxicol. 84: 438–442.

Morales, M., A. Maria, S. Munoz, P. Claudia and L. Silvia. 2013. Evaluation of natural attenuation, bioventing, bioaugmentation and bioaugmentation bioventing techniques, for the biodegradation of diesel in a sandy soil, through column experiments. Gestion. y. Ambiente. 16: 83–94.

Muthuswamy, K., S. Gopalakrishnan, T.K. Ravi and P. Sivachidambaram. 2008. Biosurfactants: properties, commercial production and application. Curr. Sci. 94: 736–747.

Narro, M.L., C.E. Cerniglia, B.C. Van and D.T. Gibson. 1992b. Metabolism of phenanthrene by the marine cyanobacterium *Agmenellum quadruplicatum* PR-6. Appl. Environ. Microbiol. 58: 1351–1359.

Neos, C. and M. Varma. 1966. The removal of phosphate by algae. Wat. Sew. Works 112: 456–459.

Neu, T.R., T. Dengler, B. Jann and K. Poralla. 1992. Structural studies of an emulsion-stabilizing exopolysaccharide produced by an adhesive hydrophobic *Rhodococcus* strain. J. Gen. Microbiol. 138: 2531–2537.

Norris, P.R. and D.P. Kelly. 1977. Accumulation of cadmium and cobalt by Saccharomyces cerevisiae. J. Gen. Microbiol. 99: 317–324.

Noue, J. and A. Basseres. 1989. Biotreatment of anaerobically digested swine manure with microalgae. Biol. Wastes 29: 17–31.

Noue, J., and D. Prouk. 1990. Waste Water Treatment by Immobilized cells. CRC Press, New York.Noue JDL, Proulx D.

Olukanni, O.D., A.A. Osuntoki and G.O. Gbenle. 2006. Textile effluent biodegradation potentials of textile effluent-adapted and non-adapted bacteria. Africa. J. Biotechnol. 5(20): 1980–1984.

Oswald, W.J., E.W. Lee, B. Adan and K.H. Yao. 1978. New wastewater treatment method yields a harvest of saleable algae. W.H.O. Chronicle.

Oswald, W.J. 1992. Microalgae and wastewater treatment. pp. 305–328. *In*: Borowitzka, L.J. (eds.). Micro-algal Biotechnology. (Cambridge University Press, Cambridge).

Painter, T.J. 1983. Algal polysaccharides. pp. 195–285. *In*: Aspinall, G.O. (ed.). The Polysaccharides. Vol. 2. Academic Press, New York.

Panoff, J.M., B. Priem, H. Morvan and F. Joset. 1988. Sulphated exopolysaccharides produced by two unicellular strains of cyanobacteria, Synechocystis PCC 6803 and 6714. Arch. Microbiol. 150: 558–563.

Philippis, R.D. and M. Vincenzini. 1998. Exocellular polysaccharides from cyanobacteria and their possible applications. FEMS Microbiol. Rev. 22: 151–175.

Pons, M.P. and M.C. Fuste. 1993. Uranium uptake by immobilized cells of Pseudomonas strain EPS 5028. Appl. Microbiol. Biotechnol. 39: 661–665.

Pouliot, Y., G. Buelna, C. Racinie and J. Noue. 1989. Culture of cyanobacteria for tertiary wastewater treatment and biomass production. Biol. Waters 29: 81–91.

Radwan, S.S. and R.H. Al-Hasan. 2001. Potential application of coastal biofilm-coated gravel particles for treating oily waste. Aquat. Microb. Ecol. 23: 113–117.

Raghukumar, C., S. Sharma and V. Lande. 1991. Distribution and biomass estimation of marine shell-boring algae in the intertidal, Goa, India. Phycologia 30: 303–309.

Raghukumar, C., V. Vipparty, J.J. David and D. Chandramohan. 1997. A process for removal of crude oil pollutants using marine cyanobacteria. Indian Patent No.1873/DEL/97; 04.07.97.

Raghukumar, C., V. Vipparty and J.J. David. 2001. Chandramohan. Degradation of crude oil by marine cyanobacteria. Appl. Microbiol. Biotechnol. 57: 433–436.

Rai, L.C. and N. Mallick. 1992. Removal and assessment of toxicity of Cu and Fe to Anabaena doliolum and Chlorella vulgaris using free and immobilized cells. World J. Microbiol. Biotechnol. 8: 110–114.

Satinder, K., M. Verma, R.V. Surampalli, K. Misra, R.D. Tyagi et al. 2006. Bioremediation of hazardous wastes—A review. Practical Periodical of Hazardous, Toxic and Radioactive Waste Management 10: 59–72.

Saxena, P.N., A. Tiwari and M.A. Khan. 1974. Effect of Anacystisnidulans on physicochemical and biological characteristics of raw sewage. Proc. Indian Acad. Sci. 79: 139–146.

Semyalo, R.P. 2009. The effects of cyanobacteria on the growth, survival, and behaviour of a tropical fish (Nile Tilapia) and zooplankton (Daphnia lumholtzi). Ph.D. Thesis, University of Bergen, Norway.

Shepherd, R., J. Rockey, I. Sutherland and S. Roller. 1995. Novel bioemulsi¢ers from microorganisms for use in foods. J. Biotechnol. 40: 207–217.

Sudo, H.J., G. Burgess, H. Takemasa, N. Nakamura, T. Matsunga et al. 1995. Sulfated polysaccharide production by the halophilic cyanobacterium *Aphanocapsa halophytica.* Curr. Microbiol. 30: 219–222.

Sutherland, I.W. 1994. Structure-function relationships in microbial exopolysaccharides. Biotech. Adv. 12: 393–448.

Tease, B., U.J. Juërgens, J.R. Golecki, U.R. Heinrich, R. Rippka et al. 1991. Fine-structural and chemical analyses on inner and outer sheath of the cyanobacterium Gloeothece sp. PCC 6909. Antonie van Leeuwenhoek 59: 27–34.

Thepenier, C., D. Chaumont and C. Gudin. 1988. Mass culture of Porphyridiumcruentum: A multiproduct strategy for the biomass valorization. pp. 413–420. *In*: Stadler, T., J. Mollion, M.C. Verdus, Y. Karamanos, H. Morvan and D. Christiaen (eds.). Algal Biotechnology. Elsevier, London.

Vijayakumar, S. 2005. Studies on cyanobacteria in industrial effluents—An Environmental and Molecular Approach. Bharathidasan University, Tiruchirapalli, Tamil Nadu, India.

Vijayakumar, S., N. Tajudden and C. Manoharan. 2005. Role of cyanobacteria in the treatment dye industry effluent. Poll. Res. 24: 69–74.

Walker, J.D., R.R. Colwell, Z. Vaituzis and S.A. Meyer. 1975. Petroleum degrading achlorophyllous alga *Protothecazopfii*. Nat. 254(5499): 423–424.

Whitfield, C. 1988. Bacterial extracellular polysaccharides. Can. J. Microbiol. 34: 415–420.

Zhang, X.W., Y.M. Zhang and F. Chen. 1999. Application of mathematical models to the determination optimal glucose concentration and light intensity for mixotrophic culture of Spirulina platensis, Process. Biochem. 34: 477–481.

Zobell, C.E. 1946. Action of microorganisms on hydrocarbons. Bacteriol. Rev. 10: 1–49.

11

Microbial Biosurfactants An Eco-friendly Approach for Bioremediation of Contaminated Environments

Shah Ishfaq,[1] *Basharat Hamid,*[1] *Muzafar Zaman,*[1] *Sabah Fatima,*[1,*] *Shabeena Farooq,*[1] *Rahul Datta,*[2] *Subhan Danish,*[3] *Noshin Ilyas*[4] and *R Z Sayyed*[5,*]

1. Introduction

With the advancement in industrialization and urbanization, the earth is experiencing drastic changes in its system (Zafar et al. 2020) due to moderate to high pollution levels that in turn have led to the various inventions and scientific discoveries varying from high tech instruments to advancement in the field of naturally occurring solutions that are available for combating the pollution menace. Bioremediation through biosurfactants is a promising technique to combat pollution in any ecosystem in an eco-friendly and economical way (Roy et al. 2018). Biosurfactants are the surface-active amphiphilic substances usually produced by microorganisms. However, their use in bioremediation processes is not well established due to high production costs (Santos et al. 2017a) that require low-cost bio-surfactants produced from waste or are of renewable origin to be usefully applied. These surfactants can be of bacterial

[1] Department of Environmental Science, University of Kashmir, Hazratbal, Srinagar-190006, Jammu and Kashmir, India.

[2] Department of Geology and Pedology, Mendel University in Brno, Czech Republic.

[3] Hainan Key Laboratory for Sustainable Utilization of Tropical Bioresource, College of Tropical Crops, Hainan University, Haikou 570228, China.

[4] Department of Botany, PMAS Arid Agricultural University, Rawalpindi, 46300, Pakistan.

[5] Department of Microbiology, PSGVP Mandal's Arts, Science and Commerce College, Shahada-425 409, Maharashtra, India.

* Corresponding authors: sabahfatima333@gmail.com; sayyedrz@gmail.com

origin, e.g., rhamnolipids and surfactin (Santos et al. 2016) or maybe extracted from or secreted by yeasts, e.g., Glycolipids (Jezierska et al. 2018) that enhance the dispersal desorption and solubilisation of contaminants that are present in the environment, especially soil and enable its reuse (Liu et al. 2017a). These naturally occurring surfactants are advantageous over chemically prepared surfactants. They can be produced from natural sources that are less toxic, possess high specificity, better foaming and emulsifying properties, resulting in survivability under extreme temperatures, salinity, pH (Busi and Rajkumari 2017, Bee et al. 2019), and are cheaper as they can be produced from the industrial waste and by-products (Sarubbo et al. 2015, Olasanmi and Thring 2018).

The use of biosurfactants in the remediation of contaminants is an emerging technology. In addition to remediation, these surface-active compounds also have their application in food production, agriculture, environmental degradation and oil recovery. In this chapter, the biosurfactants that have application in bioremediation of various contaminants will be discussed.

2. Classification of Biosurfactants

Biosurfactants are usually classified based on their chemical structure, molecular weight, mode of action, origin and physico-chemical properties (Figure 1). On the basis of molecular weight, these are classified into low and high-molecular-weight surfactants. Biosurfactants with low molecular weight like glycolipids, surfactin and rhamnolipids are mainly used in reducing interfacial and surface tensions, whereas high molecular weight surfactants like phospholipids, alasan, lipoproteins and emulsan mainly act as bioemulsifiers (Jahan et al. 2020).

Among the bacteria, *Pseudomonas* is known for rhamnolipid production (Wood et al. 2018, Soberón-Chávez et al. 2020) and *Bacillus* sp. for lipopeptides (Sharma et al. 2020a). Also, fungi and yeasts produce biosurfactants, for example, *Candida lipolytica* (Santos et al. 2017), *Yarrowia lipolytica* and *Apiotrichum loubieri* (Yalçin

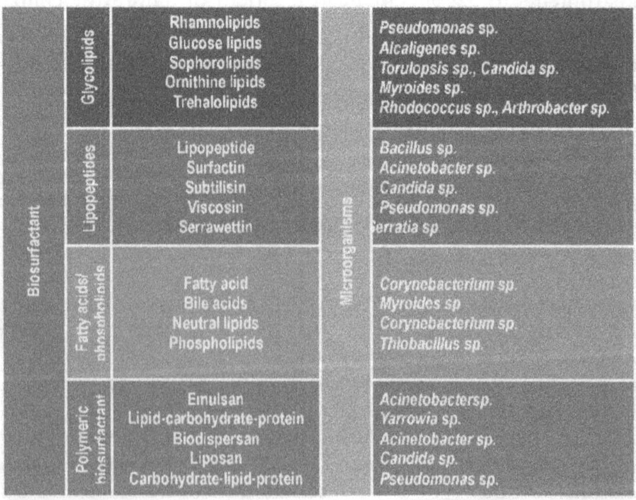

Figure 1. Different classes of biosurfactants and their associated microbes.

et al. 2018), and *Saccharomyces cerevisiae* (Kreling et al. 2020). Among all the biosurfactants, glycolipids are the most studied type comprising long-chain aliphatic acids with linked carbohydrates or an ester group linked with hydroxyl aliphatic acids. Lipopeptides are composed of a polypeptide chain with attached lipids. The other groups include fatty acids, phospholipids and neutral acids produced by both bacteria and yeasts. Polymeric biosurfacants are also termed bioemulsifiers composed of proteins, lipids and carbohydrates (Busi and Rajkumari 2017). Some of the examples that are helpful in bioremediation are depicted in Table 1.

Table 1. Classification of some biosurfactants and their producing organisms applicable for remediation of metals and oil recovery and remediation (Modified from Pacwa-Płociniczak et al. 2011).

Biosurfactant	Bacteria	Metal Remediation	Remediation of Oil
Rhamnolipids	*Pseudomonas aeruginosa, Pseudomonas* sp.	Yes	Yes
Sophorolipids	*Torulopsis bombicola, Torulopsis petrophilum, Torulopsis apicola*	Yes	Yes
Phosphati-dylethanolamine	*Acinetobacter* sp., *Rhodococcus erythropolis*	Yes	–
Surfactin	*Bacillus subtilis*	Yes	–
Lichenysin	*Bacillus licheniformis*	–	Yes

3. Bioremediation Favoring Properties of Biosurfactants

Unlike chemical surfactants, biosurfactants have advantages for several reasons. The most important and common being their eco-friendly nature. They are also less toxic, have high biodegradability and possess a wide range of physicochemical properties like pH, temperature and salinity. These properties are the reason why they have extensive potential for use in bioremediation of xenobiotics or contaminants, food processing ad pharmaceuticals (Rodrigues 2015).

3.1 *Surface Activity*

Biosurfactants allow two immiscible fluids form two different phases by reducing the tension (surface and interfacial) and repulsive force between them. The most active ones lower the surface tension of water from 72 to 30 mNm^{-1}. Such properties favor the use of biosurfactants FPR pollutant (e.g., metals) removal from liquids (Ravindran et al. 2020). The extended efficiency of surface activity is determined by the micro-organisms producing a particular type of biosurfactant. Another property that determines the activity of biosurfactants is critical micelle concentration (CMC). The lower the CMC, the more efficient will a biosurfactant be, i.e., to decrease the surface tension, lower quantities of biosurfactants are needed.

3.2 *Temperature and pH Tolerance*

The microbial surfactants are not much affected by environmental pH and temperature stresses of. For example, lichenysin produced by *B. licheniformis* shows a temperature tolerance upto 50°C, pH tolerance range of 4.5–9.0 and ionic tolerance of upto 50 g/l for NaCl and 25 g/l for Ca (Anjum et al. 2016).

3.3 *Biodegradability*

Biosurfactants are easily degraded by microorganisms, including bacteria that make them most suitable for environmental applications like remediation and more.

3.4 *Low Toxicity*

The less toxic nature of biosurfactants makes them appropriate for use in food, pharmaceuticals, cosmetics and other environmental applications. Contrary to chemically produced surfactants with mutagenic effects on various organisms, many studies depicted that biosurfactants are often less toxic and reduced mutagenic effects on organisms (Shah et al. 2016, Busi and Rajkumari 2017).

3.5 *Availability*

This property is the one that makes the production of biosurfactants economical as they can be produced using the raw materials that are available in abundance like industrial waste, starchy substances, oil mill effluents, municipal wastes, or by-products and wastewaters.

3.6 *Specificity*

A particular functional group's presence makes the biosurfactant specific for detoxification of any xenobiotic and emulsification of industrial effluents.

4. Biosurfactants and Remediation of Hydrocarbons

Hydrocarbons are the hydrophobic compounds that exhibit poor solubility in water, have limited recovery by physico-chemical treatments (Chaprão et al. 2015), are not readily available to microorganisms, thus rendered as toxic and persistent with a negative impact on living organisms. So far, detergents for emulsification of these hydrocarbons have proven to be destructive to the environment and releasing toxic byproducts. Biosurfactants being less toxic and highly biodegradable, can play an important role in these hydrocarbons' bioremediation (Silva et al. 2014). Biosurfactants can be widely employed to remediate oil spills, removal of oil residue and enhanced microbial oil recovery (Patel et al. 2015, Zhang et al. 2016, Joshi et al. 2016, Purwasena et al. 2019, Xu et al. 2020). Biosurfactants increase the bioavailability of hydrocarbons to microorganisms by reducing the surface and interfacial tensions and also increasing the hydrophobicity by interacting with the cell surface by increasing the contact angle and reducing the capillary force holding the oil and soil together (Figure 2). The main processes for removing hydrocarbon degradation are mobilization, emulsification, or solubilization (Mnif et al. 2015, Carolin et al. 2021) (Figure 3). These processes depend on the CMC values, i.e., mobilization occurs below CMC of the biosurfactant (Mańko-Jurkowska et al. 2019) while solubilisation occurs above the CMC of the biosurfactant (Li et al. 2015). Both

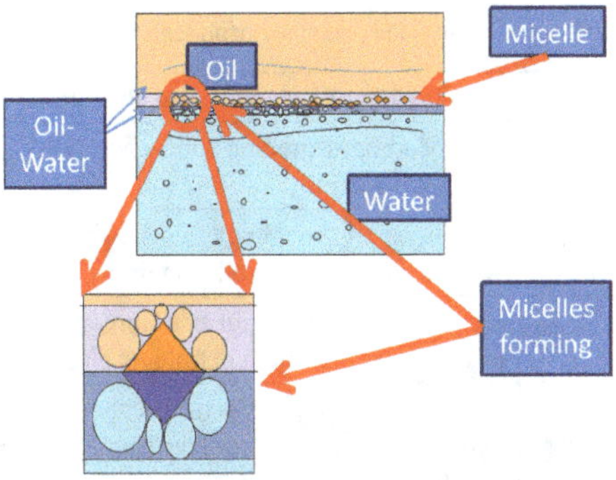

Figure 2. Complex formation by micelles at oil-water interface.

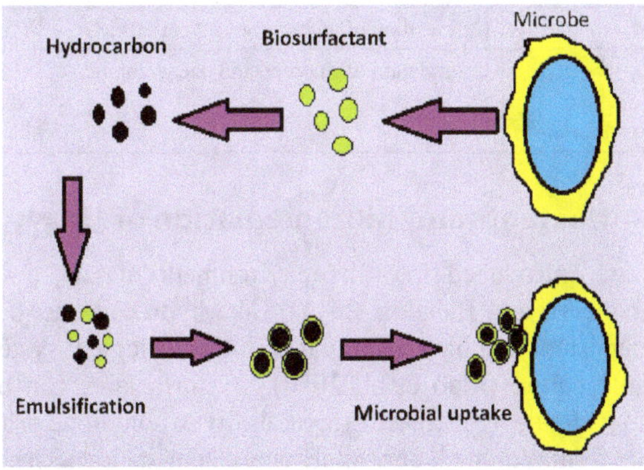

Figure 3. Biosurfactant enhanced hydrocarbon uptake by microbes.

mobilization and solubilisation are performed by low molecular mass biosurfactants (Mańko-Jurkowska et al. 2019) and they carry out emulsification with a high molecular mass (Gudiña et al. 2015).

With the advancement in biosurfactant isolation and characterization, a large number of biosurfactant producing microbes have been isolated, which can degrade hydrocarbons (Xu et al. 2018, Xu et al. 2020). Some of the biosurfactant-producing microorganisms are listed in Table 2. These microbial surfactants also can remediate soils contaminated with crude petroleum oil. In one of study, it was reported that the consortium of isolated culturable biosurfactants-producing freshwater lake bacteria showed >80% contribution in the degradation of total petroleum-hydrocarbons by the natural microbiome of the ecosystem (Phulpoto et al. 2021).

In addition, biosurfactants find application in soil washing done by applying aqueous solutions of biosurfactants to release compounds with low solubility from soil and other media. Some studies have concluded that biosurfactants like rhamnolipids have the same elimination ability for petroleum contaminated soils as synthetic surfactants like sodium dodecyl sulfate (SDS) (Olasanm et al. 2018). Microbes and their metabolites, including biosurfactants, are also involved in the oil recovery process known as microbial enhanced oil recovery (MOER).

Table 2. Application of some biosurfactants produced by various microorganisms.

Bacteria	Application of Biosurfactant Produced	References
E. fergusonii	Hydrocarbon-degrading and heavy metal-tolerant bacterium	(Raj et al. 2021)
Candida lipolytica	Exhibit an oil spreading efficiency of 75% when tested on motor oil containing seawater	(Santos et al. 2017b)
P. aeruginosa	Hexadecane degradation and bioremediation of n-alkanes present in petroleum sludge	(Zhong et al. 2016)
Pseudomonas marginalis	Solubilization of polycyclic aromatic hydrocarbons (PAHs) such as phenanthrene and enhanced biodegradation	(Ławniczak et al. 2020)
Pseudomonas alcaligenes	Naphthalene and phenanthrene degradation	(Abo-State et al. 2018)
Burkholderia cepacia	Degradation of anthracene and carbazole	(Gosh et al. 2020)
Ochrobactrum sp., *Enterobacter cloacae* and *Stenotrophomonas maltophilia*	Degradation of fluorene and phenanthrene	(Ramasamy et al. 2017)

5. Biosurfactants and Bioremediation of Heavy Metals

Heavy metals are introduced into the environment mainly through industrial processes (Ayangbenro and Babalola 2017). Like hydrocarbons, heavy metals are also toxic and persistent and bio-accumulated in the biological systems due to their non-biodegradable nature (Mao et al. 2016). Biosurfactants can be beneficial in remediating heavy metals by increasing their ability to solubilize or form complexes known as micelle formation with oppositely charged metal ion (Sarubbo et al. 2015). The biosurfactants can be useful in removing the metals from soil, wastewater, and sediments. The anionic biosurfactants remove the non-anionic metals by complex formation, while the ion exchange process removes cationic metals. For example, biosurfactants from *Burkholderia* sp. remove metals by forming a complex with them and also by reducing the metal toxicity to microbes (Usman et al. 2016). Rhamnolipids have been studied for lead removing capacity (Wan et al. 2015, Chen et al. 2017, Yang et al. 2020) and biosurfactants like surfactin remediate metals like magnesium, manganese, calcium, barium, lithium and rubidium as they contain two negative charges-1 on aspartate and the other on glutamate residues (Arutchelvi et al. 2014). Surfactin also possesses the ability to remove metals like cadmium and zinc by micelle formation. Lipoprotein biosurfactants have shown an effective ability to remove ions and metals from the aqueous solutions (Md Badrul Hisham et al. 2019). Biosufactants from marine microbes have been studied for chelate formation with heavy metals, hence pose bioremediation in wastewaters (Jimoh and Lin 2019). From the sediments, heavy metals could be removed using solutions containing biosurfactants and inorganic compounds. For example, adding an inorganic compound with an OH- group can solubilise the organic fraction of a complex, making more metals available for removal.

In soils, heavy metals are removed by a process called biosurfactant foam technology. Metals like Cd and Ni can be removed by using rhamnolipid foam.

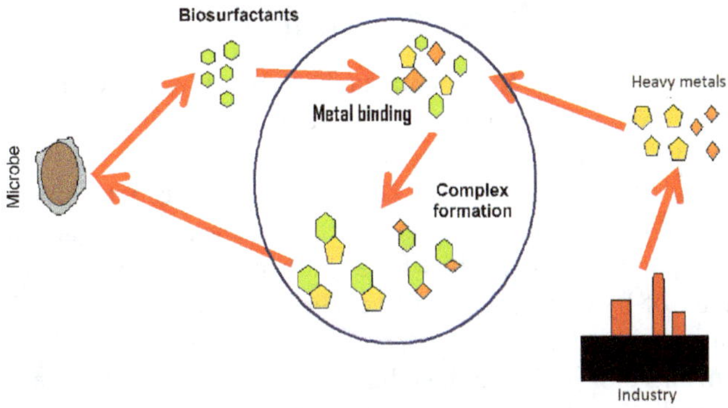

Figure 4. Biosurfactant induced bioremediation of heavy metals.

However, removing heavy metal from soils depends on their chemical composition, i.e., sand silt ratio or mineral composition. Metal recovery has been noted from quartz by using biosurfactants like rhamnolipids. On the other hand, lipopeptides act as ion collectors by adsorping the metal and then removing it by the foam floatation process (Sarubbo et al. 2015). Lipopeptides have been used to remove metals like Cu, Co, Fe, Zn, Ni and Pb from wastewater and soil (Rozas et al. 2019). Lipopeptide biosurfactants increase bioavailability and increase heavy metal solubility (Manoj et al. 2020).

6. Biosurfactants and Bioremediation of Pesticides and Insecticides

Biosurfactant use in the remediation of organic pollutants like pesticides has been of significant interest due to their ecofriendliness, diverse functionality and excellent biodegradability. Biosurfactants help in dispersing low water-soluble organic pollutants like beta-cypermethrin, increasing its bioavailibilty to the bacteria that render it to non-toxic fragments. Rhamnolipids from *P. aeroginosa* have been reported to isomerise and biodegrade the organic pollutants by enhancing their mobilization, absorption and dissolution (Busi and Rajkumari 2017). Similarly, the surfactin from *B. subtilis* has shown the ability to bioremediate the pesticide-endosulfan, which has adeleterious impact on human health and the environment (Jayaprabha and Suresh 2016).

7. Biosurfactants and Bioremediation of Dyes

Dyes are the main constituent of wastewater from the textile industry, leading to environmental degradation and, at the same time, its management. So far, microbes have been used extensively for discolorations and degradation of persistent dyes (Bhosale et al. 2019, Sharma et al. 2020b), but biosurfactant's role in the bioremediation of dyes has not been fully understood. However, rhamnolipids have shown enhanced ability by increasing the enzyme activity, and decreasing the time by half (Madsen et al. 2015). Recent studies have reported a biosurfactant role in improving dye degradation (Jiao et al. 2017, Liu et al. 2017b), though more investigation is required to understand the underlying mechanisms fully.

8. Conclusions

Biosurfactants are surface-active compounds secreted by bacteria, fungi and yeast. The implementation of these substances for remediation is a promising eco-friendly and economical tool. Industrial production, however, is a costly way to produce these biological origin surfactants. Their availability can be made cheaper by isolating these compounds from industrial wastes, wastewater and other sources. Biosurfactants possess the capability to facilitate biodegradation of oil, oil recovery, bioremediation of hydrocarbons and heavy metals, as well as pesticides dyes and co-contaminated sites. The low toxicity, biodegradability and amphilic nature make them a promising tool in the bioremediation of wastewaters, marine environments, soils and oil pollution. However, there is a need to understand these biosurfactant roles at the microlevel in the bioremediation of various pollutants and contaminants present in different environments. Also, their potential to remediate persistent organic pollutants is yet to be understood on a full scale; however, the studies regarding the remediation of PAH's has been well established. The biosurfactants' production on a large scale from wastes needs further investigations to develop cheaper technologies ad methods.

References

Abo-State, M.A.M., B.Y. Riad, A.A. Bakr and M.A. Aziz. 2018. Biodegradation of naphthalene by Bordetella avium isolated from petroleum refinery wastewater in Egypt and its pathway. Journal of Radiation Research and Applied Sciences 11(1): 1–9.

Anjum, F., G. Gautam, G. Edgard and S. Negi. 2016. Biosurfactant production through *Bacillus* sp. MTCC 5877 and its multifarious applications in food industry. Bioresource Technology 213: 262–269.

Arutchelvi, J., J. Sangeetha, J. Philip and M. Doble. 2014. Self-assembly of surfactin in aqueous solution: Role of divalent counterions. Colloids and Surfaces B: Biointerfaces 116: 396–402.

Ayangbenro, A.S. and O.O. Babalola. 2017. A new strategy for heavy metal polluted environments: A review of microbial biosorbents. International Journal of Environmental Research and Public Health 14(1): 94.

Bee, H., M.Y. Khan and R.Z. Sayyed. 2019. Microbial surfactants and their significance in agriculture. *In*: Plant Growth Promoting Rhizobacteria (PGPR): Prospects for Sustainable Agriculture. Springer, Singapore, pp. 205–215.

Bhosale, S.S., S.S. Rohiwal, L.S. Chaudhary, K.D. Pawar, P.S. Patil et al. 2019. Photocatalytic decolorization of methyl violet dye using Rhamnolipid biosurfactant modified iron oxide nanoparticles for wastewater treatment. Journal of Materials Science: Materials in Electronics 30(5): 4590–4598.

Busi, S. and J. Rajkumari. 2017. Biosurfactant: A promising approach toward the remediation of xenobiotics, a way to rejuvenate the marine ecosystem. *In*: Marine Pollution and Microbial Remediation. Springer, Singapore, pp. 87–104.

Carolina, C.F., P.S. Kumar and P.T. Ngueagni. 2021. A review on new aspects of lipopeptide biosurfactant: Types, production, properties and its application in the bioremediation process. Journal of Hazardous Materials 407: 124827.

Chaprão, M.J., I.N. Ferreira, P.F. Correa, R.D. Rufino, J.M. Luna et al. 2015. Application of bacterial and yeast biosurfactants for enhanced removal and biodegradation of motor oil from contaminated sand. Electronic Journal of Biotechnology 18(6): 471–479.

Chen, W., Y. Qu, Z. Xu, F. He, Z. Chen et al. 2017. Heavy metal (Cu, Cd, Pb, Cr) washing from river sediment using biosurfactant rhamnolipid. Environmental Science and Pollution Research 24(19): 16344–16350.

Ghosh, P. and S. Mukherji. 2020. Modeling growth kinetics and carbazole degradation kinetics of a *Pseudomonas aeruginosa* strain isolated from refinery sludge and uptake considerations during growth on carbazole. Science of the Total Environment 738: 140277.

Gudiña, E.J., A.I. Rodrigues, E. Alves, M.R. Domingues, J.A. Teixeira et al. 2015. Bioconversion of agro-industrial by-products in rhamnolipids toward applications in enhanced oil recovery and bioremediation. Bioresource Technology 177: 87–93.

Jahan, R., A.M. Bodratti, M. Tsianou and P. Alexandridis. 2020. Biosurfactants, natural alternatives to synthetic surfactants: Physicochemical properties and applications. Advances in Colloid and Interface Science 275: 102061.

Jayaprabha, K.N. and K.K. Suresh. 2016. Endosulfan contamination in water: A review on to an efficient method for its removal. J. Chem. Chem. Sci. 6: 182–191.

Jezierska, S., S. Claus and I. Van Bogaert. 2018. Yeast glycolipid biosurfactants. Febs Letters 592(8): 1312–1329.

Jiao, J., X. Xin, X. Wang, Z. Xie, C. Xia et al. 2017. Self-assembly of biosurfactant-inorganic hybrid nanoflowers as efficient catalysts for degradation of cationic dyes. RSC Advances 7(69): 43474–43482.

Jimoh, A.A. and J. Lin. 2019. Production and characterization of lipopeptide biosurfactant producing *Paenibacillus* sp. D9 and its biodegradation of diesel fuel. International Journal of Environmental Science and Technology 16(8): 4143–4158.

Joshi, S.J., Y.M. Al-Wahaibi, S.N. Al-Bahry, A.E. Elshafie, A.S. Al-Bemani et al. 2016. Production, characterization, and application of Bacillus licheniformis W16 biosurfactant in enhancing oil recovery. Frontiers in Microbiology 7: 1853.

Kreling, N.E., M. Zaparoli, A.C. Margarites, D. Zampieri, L.M. Colla et al. 2019. Produção de biossurfactantes: manoproteínas intracelulares e soforolipídios extracelulares por Saccharomyces cerevisiae. Engenharia Sanitaria e Ambiental 24(6): 1209–1219.

Ławniczak, Ł., M. Woźniak-Karczewska, A.P. Loibner, H.J. Heipieper, Ł. Chrzanowski et al. 2020. Microbial degradation of hydrocarbons—Basic principles for bioremediation: A review. Molecules 25(4): 856.

Li, S., Y. Pi, M. Bao, C. Zhang, D. Zhao et al. 2015. Effect of rhamnolipid biosurfactants on solubilization of polycyclic aromatic hydrocarbons. Marine Pollution Bulletin 101(1): 219–225.

Liu, Z., Z. Li, H. Zhong, G. Zeng, Y. Liang et al. 2017a. Recent advances in the environmental applications of biosurfactant saponins: A review. Journal of Environmental Chemical Engineering 5(6): 6030–6038.

Liu, C., Y. You, R. Zhao, D. Sun, P. Zhang et al. 2017b. Biosurfactant production from Pseudomonas taiwanensis L1011 and its application in accelerating the chemical and biological decolorization of azo dyes. Ecotoxicology and Environmental Safety 145: 8–15.

Madsen, J.K., R. Pihl, A.H. Møller, A.T. Madsen, D.E. Otzen et al. 2015. The anionic biosurfactant rhamnolipid does not denature industrial enzymes. Frontiers in Microbiology 6: 292.

Mańko-Jurkowska, D., E. Ostrowska-Ligęza, A. Górska and R. Głowacka. 2019. The role of biosurfactants in soil remediation. Zeszyty Problemowe Postępów Nauk Rolniczych 596: 33–43.

Manoj, S.R., C. Karthik, K. Kadirvelu, P.I. Arulselvi, T. Shanmugasundaram, B. Bruno, M. Rajkumar. 2020. Understanding the molecular mechanisms for the enhanced phytoremediation of heavy metals through plant growth promoting rhizobacteria: A review. Journal of Environmental Management 254: 109779.

Mao, X., Z. Yu, Z. Ding, T. Huang, J. Ma et al. 2016. Sources and potential health risk of gas phase PAHs in Hexi Corridor, Northwest China. Environmental Science and Pollution Research 23(3): 2603–2612.

Md Badrul Hisham, N.H., M.F. Ibrahim, N. Ramli and S. Abd-Aziz. 2019. Production of biosurfactant produced from used cooking oil by Bacillus sp. HIP3 for heavy metals removal. Molecules 24(14): 2617.

Mnif, I., S. Mnif, R. Sahnoun, S. Maktouf, Y. Ayedi et al. 2015. Biodegradation of diesel oil by a novel microbial consortium: Comparison between co-inoculation with biosurfactant-producing strain and exogenously added biosurfactants. Environmental Science and Pollution Research 22(19): 14852–14861.

Olasanmi, I.O. and R.W. Thring. 2018. The role of biosurfactants in the continued drive for environmental sustainability. Sustainability 10(12): 4817.

Pacwa-Płociniczak, M., G.A. Płaza, Z. Piotrowska-Seget and S.S. Cameotra. 2011. Environmental applications of biosurfactants: Recent advances. International Journal of Molecular Sciences 12(1): 633–654.

Patel, J., S. Borgohain, M. Kumar, V. Rangarajan, P. Somasundaran et al. 2015. Recent developments in microbial enhanced oil recovery. Renewable and Sustainable Energy Reviews 52: 1539–1558.

Phulpoto, I.A., B. Hu, Y. Wang, F. Ndayisenga, J. Li et al. 2020. Effect of natural microbiome and culturable biosurfactants-producing bacterial consortia of freshwater lake on petroleum-hydrocarbon degradation. Science of the Total Environment 751: 141720.

Purwasena, I.A., D.I. Astuti, M. Syukron, M. Amaniyah, Y. Sugai et al. 2019. Stability test of biosurfactant produced by Bacillus licheniformis DS1 using experimental design and its application for MEOR. Journal of Petroleum Science and Engineering 183: 106383.

Raj, D.S., S.V. Nagarajan, T. Raman, P. Venkatachalam, M. Parthasarathy et al. 2021. Remediation of textile effluents for water reuse: Decolorization and desalination using Escherichia fergusonii followed by detoxification with activated charcoal. Journal of Environmental Management 277: 111406.

Ramasamy, S., A. Arumugam and P. Chandran. 2017. Optimization of Enterobacter cloacae (KU923381) for diesel oil degradation using response surface methodology (RSM). Journal of Microbiology 55(2): 104–111.

Ravindran, A., A. Sajayan, G.B. Priyadharshini, J. Selvin, G.S. Kiran et al. 2020. Revealing the efficacy of thermostable biosurfactant in heavy metal bioremediation and surface treatment in vegetables. Frontiers in Microbiology 11: 222.

Rodrigues, L.R. 2015. Microbial surfactants: Fundamentals and applicability in the formulation of nano-sized drug delivery vectors. Journal of Colloid and Interface Science 449: 304–316.

Roy, A., A. Dutta, S. Pal, A. Gupta, J. Sarkar et al. 2018. Biostimulation and bioaugmentation of native microbial community accelerated bioremediation of oil refinery sludge. Bioresource Technology 253: 22–32.

Rozas, E.E., M.A. Mendes, M.R. Custodio, D.C.R. Espinosa, A.O. Claudio et al. 2019. Self-assembly of supramolecular structure based on copper-lipopeptides isolated from e-waste bioleaching liquor. J. Hazard. Mater. 368: 63–71.

Santos, D.K.F., J.M. Luna, R.D. Rufino, V.A. Santos and L.A. Sarubbo. 2016. Biosurfactants: Multifunctional materials of the XXI century. International Journal of Molecular Sciences 17(401): 2016.

Santos, D.K.F., H.M. Meira, R.D. Rufino, J.M. Luna, L.A. Sarubbo et al. 2017a. Biosurfactant production from Candida lipolytica in bioreactor and evaluation of its toxicity for application as a bioremediation agent. Process Biochemistry 54: 20–27.

Santos, D.K., A.H. Resende, D.G. de Almeida, R.D.C.F. Soares da Silva, R.D. Rufino et al. 2017b. Candida lipolytica UCP0988 biosurfactant: Potential as a bioremediation agent and in formulating a commercial related product. Frontiers in Microbiology 8: 767.

Sarubbo, L.A., R.B. Rocha Jr, J.M. Luna, R.D. Rufino, V.A. Santos et al. 2015. Some aspects of heavy metals contamination remediation and role of biosurfactants. Chemistry and Ecology 31(8): 707–723.

Shah, A., S. Shahzad, A. Munir, M.N. Nadagouda, G.S. Khan et al. 2016. Micelles as soil and water decontamination agents. Chemical Reviews 116(10): 6042–6074.

Sharma, D., S.S. Singh, P. Baindara, S. Sharma, N. Khatri et al. 2020a. Surfactin like broad spectrum antimicrobial lipopeptide Co-produced with sublancin from Bacillus subtilis strain A52: Dual reservoir of bioactives. Frontiers in Microbiology 11: 1167.

Sharma, B., S. Menon, S. Mathur, N. Kumari, V. Sharma et al. 2020b. Decolorization of malachite green dye from aqueous solution using biosurfactant-stabilized iron oxide nanoparticles: process optimization and reaction kinetics. International Journal of Environmental Science and Technology, 1–14.

Silva, R.C.F.S., D.G, Almeida, R.D. Rufino, J.M, Luna, V.A. Santos and L.A. Sarubbo. 2014. Applications of biosurfactants in the petroleum industry and the remediation of oil spills. International Journal of Molecular Science 15: 12523–12542.

Soberón-Chávez, G., A. González-Valdez, M.P. Soto-Aceves and M. Cocotl-Yañez. 2021. Rhamnolipids produced by Pseudomonas: from molecular genetics to the market. Microbial Biotechnology 14(1): 136–146.

Usman, M.M., A. Dadrasnia, K.T. Lim, A.F. Mahmud, S. Ismail et al. 2016. Application of biosurfactants in environmental biotechnology; remediation of oil and heavy metal. AIMS Bioengineering 3(3): 289–304.

Wan, J., D. Meng, T. Long, R. Ying, M. Ye et al. 2015. Simultaneous removal of lindane, lead and cadmium from soils by rhamnolipids combined with citric acid. Plos One 10(6): e0129978.

Wood, T.L., T. Gong, L. Zhu, J. Miller, D.S. Miller et al. 2018. Rhamnolipids from Pseudomonas aeruginosa disperse the biofilms of sulfate-reducing bacteria. NPJ Biofilms and Microbiomes 4(1): 1–8.

Xu, M., X. Fu, Y. Gao, L. Duan, C. Xu et al. 2020. Characterization of a biosurfactant-producing bacteria isolated from Marine environment: Surface activity, chemical characterization and biodegradation. Journal of Environmental Chemical Engineering 8(5): 104277.

Xu, X., W. Liu, S. Tian, W. Wang, Q. Qi et al. 2018. Petroleum hydrocarbon-degrading bacteria for the remediation of oil pollution under aerobic conditions: A perspective analysis. Front. Microbiol. 9: 2885.

Yalçın, H.T., G. Ergin-Tepebaşı and E. Uyar. 2018. Isolation and molecular characterization of biosurfactant producing yeasts from the soil samples contaminated with petroleum derivatives. Journal of Basic Microbiology 58(9): 782–792.

Yang, R., H. Wang, M. Shi, Y. Jiang, Y. Dong et al. 2020. Biosurfactant rhamnolipid affacts the desorption of sorbed As (III), As (V), Cr (VI), Cd (II) and Pb (II) on iron (oxyhydr) oxides and clay minerals. International Biodeterioration & Biodegradation 153: 105019.

Zafar, A., S. Ullah, M.T. Majeed and R. Yasmeen. 2020. Environmental pollution in Asian economies: Does the industrialisation matter? OPEC Energy Review 44(3): 227–248.

Zhang, J., Q. Xue, H. Gao, H. Lai, P. Wang et al. 2016. Production of lipopeptide biosurfactants by Bacillus atrophaeus 5-2a and their potential use in microbial enhanced oil recovery. Microbial Cell Factories 15(1): 168.

Zhong, H., Z. Wang, Z. Liu, Y. Liu, M. Yu et al. 2016. Degradation of hexadecane by Pseudomonas aeruginosa with the mediation of surfactants: Relation between hexadecane solubilization and bioavailability. International Biodeterioration & Biodegradation 115: 141–145.

12

Rhamnolipids (RLs) Green Solution for Global Hydrocarbon Pollution

Anushka Devale,[1] *Rupali Sawant*,[1] *Karishma Pardesi*,[2] *Surekha Satpute*[2] and *Shilpa Mujumdar*[1,*]

1. Introduction

The global population has increased drastically over the past few years and reflected surges in the economic output, energy demands and natural resource consumption. Currently, petroleum is one of the major energy resources that have been utilized routinely to fulfill an enormous energy demand (Ince and Ince 2019). The practices involved in petroleum production, storage, carriage, as well as dumping processes release huge amounts of hydrocarbons in the environment (Badowska and Bandzierz 2019, Xu et al. 2018). The estimate of the global crude oil seepage rate given by Kvenvolden and Cooper (2003) was 600,000 metric tons. This has led to environmental pollution including both soil and water (Ezekwe et al. 2018, Truskewycz et al. 2019). A survey carried out in Nigeria by Emoyan and coworkers in 2020, significantly indicates the severity of situation. They investigated, soils from different vehicle parks which were obvious sites of petroleum pollutants and analyzed for 16 different polycyclic aromatic hydrocarbon (PAHs) concentrations. They found that in the wet season, PAH concentrations ranged from 365.8 to 1065.8 μg/kg (top soil) and 142.9 to 5321.9 μg/kg (sub soil) whereas in the dry season 66.4 to 269.3 μg/kg (top soil) and 212.1 to 1342.5 μg/kg (wet soil). As per the Environmental Guidelines and Standards for the Petroleum Industry in Nigeria (2002), these values were remarkably more than the permissible limits. A similar study was

[1] Department of Microbiology, Modern College of Arts, Science and Commerce, Shivajinagar, Pune 411 005, Maharashtra, India.

[2] Department of Microbiology, Savitribai Phule Pune University, Ganeshkhind, Pune-411007, Maharashtra, India.

* Corresponding author: hodmicro@moderncollegepune.edu.in

carried out by Suman and coworkers in 2015 in Dhanbad, India. In this study PAH contamination in traffic soils was evaluated. The control for this experiment was a rural site soil (control) without traffic. Average total PAHs concentration of traffic site soils was 3.488 μg/g while that of rural site soils without traffic (control) was 0.640 μg/g. Further it was also revealed that PAH components in traffic soils were 6.15 μg/g which is more carcinogenic than that of control soil. Therefore, this situation demands a focus on development of innumerable strategies to prevent ecological loss as well as rehabilitation of contaminated locations. Remediation procedures involve elimination of the contaminants or reduction in their toxicity or converting them into the less harmful components (Ossai et al. 2020). Currently, both physico-chemical and biological remediation processes are being practiced (Jain et al. 2011). Physicochemical techniques include solvent extraction, soil vapor extraction, thermal desorption, stabilization, incineration and use of chemical surfactants (Ossai et al. 2020). However, the mentioned physico-chemical methods pose consequences like the cost, labors involved and toxicity (Jain et al. 2011, Makkar et al. 2003, You et al. 2018). On an average expenses for cleaning oil spills in water range from $40–400/L (Kapsalis et al. 2021). Comparatively, biological remediation techniques like bioventing, biostimulation, biodegradation, bioaugmentation and phytoremediation are found to be economic and eco-friendly to deal with challenges associated with hydrocarbon-contamination (Olawale et al. 2020, Yuniati 2018). Other critical issues in biological remediation are bioavailability and solubility of hydrocarbons. The most promising approach to resolve this problem is bioremediation processes assisted with biosurfactants (Matvyeyeva et al. 2014). Biosurfactants simply enhance bioavailability and solubility, making it accessible to microbes present in the polluted surroundings (Karlapudi et al. 2018, Rahman 2016). From the huge world of biosurfactants; rhamnolipids appear to be the major attraction of the global market. Biosurfactants are biological surface-active amphiphilic microbial compounds produced by different species of bacteria, fungi, yeast and actinomycetes (Atuanya et al. 2016, Shekhar et al. 2014). These microbial compounds are very complex in nature comprising peptides, glycolipids, fatty acids, phospholipids and glycopeptides in various combinations and proportions (Nitschke et al. 2004).

Biosurfactants like Rhamnolipids, Alasan, Surfactin, Emulsan and Sophorolipids have been frequently reported to enhance the hydrocarbon degradation process (Matvyeyeva et al. 2014). The reasons for Rhamnolipids standing apart from other surfactants is their exceptional physico-chemical properties and enormous range of applications (Tan and Li 2018). Literature depicts the usage of rhamnolipids as one of the most promising biosurfactants for several bioremediation processes. This chapter encompasses comprehensive information on rhamnolipid type biosurfactants which are explored extensively for bioremediation of hydrocarbons. We have also heightened the major challenges associated with applicability of rhamnolipids for bioremediation purposes.

2. Rhamnolipids: Successful Molecules of Green Energy and Technology

Rhamnolipids (RLs) are anionic surface-active glycolipids of microbial origin, and have been studied extensively after the first report documented by Jarvis and

Johnson (1949) from *Pseudomonas aeruginosa.* RLs consist of rhamnose and β-hydroxyalkanoic acid units linked by glycosidic bonds (Armendáriz et al. 2019). Diverse RLs with variations in chain length, unsaturated fatty acids and number of sugars have been reported till date. On the basis of the presence of one or two rhamnose sugars respectively, RLs are classified as mono-rhamnolipids and di-rhamnolipids (Chong and Li 2017). Figure 1 illustrates the structure and types of rhamnolipids.

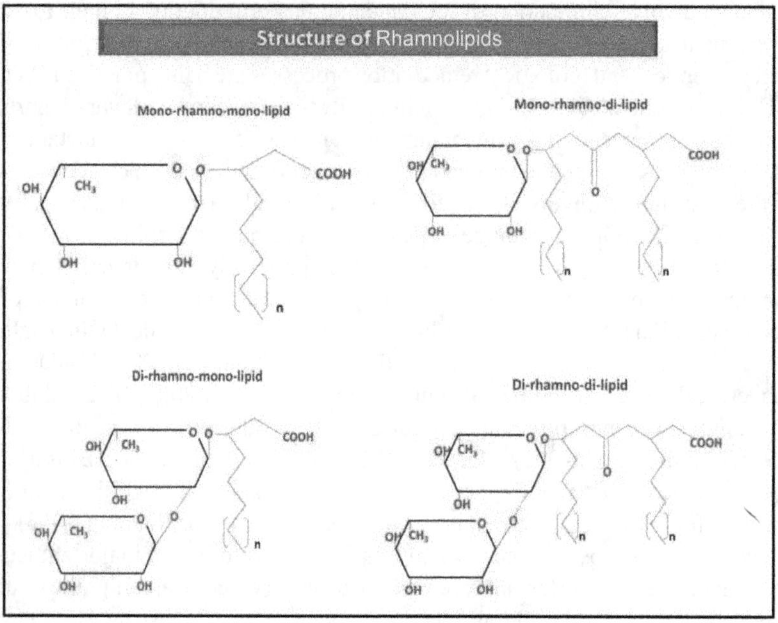

Figure 1. Structure of Rhamnolipids.

Rhamnolipids are less toxic, biodegradable and eco-friendly in nature (Arora et al. 2021, Baskaran et al. 2021). They reduce the surface tension of liquid solutions and have an emulsifying property (Leite et al. 2016). Moreover, they exhibit various activities such as foaming, solubilization, sequestration and wetting which make them an industrially valuable compound (Cheng et al. 2017, Kłosowska-Chomiczewska et al. 2021). Glycolipid biosurfactants are the best alternatives for synthetic surfactants in many industries (Sinumvayo et al. 2015). Rhamnolipids possess different promising applications in industries like agriculture, pharmaceuticals, cosmetics and medicine (Liu et al. 2017, Thakur et al. 2021). For example, Green Pyramid Biotech. Pvt. Ltd. Company located in Pune, Maharashtra produces "Evergreen and Bio Clean Washing Solutions". The product formulation is based on a biosurfactant which can sanitize fruits and vegetables successfully before their consumption (https://www.greenpyramidbiotech.com/). Although RLs have many advantages over chemical surfactants, their production at industrial level was hard until 2016. The first company, to produce RLs on a large scale was Evonik Industry, using recombinant *P. putida* and butane for the production. Another company named Biotensidon

GmbH, Germany developed the first cost-effective process to produce RLs at an industrial scale (Geissler et al. 2019). Currently, other leading companies namely TeeGene Biotech (UK), AGAE Technologies LLC (USA), Jeneil Biosurfactant (USA), Paradigm Biomedical Inc. (USA), Shaanxi Pioneer Biotech Company (China) (Eslami et al. 2020) are involved in the production of RLs commercially. Today, a huge number of patents and formulations are evident, making this market open to explore many more applications (Eslami et al. 2020).

2.1 Rhamnolipid producing Bacteria

It has been observed that mainly, hydrocarbon-degrading microorganisms produce RLs (Bezza and Chirwa 2014, Sharma et al. 2015, Wang et al. 2014). The amphipathic nature of RLs helps the bacteria to access the hydrophobic nutrients in the form of hydrocarbons. Some strains may produce RLs as a virulence factor (Zulianello et al. 2006). *Pseudomonas* strains are the maximally reported and well-studied producers of RL biosurfactants (Carrazco-Palafox et al. 2021). The list of efficient RL producing *Pseudomonas* species reported till today includes *P. aeruginosa, P. chlororaphis, P. fluorescens, P. indica, P. luteola, P. nitroreducens, P. putida, P. stutzeri, P. alcaligenes, P. clemancea, P. collierea, P. nitroreducens* and *P. teessidea* (Abdel-Mawgoud et al. 2010, Arora et al. 2021, Chong et al. 2017, Muller et al. 2012, Onwosi and Odibo 2012). Studies suggest that among the literature reported on various *Pseudomonas* strains; *P. aeruginosa* appears to be one of the most prospective candidates for RL production (Liu et al. 2017). Although *P. aeruginosa* was reported as the best candidate for RLs production, its pathogenic nature to humans is a major hindrance in its usage for various applications and industrial scale RL production. Hence, search for novel non-pathogenic RL producers was inevitable (Abdel-Mawgoud et al. 2010, Toribio et al. 2010). Recently, di-rhamnolipid producing *Burkholderia thailandensis* and *Burkholderia kururiensis* have been reported (Funston et al. 2016, Tavares et al. 2013). Nonpathogenic strains of *Acinetobacter* and *Enterobacter* are also found to be efficient RL producers (Dong et al. 2016, Hošková et al. 2013, Hošková et al. 2015a). Moreover recently, for the first time, *Shewanella* sp. was reported for the production of RLs which is isolated from contaminated soils of Tamil Nadu, India (Joe et al. 2019). A novel strain *Serratia rubidaea* SNAU02 was also used in the studies on Mahua oil cake as a substrate for the production of RL biosurfactants (Nalini and Parthasarathi 2014).

Another popular approach is 'heterologous recombination' to achieve safer and maximum production of RLs. In heterologous recombination, a safe host strain is exploited for expression of the RL gene and its large-scale production (Kryachko et al. 2013, Tiso et al. 2015). Table 1 summarizes few examples of recombinant mono-rhamnolipid expression studies done in the last two decades. In recent studies, *Pantoea* sp. P37 was suggested to be a novel non-pathogenic suitable host and its efficacy was evaluated for the heterologous mono-rhamnolipid production. The non-pathogenic nature of *Pantoea* species was confirmed using the virulence factor database (VFDB) of common bacterial pathogens and the VF analyzer pipeline. The said strain was transformed with different vectors namely pETrhlAB8, pETrhlAB100, pET3110 and pET2711 carrying mono-rhamnolipid biosynthesis

Table 1. Recombinant expression of mono-rhamnolipid.

***Pseudomonas aeruginosa* Genes**	**Gene Product**	**Heterologous Expression in Safe Host**	**References**
rhlAB operon	Rhamnosyltransferase	*Pantoea* sp. P37	Nawrath et al. 2020
rmlA, rmlB, rmlC, rmlD, rhlA and *rhlB* and *Gft* gene from *Pelomonas saccharophila*	Mono-rhamnolipid biosynthesis enzymes and sucrose phosphorylase	*Saccharomyces cerevisiae*	Bahia et al. 2018
rhlAB with *rhlRI*	Rhamnosyltransferase with quorum sensing system	*P. putida* 1067	Cha et al. 2008
rhlAB	Enzymatic complex rhamnosyltransferase 1	*E. coli*	Cabrera-valladares et al. 2006

enzyme gene (rhamnosyltransferase) derived from *P. aeruginosa.* The highest production of rhamnolipid (409.4 mg/L) was obtained by pET3110 transformants while the other transformants showed very little or no production (Nawrath et al. 2020). This study also depicts the best examples of applications and advantages of modern bioinformatics tools and software for quick and precise genome analysis and characterization. Figure 2 illustrates the generalized protocol followed for a heterologous expression of mono-rhamnolipids in novel safe host bacteria.

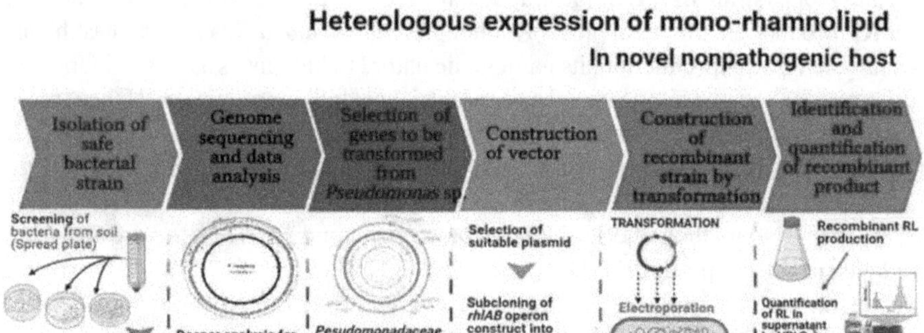

Figure 2. The generalized protocol followed for heterologous expression of mono-rhamnolipids in novel safe host bacteria.

3. Application of Rhamnolipids in Hydrocarbon Bioremediation

Hydrocarbons such as petroleum are complex mixtures of alkanes, aromatics and resins. Polycyclic aromatic hydrocarbons (PAHs) are naturally occurring compounds with benzene rings, in coal, crude oil, and gasoline. Burning of coal, oil, and garbage also releases PAHs. These are listed as major pollutants because of their properties like toxicity, carcinogenicity, mutagenicity and recalcitrance (Li et al. 2020). In 2010, in one research done on the Tamil Nadu coast, Bay of Bengal, India, it was revealed that coastal waters contain very high petroleum hydrocarbon concentrations. Not only waters but a commercial fish species *Sardinella longiceps* was also found to have very high PHC content (Venkatachalapathy et al. 2010). This indicates biomagnification of petroleum pollutants. This is an alarming scenario with regard to ecosystem sustainability. Hence it is essential to develop and employ various environmentally friendly, economic and effective remediation processes to reduce and/or eliminate hydrocarbon contamination. Application of RLs through various techniques like bacterial degradation of PAHs, soil washing and phytoremediation have been extensively evaluated and found to be effective to enhance bioremediation of PAHs (Liu et al. 2018). Figure 3 describes the applications and key roles played by RLs in various soil bioremediation methods used for removal of hydrocarbon contamination.

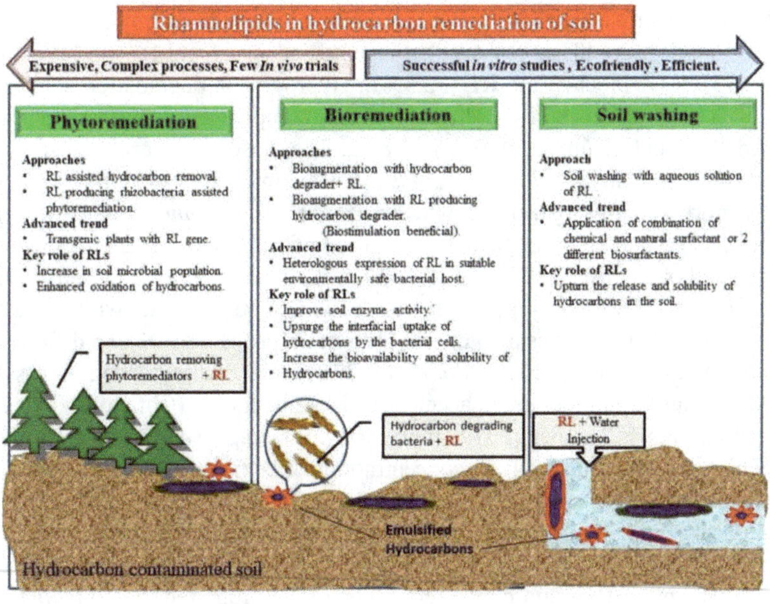

Figure 3. Applications of rhamnolipids in remediation of hydrocarbon contaminated soil.

3.1 *Rhamnolipid Assisted Hydrocarbon Degradation*

Bioremediation of hydrocarbon polluted environments becomes challenging due to lack of solubility and bioavailability. Hydrocarbon degrading microorganisms generally access the hydrocarbons by producing biological surfactants. Hence augmentation of biosurfactant producing hydrocarbon degraders at the polluted sites is a popular and practical approach to achieve successful bioremediation (Musale et al. 2015). RLs have been expansively studied and are found to be highly applicable in the biodegradation of organic contaminants and even organic and metal contaminants together. Researchers also found that the ability to produce RL biosurfactants was an important feature of many petroleum degraders investigated till today (You et al. 2018). A comparative study on two efficient hexadecane degraders namely *P. aeruginosa* ATCC 9027 (RL producing strain) and *P. putida* CICC 20575 (RL degrading strain) was performed by Liu et al. (2018). In this study *P. aeruginosa* ATCC 9027 showed an enhanced uptake of RL solubilized hexadecane as compared to mass hexadecane. In contrast, *P. putida* CICC20575, could not uptake hexadecane at all. This highlights the role of RLs in enhancing the bioavailability of hydrocarbons by increasing their solubility. Patowary and coworkers (2017), reported RL production from *Pseudomonas aeruginosa* PG1 isolated from a hydrocarbon polluted site and suggested its usefulness in crude oil degradation. The coastal wetland soil oil spill clearance is a challenging venture. Recently in 2020, efficiency in *in vitro* RL mediated remediation of Louisiana coastal saline marshes by indigenous microflora was demonstrated by Wei and colleagues (USA). The study also demonstrated removal of Total petroleum hydrocarbons (TPH) by diverse indigenous flora in the test soil was enhanced after supplementation with biochar, Nitrogen (N) and RL. However maximum TPH removal was achieved after RL application. [TPH removal achieved: Biochar, N and L (80.9%) > biochar and N (73.2%) > biochar and RL (32.3%)]. Hydrocarbon degradation studies were done by You et al. 2018 on 14 different diesel components using bacterial strains isolated from petroleum refineries in China. These isolates were identified to be *Klebsiella pneumoniae* and *P. aeruginosa*. It was observed that *P. aeruginosa* was a 15% more superior hydrocarbon degrader than *K. pneumoniae*. However, the authors also commented that *in situ* application of *P. aeruginosa* is restricted due to possible pathogenicity. In 2004, a novel rhamnolipid producing strain *Renibacterium salmoninarum* 27BN was reported for utilization of hexadecane, which is a major component of diesel. It was also the first report on production of two different RLs from one strain. Hence, *R. salmoninarum* 27BN was recommended as a suitable bacterial strain for application in accelerated bioremediation (Christova et al. 2004). Martino and his colleagues (2019) found RL producing *Pseudomonas* sp. KA and KB, which not only tolerate high concentrations of benzene, toluene, xylene (BTX) but also degrade them efficiently. Further it was revealed that both the strains possess the important BTX degrading genes namely *xylA* and *xylE*. The first report on the aerobic utilization of naphthalene which is made up of crude oil or coal tar by bacteria was by Tuleva and coworkers (2005). They reported that *Bacillus cereus* utilizes naphthalene aerobically and the strain has the potential for hydrocarbon bioremediation (Tuleva et al. 2005). Rocha et al. (2011) reported biodegradation of methyl-branched alkanes by RL producing *P. aeruginosa* ATCC 55925.

In several studies, RLs have shown to enhance the interfacial uptake of hydrocarbons by the bacterial cells. This is achieved by modifying cell surface properties like membrane permeability, cell surface hydrophobicity and cell surface zeta potential (Liu et al. 2017, Zeng et al. 2018). In the studies of the application of *P. stutzeri*, isolated from crude oil-contaminated soil, the effect of three surfactants namely rhamnolipids, saponins and Triton X-100 on diesel oil biodegradation were investigated. RLs were found to be the most effective, achieving 88% diesel oil removal. Studies showed that RLs efficiently enhance cell surface hydrophobicity of *P. stutzeri*

Table 2. *In vitro* studies on rhamnolipid assisted hydrocarbon removal from soil.

Pollutant	Bioremediator	Maximum Removal (%)	Reference
Crude oil	*Pseudomonas aeruginosa*	58	Zhang et al. 2005
TPH*	*Pseudomonas putida* CB-100	40.6	Ángeles et al. 2013
Octane	*L. pentosus*	76	Moldes et al. 2011
Chlorpyrifos organo-phosphate pesticides (OPs)	*Pseudomonas* sp. (ChlD)	98	Singh et al. 2009
TPH*	Indigenous Microbial population	60	Millioli et al. 2009
Crude oil	*P. stutzeri* G11	96	Celik et al. 2008
Phenanthrene (co-contaminant cadmium)	*Pseudomonas aeruginosa* strains	58	Maslin and Maier 2000

Note*: TPH: Total petroleum hydrocarbon.

compared to saponins and Triton X-100 (Kaczorek et al. 2012). Similar studies on RL enhanced pyrene biodegradation by *K. oxytoca* PYR-1 showed RL media uptake of pyrene by increasing the cell surface permeability (Zhang et al. 2013). Moreover similar RL assisted bioremediation experiments were performed in contaminated farmlands located in China in which the biodegradation ability of a novel strain of *A. globiformis* was found to be noticeably enhanced by the application of an optimal RL concentration [5 mg/kg soil]. Furthermore, the proportions of removal of pollutants such as DDT and PAH achieved from the farmland soils were 64.3% and 35.6%, respectively. Similarly, the soil analysis studies of farmlands revealed that optimal RL application enhanced *A. globiformis* reproduction and oil enzymes activity (Wang et al. 2017). Table 2 summarizes additional studies on RL producing bacteria in bioremediation of hydrocarbon contaminated soils.

3.2 *Rhamnolipid Assisted Soil Washing Technology*

Rhamnolipids are applied in hydrocarbon removal by soil-washing technology- (Pacwa-Płociniczak et al. 2011). It is a physicochemical method for removing contaminants from soils generally using liquids. The soil washing technology is economic, time consuming, environment friendly and efficient in contaminant removal (Ossai et al. 2020). It is thought to be a promising way of hydrocarbon elimination from polluted sites. Urum et al. (2006) reported that two biosurfactants namely, rhamnolipids and saponins and one chemical surfactant sodium dodecyl sulphate, were the most effective candidates used as an aid in soil washing. Out of the biosurfactants studied RLs were found to be the superior compared to saponins in the removal of crude oil. Researchers also observed that the factors such as RL concentration, temperature, washing time and stirring speed were essential for designing the soil washing process (Olasanmi et al. 2020). Recently, Fanaei et al. (2020) also underlined the effectiveness of RLs in the soil washing process. They reported 86% removal of total petroleum hydrocarbons (TPH) from an oil-contaminated soil by application of a mixture of surfactin and RL in soil washing.

Table 3. Recent studies on rhamnolipids-assisted soil washing for removal of hydrocarbons.

Hydrocarbon Pollutant	Soil Type	Hydrocarbon Removal (%)	Reference
TPH*	Soil contaminated with oily-sludge	71.8	Pourfadakari et al. 2020
TPH*	A gravel-sandy soil (from the former refinery)	80	Torres et al. 2018
Petroleum, diesel oil	Sandy soil	44.75	da Rosa et al. 2015
Lindane BSR	Garden soils	85.4	Zhang 2015
Oil	Sand and clay loam	99 40–60	Hallmann and Mędrzycka 2015

Note*: TPH: Total petroleum hydrocarbon.

They also reported removal of phenanthrene (Phe) and lead (Pb) by RL alone and a mixture of Rhamnolipid-sophorolipid biosurfactants as well. WU Liang and colleagues (2020), stated that RL enhanced-flushing systems can be a key approach for successful biological remediation of hydrocarbons. Table 3 shows recent studies on rhamnolipids-assisted soil washing for the removal of hydrocarbons.

3.3 *Rhamnolipid Assisted Phytoremediation*

Phytoremediation is an economic and effective technology in which plants aid in pollutant removal and degradation. Plants not only accumulate pollutants, heavy metals and hydrocarbons from contaminated soil and water but also absorb them into their roots and shoots to facilitate their degradation into simpler and nontoxic components (Sandermann 1992). The only limiting factor in the application of this technique is low bioavailability and solubility of hydrocarbon pollutants in petroleum bioremediation (Agnello 2015). Hence, biosurfactants can play a key role in the process to make the pollutant available for absorption, assimilation and degradation by plants and thus assist in achieving successful hydrocarbon bioremediation (Lai et al. 2009). Thus, literature depicts the role of Rhamnolipids in enhancing phytoremediation of hydrocarbon contaminated soils (Liao et al. 2016 and Liduino et al. 2018). Figure 4 illustrates the role of RLs in phytoremediation.

In 2016, Liao and coworkers, carried out an *in vitro* experiment for evaluating the impact and role of various surfactants in the phytoremediation of hydrocarbon contaminated soils by maize plant. The other objectives of this study were to compare the hydrocarbon removal efficacies of chemical surfactants and biosurfactants of microbial (RL) and plant origin (Soybean lecithin). The outcome of the study, suggested that the surfactants do not pose any negative impact on plant growth. Rather RL amended soils showed improved physical soil quality and plant growth promoting microbial flora. A noticeable rise in microbial flora was observed in biosurfactant treated soils as compared to chemical surfactants. This may be due to the biodegradable nature of the natural surfactants. The percent TPH removal obtained during three months of the phytoremediation process using Maize plant with surfactant (52%) is significantly commendable compared to that without Maize plant (13%). This proves that the Maize plant is an effective phytoremediator. The TPH

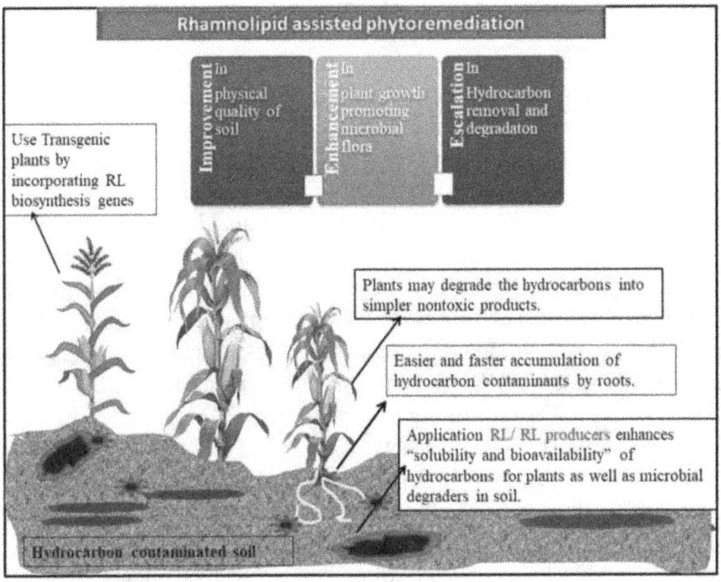

Figure 4. The role of rhamnolipids (RLs) in phytoremediation.

removal efficacy values of maize plant mediated phytoremediation with assistance of soybean lecithin (62%), rhamnolipid (58%), Tween 80 (47%) are evident to conclude that biological surfactants influence the process positively. Further studies on RL assisted phytoremediation of co-contaminated soils with heavy metals and petroleum hydrocarbons using sunflower (*Helianthus annuus* L.) indicated positive outcomes. The supplementation of 4 mg/kg RL in sunflower cultivated contaminated soil showed 58% and 48% reduction in TPH and PAH concentrations respectively (Liduino et al. 2018). Bioremediation of hydrocarbon contaminated soil was successfully achieved in the studies based on RL supplemented phytoremediation using vetiver grass. In these experiments, the augmented novel strain *Shewanella seohaensis*, served as a source of RL. Thus, it is proved from these studies that such combined tactics may serve as the novel and promising approaches in the hydrocarbon bioremediation process (Ram et al. 2019).

Researchers also have attempted to develop transgenic plants by introducing genes for biosurfactant biosynthesis. Several trials have been made to create transgenic plants by incorporating RL biosynthesis genes in plants for efficient hydrocarbon phytoremediators. One such successful genetic engineering attempt was made by Stepanova and coworkers to develop a transgenic alfalfa plant (*Medicago sativa* L.) by introducing RL biosynthesis enzyme coding gene (rhlA gene). Transformation was accomplished using *Agrobacterium tumefaciens* strain LBA4404 having a binary pBI121 vector. *In vitro* investigations about oil removal efficacies revealed that the transgenic alfalfa plants showed higher crude oil removal (71%) than that of non-transgenic alfalfa plants (50%). It has been inferred in various such investigations that transgenic plants with rhamnolipid genes improve and fasten soil reclamation by eliminating the pollutant.

4. Market for Rhamnolipids

To establish any commercial product in the market, the recommended strategies include excellent featured product or technology, convincing consumers about the significance and advantages of the product, reducing production cost, increasing profit margins, and ease and safety of application. RLs are very popular biomolecules amongst the global consumers because of their inimitable features, biological activities and applications. The environmental sustainability and safety add on to the reasons why the consumers would be attracted towards RL biosurfactants compared to synthetic surfactants. There are around 50 companies which are involved in commercial production and supply of biosurfactants, out of which around 10–12 companies are involved in the production of RL biosurfactants. For instance, EC 601 is a commercially produced RL by Ecochem Ltd., Canada. Currently, the market cost of RL (R-90%) is 98.80 USD/10 mg (Sigma-aldrich 2021).

Although ample research has been carried out in the area of production and applications of RLs; commercial large-scale production is still awaited. The high production cost is the foremost reason for the high selling cost of RL which limits the replacement of chemical surfactants. The microbial strain improvement for safer and better yields of RLs involves a lot of labor and investments, hence ultimately increasing the cost (Nawrath et al. 2020). It has been understood that downstream processing alone costs 70–80% of the total production costs (Randhawa et al. 2014).

Overview of literature suggested that bioaugmentation is one of the most promising approaches for the remediation of petroleum hydrocarbon contaminated soils. There are several successful *in vitro* trials on soil hydrocarbon removal using RL producing bacteria but actual contaminated field application studies are rare. This might be due to the following reasons:

The first challenge is about probable pathogenic nature of producer microbes and a limited number of well-known RL producing resources (Irorere et al. 2017). Researchers are now exploring novel non-pathogenic RL producers. Recombinant non-pathogenic microorganisms have been proved to be a good option for RL production but the technology is time consuming and demands monetary investments. Second, it has been observed that, the process of bioremediation by augmenting RL producing hydrocarbon degraders in aged and highly contaminated soil is generally hindered or less effective at field level. The reason might be, strong sorption capacity of organic and metal pollutants which makes RL concentrations insufficient to achieve desorption (Mohanty et al. 2013). Third, the hydrocarbon contaminated sites are generally co-contaminated with pollutants like heavy metals, pesticides (Olaniran et al. 2009 and Wan et al. 2015), which might act as inhibitors to the augmented microbes metabolism and expected microbial activities. This warrants further investigation. Fourth, there is a need for bio-stimulation of the augmented microorganisms (Tahseen et al. 2016). Though hydrocarbon contaminated sites are rich in organic content, periodic supply of inorganic nutrients is a must to retain good metabolic activity of the microbes (Adams et al. 2015). This might be contributing to the expenses of the remediation process. Fourth, it's an entrenched fact that microbial surfactants like RLs also have the potential to restore the soil quality (Tahseen et al. 2016). However, if high concentrations of RLs are produced by augmented microbes at the application site, they may cause adverse effects on the process. One of the experiments by Millioli and coworkers (2009), suggested that an increase in the RL concentration to a certain limit reflected an increase in the degradation activity as well as the seed germination rate of the *Lactuca sativa* (lettuce) species. However, a further increase in concentration showed reduction in the TPH degradation and seed germination rates. It requires further investigation for the interactions between biosurfactants, hydrocarbon pollutants (Kaczorek et al. 2018), soil constituents and soil microorganisms (Millioli et al. 2009). Fifth, the direct application of RLs at the hydrocarbon polluted sites is an ultimate but costly strategy. The cost of RL production is very high because of the specific, purified and costly raw material as well as critical downstream procedures involved in fermentations (Moldes et al. 2011). The high cost issue can be addressed by employing easily available low-cost substrates for the production of RLs (Conceição et al. 2020 and Tan et al. 2018). Strategies like solid state fermentation can also be advantageous to cut down the efforts and cost involved in downstream processing of the product (Cerda et al. 2019).

5. Concluding Remarks

Microbial rhamnolipids are a great alternative to synthetic surfactants for bioremediation purposes. *P. aeruginosa* appeared to be one of the most frequently investigated bacteria for the production of rhamnolipids. The lab scale experiments suggested that rhamnolipid assisted bioremediation using methods like biodegradation; soil washing and phytoremediation are truly promising methods to reduce environmental hydrocarbon pollution. However, sparse literature discusses combinations of rhamnolipids with various other biosurfactants and successful attempts for field trials. Impact of surfactants on the environment also have to be investigated and evaluated critically. In future, researchers need to focus on challenges such as low cost production processes, non-pathogenic and engineered microbial resources of rhamnolipid production. Additional efforts are needed to be carried out for *in situ* rhamnolipid assisted bioremediation field experiments on hydrocarbon contaminated and co-contaminated sites. Marketing for generating global consumers also has to be motivated. It is the need of time now to create knowledge and awareness about significant usage of biosurfactants over synthetic surfactants for environmental remediation in society.

References

Abdel-Mawgoud, A.M., F. Lépine and E. Déziel. 2010. Rhamnolipids: Diversity of structures, microbial origins and roles. Applied Microbiology and Biotechnology 86(5): 1323–1336. https://doi.org/10.1007/s00253-010-2498-2.

Adams, G.O., P.T. Fufeyin, S.E. Okoro and I. Ehinomen. 2015. Bioremediation, biostimulation and bioaugmention: A review. Int. J. Environ. Bioremed. Biodegr. 3(1): 28–39. https://doi.org/10.12691/ijebb-3-1-5.

Agnello, C.A. 2015. Potential of alfalfa for use in chemically and biologically assisted phytoremediation of soil co-contaminated with petroleum hydrocarbons and metals. Earth Sciences. Université Paris-Est, 2014. English. ffNNT: 2014PEST1087ff. fftel-01131249f.

Ángeles, M.T. and R.V. Refugio. 2013. *In situ* biosurfactant production and hydrocarbon removal by *Pseudomonas putida* CB-100 in bioaugmented and biostimulated oil-contaminated soil. Brazilian

Journal of Microbiology: [Publication of the Brazilian Society for Microbiology] 44(2): 595–605. https://doi.org/10.1590/S1517-83822013000200040.

Armendáriz, B.P., C. Cal, E. Girgis, E. Kassis, L. Daniel et al. 2019. Use of waste canola oil as a low cost substrate for rhamnolipid production using *Pseudomonas aeruginosa*. AMB Express. https://doi.org/10.1186/s13568-019-0784-7.

Arora, A., S.S. Cameotra and C. Balomajumder. 2021. Rhamonolipids produced by *Pseudomonas aeruginosa* promotes methane hydrates formation in fixed bed silica gel medium. Geophys. Res. 42: 5. https://doi.org/10.1007/s11001-020-09426-6.

Atuanya, E.I., D. Okafor and U. Udochukwu. 2016. Production of biosurfactants by Actinomycetes isolated from hydrocarbon contaminated soils and Ikpoba River sediments in Benin-City, Nigeria, Nigerian. Journal of Basic and Applied Sciences 24(2): 45–52. Doi: 10.4314/njbas.v24i2.7.

Badowska, E. and D. Bandzierz. 2019. The analysis of petroleum hydrocarbons in soils deriving from areas of various development. E3S Web of Conferences. 100. 00002. EDP Sciences. https://doi.org/10.1051/e3sconf/201910000002.

Bahia, F.M., G.C. de Almeida, L.P. de, Andrade, C.G. Campos, L.R. Queiroz et al. 2018. Rhamnolipids production from sucrose by engineered *Saccharomyces cerevisiae*. Scientific Reports 8(1): 1–10.

Baskaran, S.M., M.R. Zakaria and A.S. Mukhlis Ahmad Sabri. 2021. Valorization of biodiesel side stream waste glycerol for rhamnolipids production by *Pseudomonas aeruginosa* RS6. Environmental Pollution (Barking, Essex: 1987) 276: 116742. Doi: 10.1016/j.envpol.2021.116742.

Bezza, F. and E. Chirwa. 2014. Optimization strategy of polycyclic aromatic hydrocarbon contaminated media bioremediation through biosurfactant addition. Chemical Engineering Transactions 39: 1597–1602. https://doi.org/10.3303/CET1439267.

Cabrera-Valladares, N., A.P. Richardson, C. Olvera, L.G. Treviño, E. Déziel, F. Lépine and G. Soberón-Chávez. 2006. Mono-rhamnolipids and 3-(3-hydroxyalkanoyloxy) alkanoic acids (HAAs) production using *Escherichia coli* as a heterologous host. Applied Microbiology and Biotechnology 73(1): 187–194. https://doi.org/10.1007/s00253-006-0468-5.

Carrazco-Palafox, J., B.E. Rivera-Chavira, J.R. Adame Gallegos, L.M. Rodríguez-Valdez, E. Orrantia-Borunda et al. 2021. Rhamnolipids from *Pseudomonas aeruginosa* Rn19a modifies the biofilm formation over a borosilicate surface by clinical isolates. Coatings 11: 136. https://doi.org/10.3390/coatings11020136.

Celik, G.Y., B. Aslim and Y. Beyatli. 2008. Enhanced crude oil biodegradation and rhamnolipid production by *Pseudomonas stutzeri* strain G11 in the presence of Tween-80 and Triton X-100. Journal of Environmental Biology 29(6): 867–870. PMID: 19297982.

Cerda, A., A. Artola, R. Barrena, X. Font, T. Gea et al. 2019. Innovative production of bioproducts from organic waste through solid-state fermentation. Frontiers in Sustainable Food Systems (3). https://doi.org/10.3389/fsufs.2019.00063.

Cha, M., N. Lee, M. Kim, M. Kim and S. Lee. 2008. Heterologous production of *Pseudomonas aeruginosa* EMS1 biosurfactant in *Pseudomonas putida*. Bioresource Technology 99(7): 2192–2199. https://doi.org/10.1016/j.biortech.2007.05.035.

Cheng, T., J. Liang, J. He, X. Hu, Z. Ge et al. 2017. A novel rhamnolipid producing *Pseudomonas aeruginosa* ZS1 isolate derived from petroleum sludge suitable for bioremediation. AMB Express. https://doi.org/10.1186/s13568-017-0418-x.

Chong, H. and Q. Li. 2017. Microbial production of rhamnolipids: Opportunities, challenges and strategies. Microbial Cell Factories, 1–12. https://doi.org/10.1186/s12934-017-0753-2.

Christova, N., B. Tuleva, Z. Lalchev, A. Jordanova, B. Jordanov et al. 2004. Rhamnolipid biosurfactants produced by *Renibacterium salmoninarum* 27BN during growth on n-hexadecane. Zeitschrift fur Naturforschung. C, Journal of Biosciences 59(1-2): 70–74. https://doi.org/10.1515/znc-2004-1-215.

Conceição, K.S., M. de Alencar Almeida, I.C. Sawoniuk, G.D. Marques, P.C. de Sousa Faria-Tischer, C.A. Tischer and D. Camilios-Neto. 2020. Rhamnolipid production by *Pseudomonas aeruginosa* grown on membranes of bacterial cellulose supplemented with corn bran water extract. Environ. Sci. Pollut. Res. 27: 30222–30231. doi: 10.1007/s11356-020-09315-w.

da Rosa, C.F., D.M. Freire and H.C. Ferraz. 2015. Biosurfactant microfoam: Application in the removal of pollutants from soil. Journal of Environmental Chemical Engineering 3(1): 89–94.

Dong, H., W. Xia, H. Dong, Y. She, P. Zhu et al. 2016. Rhamnolipids produced by indigenous *Acinetobacter junii* from Petroleum Reservoir and its Potential in Enhanced Oil Recovery 7: 1–13. https://doi.org/10.3389/fmicb.2016.01710.

Emoyan, O.O., E.O. Onocha and G.O. Tesi. 2020. Concentration assessment and source evaluation of 16 priority polycyclic aromatic hydrocarbons in soils from selected vehicle-parks in southern Nigeria. Scientific African 7: e00296, ISSN 2468–2276. https://doi.org/10.1016/j.sciaf.2020.e00296.

Environmental Guidelines and Standard For the Petroleum Industry in Nigeria (Revised Edition) Department of Petroleum Resources (DPR) Department of Petroleum Resources. Ministry of Petroleum and Mineral Resources, Abuja Nigeria, 2002.

Eslami, P., H. Hajfarajollah and S. Bazsefidpar. 2020. Recent advancements in the production of rhamnolipid biosurfactants by *Pseudomonas aeruginosa.* RSC Adv. 10: 34014–34032. Doi: 10.1039/D0RA04953K.

Ezekwe, I.C., E.O. Oshionya and L.D. Demua. 2018. Ecological and potential health effects of hydrocarbon and heavy metal concentrations in the KoloCreek wetlands, South-South, Nigeria. International Journal of Environmental Sciences & Natural Resources, Juniper Publishers Inc. 11(1): 01–15. Doi: 10.19080/IJESNR.2018.11.555801.

Fanaei, F., G. Moussavi and S. Shekoohiyan. 2020. Enhanced treatment of the oil-contaminated soil using biosurfactant-assisted washing operation combined with H_2O_2-stimulated biotreatment of the effluent Enhanced treatment of the oil-contaminated soil using biosurfactant-assisted washing operation combined with H_2O_2-stimulated biotreatment of the effluent. Journal of Environmental Management 271: 110941. https://doi.org/10.1016/j.jenvman.2020.110941.

Funston, S.J., K. Tsaousi, M. Rudden, T.J. Smyth, P.S. Stevenson et al. 2016. Characterising rhamnolipid production in *Burkholderia thailandensis* E264, a non-pathogenic producer. Appl. Microbiol. Biotechnol. 100: 7945–7956.

Geissler, M., K.M. Heravi, M. Henkel and R. Hausmann. 2019. Lipopeptide biosurfactants from *Bacillus* species. Biobased Surfactants, 205–240. Doi:10.1016/b978-0-12-812705-6.00006-x.

Hallmann, E. and K. Mędrzycka. 2015. Wetting properties of biosurfactant (rhamnolipid) with synthetic surfactants mixtures in the context of soil remediation. Annales Universitatis Mariae Curie-skłodowska, Sectio Aa, Chemia 70(1): 29–39. https://doi.org/10.1515/umcschem-2015-0003.

Hošková, M., O. Schreiberová, R. Ježdík et al. 2013. Characterization of rhamnolipids produced by non-pathogenic *Acinetobacter* and *Enterobacter* bacteria. Bioresource Technology 130: 510–516. Doi: 10.1016/j.biortech.2012.12.085.

Hošková, M., R. Ježdík, O. Schreiberová, J. Chudoba, M. Šír et al. 2015. Structural and physiochemical characterization of rhamnolipids produced by *Acinetobacter calcoaceticus*, *Enterobacter asburiae* and *Pseudomonas aeruginosa* in single strain and mixed cultures. J. Biotechnol. 193: 45–51.

https://www.sigmaaldrich.com/IN/en/product/sigma/r90?gclid=Cj0KCQjw-NaJBhDsARIsAAja6dOou-Egtdxc0eKHTP-XwRWaSFsTDDIa90B8oZmNrfSmcebkWuyZdSEaAsbhEALw_wcB.

Ince, M. and O.K. Ince (eds.). 2019. Hydrocarbon Pollution and its Effect on the Environment. BoD–Books on Demand.

Irorere, V.U., L. Tripathi, R. Marchant, S. McClean, I.M. Banat et al. 2017. Microbial rhamnolipid production: A critical re-evaluation of published data and suggested future publication criteria. Applied Microbiology and Biotechnology 101(10): 3941–3951. https://doi.org/10.1007/s00253-017-8262-0.

Jain, P.K., V.K. Gupta, R.K. Gaur, M. Lowry, D.P. Jaroli et al. 2011. Bioremediation of petroleum oil contaminated soil and water. Res. J. Environ. Toxicol. 5: 1–26. Doi: 10.3923/rjet.2011.1.26.

Jarvis, F.G. and M.J. Johnson. 1949. A glyco-lipid produced by *Pseudomonas aeruginosa*. J. Am. Chem. Soc. 71: 4124–4126. https://doi.org/10.1021/ja01180a073.

Joe, M.M., R. Gomathi, A. Benson, D. Shalini, P. Rengasamy et al. 2019. Simultaneous application of biosurfactant and bioaugmentation with rhamnolipid-producing *Shewanella* for enhanced bioremediation of oil-polluted soil. Applied Sciences 9(18): 3773. https://doi.org/10.3390/app9183773.

Kaczorek, E., T. Jesionowski, A. Giec and A. Olszanowski. 2012. Cell surface properties of *Pseudomonas stutzeri* in the process of diesel oil biodegradation. Biotechnology Letters 34(5): 857–862. https://doi.org/10.1007/s10529-011-0835-x.

Kaczorek, E., A. Pacholak, A. Zdarta and W. Smułek. 2018. The impact of biosurfactants on microbial cell properties leading to hydrocarbon bioavailability increase. Colloids Interfaces 2: 35. Doi: 10.3390/COLLOIDS2030035.

Kapsalis, K., M. Kavvalou and I. Damikouka. 2021. Investigation of petroleum hydrocarbon pollution along the coastline of South Attica, Greece, after the sinking of the AgiaZoni II oil tanker. SN Appl. Sci. 3: 48. https://doi.org/10.1007/s42452-020-04114-x.

Karlapudi, A.P., T.C. Venkateswarulu, J. Tammineedi, L. Kanumuri, B.K. Ravuru et al. 2018. Role of biosurfactants in bioremediation of oil pollution—A review. Petroleum 4(3): 241–249. https://doi.org/10.1016/j.petlm. 2018.03.007.

Kłosowska-Chomiczewska, I.E., A. Kotewicz-Siudowska, W. Artichowicz, A. Macierzanka, A. Głowacz-Różyńska et al. 2021. Towards rational biosurfactant design—Predicting solubilization in rhamnolipid solutions. Molecules 26: 534. https://doi.org/ 10.3390/molecules26030534.

Kryachko, Y., S. Nathoo, P. Lai, J. Voordouw, E.J. Prenner et al. 2013. Prospects for using native and recombinant rhamnolipid producers for microbially enhanced oil recovery. Int. Biodeterior. Biodegrad. 81: 133– 140. https://doi.org/10.1016/j.ibiod.2012.09.012.

Kvenvolden, K.A. and C.K. Cooper. 2003. Natural seepage of crude oil into the marine environment. Geo-Mar. Lett. 23: 140–146. https://doi.org/10.1007/s00367-003-0135-0.

Lai, C.C., Y.C. Huang, Y.H. Wei, J.S. Chang et al. 2009. Biosurfactant-enhanced removal of total petroleum hydrocarbons from contaminated soil. Journal of Hazardous Materials 167(1-3): 609–614. https://doi.org/10.1016/j.jhazmat.2009.01.017.

Leite, G.G., J.V. Figueirôa, T.C. Almeida, J.L. Valões, W.F. Marques et al. 2016. Production of rhamnolipids and diesel oil degradation by bacteria isolated from soil contaminated by petroleum. Biotechnology Progress 32(2): 262–270. https://doi.org/10.1002/btpr.2208.

Li, Q., J. Liu and G.M. Gadd. 2020. Fungal bioremediation of soil co-contaminated with petroleum hydrocarbons and toxic metals. Applied Microbiology and Biotechnology 104(21): 8999–9008. https://doi.org/10.1007/s00253-020-10854-y.

Liao, C., W. Xu, G. Lu, F. Deng, X. Liang et al. 2016. Biosurfactant-enhanced phytoremediation of soils contaminated by crude oil using maize (*Zea mays* L.). Ecological Engineering 92: 10–17. https://doi.org/10.1016/j.ecoleng.2016.03.041.

Liduino, V.S., E. Servulo and F. Oliveira. 2018. Biosurfactant-assisted phytoremediation of multi-contaminated industrial soil using sunflower (*Helianthus annuus* L.). Journal of Environmental Science and Health. Part A, Toxic/hazardous Substances & Environmental Engineering 53(7): 609–616. https://doi.org/10.1080/10934529.2018.1429726.

Liu, G., H. Zhong, X. Yang, Y. Liu, B. Shao and Z. Liu. 2018. Advances in applications of rhamnolipids biosurfactant in environmental remediation: A review. Biotechnology and Bioengineering 115(4): 796–814.

Liu, Y., G. Zeng, H. Zhong, Z. Wang, Z. Liu et al. 2017. Effect of rhamnolipid solubilization on hexadecane bioavailability: Enhancement or reduction? Journal of Hazardous Materials 322(Pt B): 394–401. https://doi.org/10.1016/j.jhazmat.2016.10.025.

Makkar, R.S. and K.J. Rockne. 2003. Comparison of synthetic surfactants and biosurfactants in enhancing biodegradation of polycyclic aromatic hydrocarbons. Environmental Toxicology and Chemistry 22(10): 2280–2292. https://doi.org/10.1897/02-472.

Martino, C., N. López and L. Raiger Iustman. 2012. Isolation and characterization of benzene, toluene and xylene degrading *Pseudomonas* sp [SM1] selected as candidates for bioremediation. International Biodeterioration & Biodegradation [SM2] 67: 15–20. https://doi.org/10.1016/j.ibiod.2011.11.004 https://doi.org/10.1002/bit.26517.

Maslin, P. and R.M. Maier. 2000. Rhamnolipid-enhanced mineralization of phenanthrene in organic-metal co-contaminated soils. Bioremediation Journal 4(4): 295–308. https://doi.org/10.1080/10889860091114266.

Matvyeyeva, O.L., O.A. Vasylchenko and O.R. Aliieva. 2014. Microbial biosurfactants role in oil products. Biodegradation 2(2): 69–74. https://doi.org/10.12691/ijebb-2-2-4.

Millioli, V., E. Servulo and L. Sobral. 2009. Bioremediation of crude oil-bearing soil: Evaluating the effect of rhamnolipid addition to soil toxicity and to crude oil biodegradation efficiency. Global NEST Journal 11: 181–188.

Mohanty, S., J. Jasmine and S. Mukherji. 2013. Practical considerations and challenges involved in surfactant enhanced bioremediation of oil. BioMed Research International 2013: 328608. https://doi.org/10.1155/2013/328608.

Moldes, A.B., R. Paradelo, D. Rubinos, R. Devesa-Rey, J.M. Cruz et al. 2011. *Ex situ* treatment of hydrocarbon-contaminated soil using biosurfactants from *Lactobacillus pentosus*. Journal of Agricultural and Food Chemistry 59(17): 9443–9447. https://doi.org/10.1021/jf201807r.

Müller, M.M., J.H. Kügler, M. Henkel, M. Gerlitzki, B. Hörmann et al. 2012. Rhamnolipids—next generation surfactants? Journal of Biotechnology 162(4): 366–380. https://doi.org/10.1016/j.jbiotec.2012.05.022.

Musale, V. and S.B. Thakar. 2015. Biosurfactant and Hydrocarbon Degradation. doi: 10.26479/2015.0101.01.

Nalini, S. and R. Parthasarathi. 2014. Bioresource technology production and characterization of rhamnolipids produced by *Serratia rubidaea* SNAU02 under solid-state fermentation and its application as biocontrol agent. Bioresource Technology 173: 231–238. https://doi.org/10.1016/j.biortech.2014.09.051.

Nawrath, M.M., C. Ottenheim, J.C. Wu and W. Zimmermann. 2020. *Pantoea* sp. P37 as a novel nonpathogenic host for the heterologous production of rhamnolipids. Microbiology Open 9(5): e1019. https://doi.org/10.1002/mbo3.1019.

Nitschke, M., C. Ferraz and G.M. Pastore. 2004. Selection of microorganisms for biosurfactant production using agroindustrial wastes. Brazilian Journal of Microbiology 35(1-2): 81–85. https://doi.org/10.1590/S1517-83822004000100013.

Olaniran, A.O., A. Balgobind and B. Pillay. 2009. Impacts of heavy metals on 1, 2-dichloroethane biodegradation in co-contaminated soil. Journal of Environmental Sciences 21(5): 661–666. https://doi.org/10.1016/S1001-0742(08)62322-0.

Olasanmi, I.O. and R.W. Thring. 2020. Evaluating rhamnolipid-enhanced washing as a first step in remediation of drill cuttings and petroleum-contaminated soils. Journal of Advanced Research 21: 79–90. https://doi.org/10.1016/j.jare.2019.07.003.

Olawale, O., K.S. Obayomi, S.O. Dahunsi and O. Folarin. 2020. Bioremediation of artificially contaminated soil with petroleum using animal waste: cow and poultry dung bioremediation of artificially contaminated soil with petroleum using animal waste: cow and poultry dung. Cogent Engineering 7(1). https://doi.org/10.1080/23311916.2020.1721409.

Onwosi, C.O. and F.J. Odibo. 2012. Effects of carbon and nitrogen sources on rhamnolipid biosurfactant production by *Pseudomonas nitroreducens* isolated from soil. World Journal of Microbiology & Biotechnology 28(3): 937–942. https://doi.org/10.1007/s11274-011-0891-3.

Ossai, C., A. Ahmed, A. Hassan and F. Shahul. 2020. Environmental technology & innovation remediation of soil and water contaminated with petroleum hydrocarbon: A review. Environmental Technology & Innovation 17: 100526. https://doi.org/10.1016/j.eti.2019.100526.

Pacwa-Płociniczak, M., G.A. Płaza, Z. Piotrowska-Seget and S.S. Cameotra. 2011. Environmental applications of biosurfactants: Recent advances. International Journal of Molecular Sciences 12(1): 633–654. https://doi.org/10.3390/ijms12010633.

Patowary, K., R. Patowary, M.C. Kalita and S. Deka. 2017. Characterization of biosurfactant produced during degradation of hydrocarbons using crude oil as sole source of carbon. Frontiers in Microbiology 8: 279. https://doi.org/10.3389/fmicb.2017.00279.

Pourfadakari, S., S. Jorfi and S. Ghafari. 2020. An efficient biosurfactant by *Pseudomonas stutzeri* Z12 isolated from an extreme environment for remediation of soil contaminated with hydrocarbons. Chemical and Biochemical Engineering Quarterly 34(1): 35–48. https://doi.org/10.15255/CABEQ.2019.1718.

Rahman, P.K.S.M. (ed.). 2016. Microbiotechnology based surfactants and their applications. Lausanne: Frontiers Media. Doi: 10.3389/978-2-88919-752-1.

Ram, G., M. Melvin Joe, S. Devraj and A. Benson. 2019. Rhamnolipid production using *Shewanella seohaensis* BS18 and evaluation of its efficiency along with phytoremediation and bioaugmentation for bioremediation of hydrocarbon contaminated soils. International Journal of Phytoremediation 21(13): 1375–1383. https://doi.org/10.1080/15226514.2019.1633254.

Randhawa, S.K.K. and P.K. Rahman. 2014. Rhamnolipid biosurfactants-past, present, and future scenario of global market. Frontiers in Microbiology 5: 454. https://doi.org/10.3389/fmicb.2014.00454.

Rocha, C.A., A.M. Pedregosa and F. Laborda. 2011. Biosurfactant-mediated biodegradation of straight and methyl-branched alkanes by *Pseudomonas aeruginosa* ATCC 55925. AMB Express 1(1): 9. https://doi.org/10.1186/2191-0855-1-9.

Sandermann, H. Jr 1992. Plant metabolism of xenobiotics. Trends in Biochemical Sciences 17(2): 82–84. https://doi.org/10.1016/0968-0004(92)90507-6.

Sharma, D., M.J. Ansari, A. Al-Ghamdi, N. Adgaba, K.A. Khan et al. 2015. Biosurfactant production by *Pseudomonas aeruginosa* DSVP20 isolated from petroleum hydrocarbon-contaminated soil and its physicochemical characterization. Environmental Science and Pollution Research International 22(22): 17636–17643. https://doi.org/10.1007/s11356-015-4937-1.

Shekhar, S., A. Sundaramanickam and T. Balasubramanian. 2015. Biosurfactant producing microbes and their potential applications: A review. Critical Reviews in Environmental Science and Technology 45: 14, 1522–1554. Doi: 10.1080/10643389.2014.955631.

Singh, P.B., S. Sharma, H.S. Saini and B.S. Chadha. 2009. Biosurfactant production by *Pseudomonas* sp. and its role in aqueous phase partitioning and biodegradation of chlorpyrifos. Lett. Appl. Microbiol. Sep 49(3): 378–83. Doi: 10.1111/j.1472-765X.2009.02672.x.Epub 2009 Jul 10. PMID: 19627480.

Sinumvayo, J.P. and N. Ishimwe. 2015. Chemical engineering & process technology agriculture and food applications of rhamnolipids and its production by *Pseudomonas aeruginosa.* Journal of Chemical Engineering & Process Technology 6(2): 2–9. https://doi.org/10.4172/2157-7048.1000223.

Srivastava, M., A. Srivastava, A. Yadav and V. Rawat. 2019. Source and control of hydrocarbon pollution, hydrocarbon pollution and its effect on the environment. Muharrem Ince and Olcay Kaplan Ince, Intech Open, Doi: 10.5772/intechopen.86487.

Stepanova, A.Y., E.V. Orlova and D.V. Teteshonok. 2016. Obtaining transgenic alfalfa plants for improved phytoremediation of petroleum-contaminated soils. Russ. J. Genet. Appl. Res. 6: 705–711. https://doi.org/10.1134/S2079059716060083.

Suman, S., A. Sinha and A. Tarafdar. 2016. Polycyclic aromatic hydrocarbons (PAHs) concentration levels, pattern, source identification and soil toxicity assessment in urban traffic soil of Dhanbad, India. The Science of the Total Environment 545-546: 353–360. https://doi.org/10.1016/j.scitotenv.2015.12.061.

Tahseen, R., M. Afzal, S. Iqbal, G. Shabir, Q. Khan et al. 2016. Rhamnolipids and nutrients boost remediation of crude oil contaminated soil by enhancing bacterial colonization and metabolic activities. International Biodeterioration & Biodegradation (115): 192–198. https://doi.org/10.1016/j.ibiod.2016.08.010.

Tan, Y.N. and Q. Li. 2018. Microbial production of rhamnolipids using sugars as carbon sources. Microbial Cell Factories 17(1): 89. https://doi.org/10.1186/s12934-018-0938-3.

Tavares, L.F., P.M. Silva, M. Junqueira, D.C. Mariano, F.C. Nogueira et al. 2013. Characterization of rhamnolipids produced by wild-type and engineered *Burkholderia kururiensis.* Applied Microbiology and Biotechnology 97(5): 1909–1921. https://doi.org/10.1007/s00253-012-4454-9.

Thakur, P., N.K. Saini and V.K. Thakur. 2021. Rhamnolipid the glycolipid biosurfactant: Emerging trends and promising strategies in the field of biotechnology and biomedicine. Microb. Cell Fact. 20: 1. https://doi.org/10.1186/s12934-020-01497-9.

Tiso, T., A. Germer, B. Küpper, R. Wichmann, L.M. Blank et al. 2015. Methods for recombinant rhamnolipid production. *In*: McGenity, T., K. Timmis and B. Nogales Fernández (eds.). Hydrocarbon and Lipid Microbiology Protocols. Springer Protocols Handbooks. Springer, Berlin, Heidelberg. https://doi.org/10.1007/8623_2015_60.

Toribio, J., A.E. Escalante and G. Sobero. 2010. Review article rhamnolipids: Production in bacteria other than *Pseudomonas aeruginosa.* European Journal of Lipid Science and Technology 112(10): 1082–1087. https://doi.org/10.1002/ejlt.200900256.

Torres, L.G., R. González and J. Gracida. 2018. Production and application of no-purified rhamnolipids in the soil-washing of TPHs contaminated soils. Asian Soil Research Journal 1(1): 1–12. https://doi.org/10.9734/asrj/2018/v1i1618.

Truskewycz, A., T.D. Gundry, L.S. Khudur, A. Kolobaric, M. Taha et al. 2019. Petroleum hydrocarbon contamination in terrestrial ecosystems-fate and microbial responses. Molecules (Basel, Switzerland) 24(18): 3400. https://doi.org/10.3390/molecules24183400.

Tuleva, B., N. Christova, B. Jordanov, B. Nikolova-Damyanova and P. Petrov et al. 2005. Naphthalene degradation and biosurfactant activity by *Bacillus cereus* 28BN. Zeitschrift fur Naturforschung C, Journal of Biosciences 60(7-8): 577–582. https://doi.org/10.1515/znc-2005-7-811.

Urum, K., S. Grigson, T. Pekdemir and S. Mcmenamy. 2006. A comparison of the efficiency of different surfactants for removal of crude oil from contaminated soils. Chemosphere 62(9): 1403–1410. https://doi.org/10.1016/j.chemosphere.2005.05.016.

Venkatachalapathy, R., S. Veerasingam and T. Ramkumar. 2010. Petroleum hydrocarbon concentrations in marine sediments along Chennai Coast, Bay of Bengal, India. Bulletin of Environmental Contamination and Toxicology 85(4): 397–401. https://doi.org/10.1007/s00128-010-0097-7.

Wan, J., D. Meng, T. Long, R. Ying, M. Ye et al. 2015. Simultaneous removal of lindane, lead and cadmium from soils by rhamnolipids combined with citric acid. PloS One 10(6): e0129978. https://doi.org/10.1371/journal.pone.0129978.

Wang, C.P., J. Li, Y. Jiang and Z. Zhang. 2014. Enhanced bioremediation of field agricultural soils contaminated with PAHs and OCPs. International Journal of Environmental Research 8: 1271–1278.

Wang, X., L. Sun, H. Wang, H. Wu, S. Chen et al. 2017. Surfactant-enhanced bioremediation of DDTs and PAHs in contaminated farmland soil. Environmental Technology. https://doi.org/10.1080/09593330.2017.1337235.

Wei, Z., J.J. Wang, L.A. Gaston, J. Li, L.M. Fultz et al. 2020. Remediation of crude oil-contaminated coastal marsh soil: Integrated effect of biochar, rhamnolipid biosurfactant and nitrogen application. Journal of Hazardous Materials 396: 122595. https://doi.org/10.1016/j.jhazmat.2020.122595.

Wu, L., D. Song and L. Yan. 2020. Simultaneous desorption of polycyclic aromatic hydrocarbons and heavy metals from contaminated soils by rhamnolipid biosurfactants. J. Ocean Univ. China 19: 874–882. https://doi.org/10.1007/s11802-020-4266-y.

Xu, X., W. Liu, S. Tian, W. Wang, Q. Qi et al. 2018. Petroleum hydrocarbon-degrading bacteria for the remediation of oil pollution under aerobic conditions: A perspective analysis. Frontiers in Microbiology 9: 2885. https://doi.org/10.3389/fmicb.2018.02885.

You, Z., H. Xu, S. Zhang, H. Kim, P. Chiang et al. 2018. Comparison of petroleum hydrocarbons degradation by *Klebsiella pneumoniae* and *Pseudomonas aeruginosa.* Appl. Sci. 8(12): 2551. https://doi.org/10.3390/app8122551.

Yuniati, M. 2018. Bioremediation of petroleum-contaminated soil: A review. IOP Conference Series: Earth and Environmental Science 118(012063). https://doi.org/10.1088/1755-1315/118/1/012063.

Zeng, Z., Y. Liu, H. Zhong, R. Xiao, G. Zeng et al. 2018. Mechanisms for rhamnolipids-mediated biodegradation of hydrophobic organic compounds. The Science of the Total Environment 634: 1–11. https://doi.org/10.1016/j.scitotenv.2018.03.349.

Zhang, D., L. Zhu and F. Li. 2013. Influences and mechanisms of surfactants on pyrene biodegradation based on interactions of surfactant with a *Klebsiella oxytoca* strain. Bioresource Technology 142: 454–461. https://doi.org/10.1016/j.biortech.2013.05.077.

Zhang, G.L., Y.T. Wu, X.P. Qian and Q. Meng. 2005. Biodegradation of crude oil by *Pseudomonas aeruginosa* in the presence of rhamnolipids. Journal of Zhejiang University Science B 6(8): 725–730. https://doi.org/10.1631/jzus.2005.B0725.

Zhang, W. 2015. Batch washing of saturated hydrocarbons and polycyclic aromatic hydrocarbons from crude oil contaminated soils using biosurfactant. Journal of Central South University 22(3): 895–903. https://doi.org/10.1007/s11771-015-2599-2.

Zulianello, L., C. Canard, T. Köhler, D. Caille, J.S. Lacroix et al. 2006. Rhamnolipids are virulence factors that promote early infiltration of primary human airway epithelia by *Pseudomonas aeruginosa.* Infection and Immunity 74(6): 3134–3147. https://doi.org/10.1128/IAI.01772-05.

13

Bioremediation of Petroleum Hydrocarbons (PHC) using Biosurfactants

Veeranna Channashettar,[1,2] *Shaili Srivastava*,[1]
Banwari Lal,[2] *Anoop Singh*[3] and *Dheeraj Rathore*[4,*]

1. Introduction

Energy from oil is considered as a major source among others and its demand has kept increasing every year. International Energy Agency (IEA 2017) estimated that around 97 million barrels of oil equivalent (MBOE) were consumed every day as on 2015 and 100 MBOE will be required per day up to 2021. India stands at the third position in the world for consumption of crude oil, after the United States and China, and has a share of about 4.81% of the total world oil consumption in the year 2016–17. India has limited oil resources; therefore, it depends on its imports. The crude oil imports to India rose from 121.7 MTOE (million tonnes of oil equivalent) in 2007–08 to 213.9 MTOE in 2016–17. Pollution due to combustion and spillage, during the transportation and storage, of petroleum oil damaged the environment which prevails as an ecological hazard.

Petroleum hydrocarbon (PHC) also known as crude oil is a flammable liquid fossil fuel, produce naturally and originating from geological formations below the earth's surface. It is formed by the decomposition of organic matter under intense heat and

[1] Amity School of Earth and Environmental Sciences, Amity University, Gurgaon, Haryana-122413, India.
[2] Environmental and Industrial Biotechnology Division, The Energy and Resources Institute (TERI), New Delhi-110003 India.
[3] Department of Scientific and Industrial Research (DSIR), Ministry of Science and Technology, Government of India, Technology Bhawan, New Mehrauli Road, New Delhi-110016 India.
[4] School of Environment and Sustainable Development, Central University of Gujarat, Gandhinagar-382030, Gujarat, India.
* Corresponding author: dheeraj.rathore@cug.ac.in

pressure in sedimentary rocks over millions of years. Petroleum consists of various types of chemical compounds and fractions. Petroleum is used as a fuel to power transport vehicles, farm machineries, heating units, industries and other machines as well for being converted into plastics and other materials. The PHC constitutes a complex mixture of hydrocarbons like aliphatic, alkanes and aromatic compounds, nitrogen, sulphur and oxygen containing compounds, and tar like asphaltene fractions (Bhattacharya et al. 2003). After release of oil into the environment its properties get changed by exposure to physical, chemical and biological processes, which alter its physical forms, accelerate or decelerate its degradation and destruction.

Various wastes like petroleum oily sludge from effluent treatment plants and storage tanks, drilling fluids and petroleum wastewater are generated during exploration, drilling and oil refining activities. Ling and Isa (2006) estimated that about 50 tons of oily sludge is produced every year from a petroleum refinery having a production capability of 105,000 drums per day. Oily sludge is a mixture of hydrocarbons (asphaltenes, paraffin, and more) and inorganic solids such as, sand, iron sulfides and iron oxides, wherein the hydrocarbon is the major component, which forms due to changes in external conditions (Johnson and Affam 2019). Generally, the formation of petroleum sludge is occurs by cooling below the cloud point, evaporation, mixing with inappropriate materials and emulsification with water. Environmental Protection Act 1986 and Resources Conservation and Recovery Act (RCRA) have regarded sludge as a hazardous waste. Asia et al. (2006) has reported that oily sludge contains different elements such as calcium, magnesium, nitrogen, Phosphorous, potassium, iron, copper, cadmium, chromium, zinc, sodium and lead. In addition, oil transportation by road, sea and pipelines are the main means of environmental (mainly land and water) pollution due to oil spills, transport accidents, storage tank rupture, ship breakage, leakage of oil pipelines, wars, earthquakes and floods.

The increasing demand for oil, prompting expansion of oil refineries, and increased offshore and deep water drilling are leading to further reasons for oil spills (Ferguson et al. 2020). Petroleum industries related to exploration, production and refining activities generate unavoidably large amounts of oily sludge and oil contaminated soil as waste. The BP oil spill occurred in April, 2010 at the Gulf of Mexico caused spillage of 12,000 to 100,000 barrels of oil per day, which is one of the largest oil spills in US history (Repanich 2010). There was a collision of two cargo ships in the Chennai shoreline in India and spilt 75 metric tons of heavy oil into the Bay of Bengal in January 2017 (Han et al. 2018). Dabbs (1999) reported that during the years1982 to 1992 in Nigerian operations of the Shell company was responsible for 1,626,000 gallons of oil discharge in 27 different oil spill incidents. During the Gulf War in 1991 massive areas of land and sea water got polluted by oil spills (Enzler 2006).

Oily wastes have been designated as hazardous wastes by the Indian Government, US EPA and OECD countries (Zhu et al. 2001, Ministry of Environment and Forests, Government of India 2000) and the management of such oily hazardous waste is a mandatory requirement to protect the environment (Arpia et al. 2021).

2. Sources of PHC Pollution

The major sources of petroleum hydrocarbon (PHC) pollution include drilling and transportation accidents, leakages from refineries, storage tanks and pipelines, natural calamities, cleaning of oil tanks, sludge generation through refining processes, and more. The major incidents of oil spills are summarized in Table 1, which represents the data of thousands of million gallons of oil spilt since 1967. Onshore and offshore exploration of natural oil reserves is causing contamination in soil and marine environments (Ron and Rosenberg 2014, Varjani 2017). Globally, every year about 1.7–8.8 million metric tonnes of PHCs are being spilt into the marine environment, and 90% of the spills occurred due to accidents caused by human errors (Zhu et al. 2001, Dadrasnia et al. 2013). In the year 2002, NRC (National Research Council of the U.S. National Academy of Science) reported that the main sources responsible for PHC pollution include 46% from natural seeps, 37% from land-based sources,

Table 1. Major oil spill incidents in the World (Hoffman and Jennings 2010, Han et al. 2018, Prasad et al. 2014).

Year	Spill Name	Location	Volume (M gallons)
1967	Torrey Canyon	UK	2536
1968	World Gory	South Africa	13.5
1970	Othello	Sweden	18–29
1972	Seastar	Oman	35.3
1976	Urquiola	Spain	29.4
1977	Hawaiian Patroit	Northern Pacific	29.1
1978	Amoco Cadiz	France	69
1979	Atlantic Impress	Trinidad and Tobago	90
1979	Burmah Agate	USA	10.7
1980	Ixtoc 1	Mexico	140
1980	Irenes Serenad	Greece	30
1983	Nowruz	Iran	80
1983	Castillo-de- bellver	South Africa	79
1988	Oddyssey	Canada	40.7
1989	Exxon Valdez	USA	10.8
1991	Arabian Gulf	Kuwait	380–520
1991	Fergana Valley	Uzbekistan	88
1991	ABT Summer	Angola	51–81
1991	M/T Haven	Italy	45
1993	Braer	UK	26
1994	Kolva River	Russia	84
1996	Seaimpress	Wales	19
2010	Deep Water Horizon	USA	205
2010	Chitra and MV Khalijia3	India	0.13–0.25
2017	Ennore	India	0.08

12% from accidental spills and 3% from oil extraction. According to Oil tanker spill statistics (2016) around 6 MT PHC has been released into the marine environment during the years 1970–2016 and leakage of 600,000 MT natural petroleum oil per annum was recorded (Kvenvolden and Cooper 2003). Although, ITOPF statistics (2020) suggested a declining trend of oil spills since 1970, the volume lost to the environment is still high. Spillages due to accidents are unavoidable and contribute to 10–15% of the total oil spills (Clark 1999).

3. Effects of PHC Pollution

Oil contamination affects plant and animal ecosystems along with human health (Mandal et al. 2007). Oil entry to the ecosystem is a matter of environmental concerns. Water and soil contamination from accidents occurs in different environmental compartments like groundwater (Liu et al. 2015), soils (Li et al. 2015), rocky shores (Kankara et al. 2016), sediments (Burnes and Jones 2016) and oceans (Broszeit et al. 2016), causing impacts mainly from a toxicological point of view. The relationship among oil, environment and the ecosystem is displayed in Figure 1.

The chemical composition, physical state, molecular mass, mode, level and time of exposure of PHC compounds determine their toxicity and bioavailability (Ossai et al. 2020). The pollutants are responsible for varied acute health problems like flu, nausea, headaches, skin and eye irritation and chronic health problems such as melanoma, cancers, leukemia, and mutagenicity to humans and animals (Lawal 2017, Gutzkow 2015, Ossai et al. 2020). The acute and chronic health effects of petroleum hydrocarbons are presented in Figure 2.

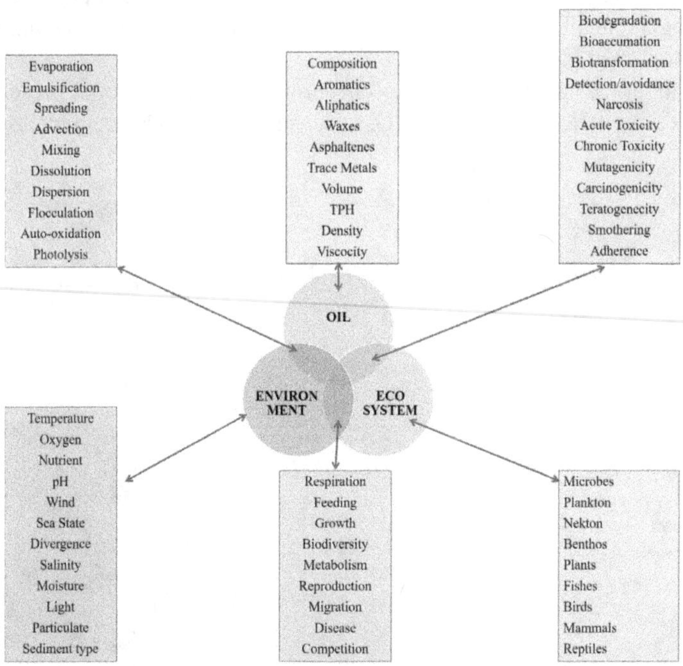

Figure 1. Oil-environment-ecosystem triad.

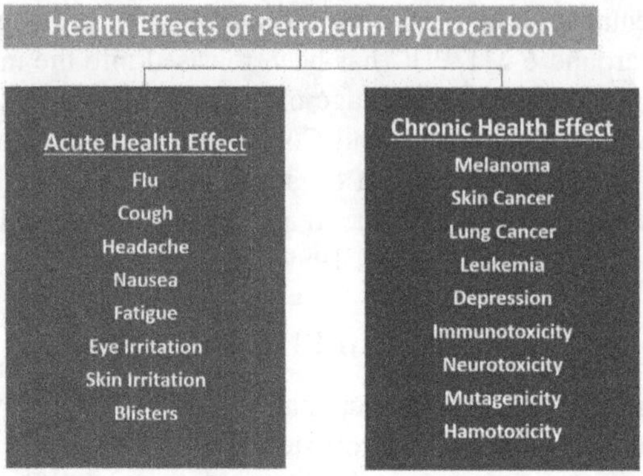

Figure 2. Health effects of petroleum hydrocarbon pollution.

4. Remediation of PHC Pollution in Soil

The petroleum hydrocarbons after entering the environment undergo a weathering process because of physical, chemical and biological changes occurring through interaction with microorganisms and metabolic pathways (Abdel-Shafy and Mansour 2016, Hassanshahian and Cappello 2013). Volatilization, adsorption to soil and dissolution in water are responsible for the weathering process (Esbaugh et al. 2016, Mishra and Kumar 2015). The aliphatic hydrocarbons are more volatile than aromatics hydrocarbons because of their molecular nature. Air pollution at the contaminated site is mainly caused by the loss of lower molecular weight aliphatic hydrocarbons through volatilization (Maletic et al. 2013, Peter 2011). Various physical, chemical and biological methods for remediation of petroleum hydrocarbon contamination in soil are presented in Figure 3.

The nature, composition and the concentration of hydrocarbons present in PHCs determines their degradation rate (Ossai et al. 2020). PHC degradation involves

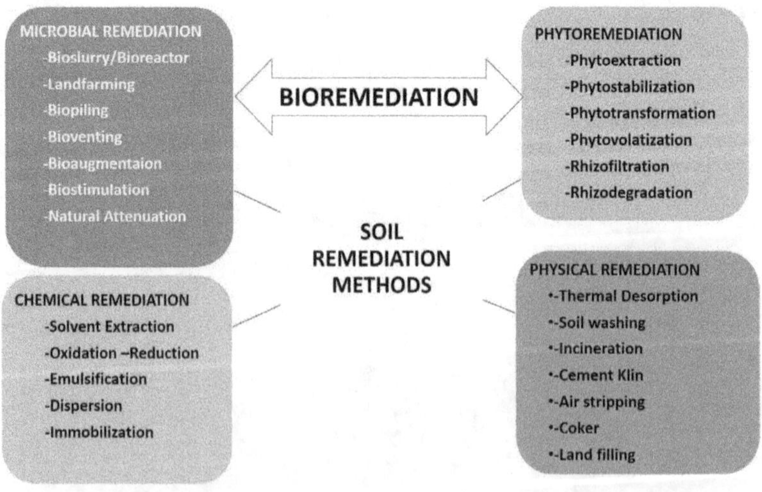

Figure 3. Remediation methods for petroleum hydrocarbon contaminated soils.

chemical transformation and mineralization by metabolic and enzymatic activities and convert PHCs into less harmful and non-hazardous substances (Maletic et al. 2013). Joutey et al. (2013) in a study concluded that the results of petroleum hydrocarbon degradation can be enhanced by combination of chemical transformation and biodegradation. The biodegradation mechanisms are enzyme-specific activities involving aerobic or anaerobic metabolisms with the production of biosurfactants and emulsifiers (Hu 2016, Varjani and Upasani, 2017).

5. Bio-surfactants

Biosurfactants are surface-active, amphiphilic cell ingredients produced by microorganisms like bacteria, fungi, actinomycetes and yeasts (Banat et al. 2010). Biosurfactants are produced intracellularly and/or extracellularly. The hydrophobic and hydrophilic characteristic of biosurfactants is responsible for reduction in surface and interfacial tensions (Cunha et al. 2004). Most biosurfactants are either anionic or neutral. Hydrophilic moieties are generally represented by carbohydrates, phosphates and amino acid groups while, hydrophobic moieties are represented by long carbon chain fatty acids (Singh et al. 2018). The amphiphilic molecules make them interfere at different interfaces between the fluids having unlike polarities like hydrocarbons and water. They possess the characteristic property of reducing the surface and interfacial tensions (Singh et al. 2007). Most of the biosurfactants are produced from hydrocarbons which are used as the main carbon source. They are secondary metabolites, which are created at the logarithmic or stationary growth phase of microbial growth (Dhiman et al. 2016). Microorganisms are capable of producing low molecular weight to high molecular weight biosurfactants. Such microbes include bacteria like *Pseudomonas, Acinetobacter, Bacillus, Clostridium* and fungi like *Penicillium, Ustilago, Aspergillus*, among others (Li et al. 2016, Shekhar et al. 2015). The biosurfactants are classified as glycolipids, phospholipids, lipopeptides, flavolipid, polysaccharide proteins, fatty acids, polymeric surfactants and lipids (de França et al. 2015, Martins and Martins 2018).

5.1 Properties of Biosurfactant

The properties of biosurfactants such as a reduction in surface tension, foaming capacity, resilience to pH, temperature and ionic quality, emulsifying and demulsifying capacity, specificity, low toxicity, solubility and detergency, are important in the evaluation of the performance and selection of microorganisms with the potential to produce the biosurfactant (Deleu and Paquot 2004).

5.1.1 Surface and Interface Activity

Surface tension is the most important property of tensioactive agents and is the attractive force between the molecules in liquids (Pacwa-Plociniczak et al. 2011). Biosurfactants help in decreasing surface tension and the interfacial pressure. The tension between the air/water and oil/water phases are respectively known as the surface tension and interfacial tension (Banat et al. 2010). A tensiometer is used for the quantitative measurement of the surface tension of liquids. Surface tension is the basis of most initial evaluations for the identification of the presence of a surfactant in the medium. The air/water surface tension for distilled water is about 72 mN/m (or dynes/cm). The solubility limit of surfactants is described by the critical micelle concentration (CMC) and CMC of biosurfactants ranges from 1 to 2000 mg/L, whereas interfacial (oil/water) and surface tensions are respectively approximately 1 and 25–30 mN/m. Biosurfactant, surfactin produced by *Bacillus subtilis* reduces the surface tension of water to 25 mN m^{-1} and interfacial strain for water/hexadecane to under 1 mN m^{-1} (Cavalero and Cooper 2003). *Pseudomonas aeruginosa* produces rhamnolipids which reduced the surface tension of water to 26 mN m^{-1} and interfacial strain of water/hexadecane to under 1 mN m^{-1} (Chakrabarti 2012).

5.1.2 *Tolerance to Temperature, pH and Ionic Strength*

Biosurfactants are capable of withstanding high temperatures, broad pH ranges and ionic concentrations. Biosurfactant, lichenysin from *Bacillus licheniformis* showed resistant to temperatures up to 50°C, pH in the vicinity of 4.5 and 9.0 and NaCl and Ca concentrations up to 50 and 25 g L^{-1} and biosurfactant produced by *Arthrobacter protophormiae* was observed to be both thermostable (30–100°C) and pH (2 to 12) stable (Roy 2017).

5.1.3 *Biodegradability*

Biosurfactants are easily degraded by bacteria and other microorganisms in water or soil compared to chemical surfactants, which makes them suitable for usage in bioremediation applications and waste water treatment (Desai and Banat 1997).

5.1.4 *Emulsifying and Demulsifying Capacity*

Biosurfactants are both emulsifiers and de-emulsifiers. An emulsion is a heterogeneous system consisting of an immiscible liquid dispersed in another liquid in the form of droplets generally measuring 0.1 mm in diameter. Two basic types of emulsion are, oil-in-water (o/w) and water-in-oil (w/o). Emulsions have minimal stability, but the addition of biosurfactants can lead to an emulsion that remains stable for a longer time (Velikonja and Kosaric 1993, Hu and Ju 2001).

5.1.5 *Specificity*

Biosurfactants as complex molecules having specific functional groups are very specific in their actions. Because of their specificity property biosurfactants are of considerable interest regarding the detoxification of specific pollutants as well as in particular applications in the oil, food, cosmetic and pharmaceutical industries (Sobrinho et al. 2014).

5.1.6 *Low Toxicity*

A low degree of toxicity allows the use of biosurfactants in oils, foods, cosmetics and pharmaceuticals. Low toxicity is also of fundamental importance to environmental applications (Singh and Rathore 2019). Biosurfactants can be produced from largely available raw materials as well as industrial wastes (Sobrinho et al. 2014).

Biosurfactants are mainly classified on the basis of chemical structure and their origin. The amphiphilic molecules of biosurfactants are constituted with hydrophilic moieties (e.g., peptides, acids, saccharides and others) and hydrophobic moieties (e.g., fatty acids, saturated hydrocarbon chains or unsaturated hydrocarbon chains) (Raza et al. 2006, Yan et al. 2012, Hoskova et al. 2013, Moya et al. 2015). The various biosurfactant producing microorganism are tabulated along with the type of biosurfactant produced by them in Table 2.

Table 2. Microorganism producing different type of biosurfactant.

Biosurfactant	Type of Surfactant	Producing Organism	Author
Rhamnolipids	Glycolipids	*Pseudomonas aeruginosa*	Mulligan et al. (2001), Abdel-Mawgoud et al. (2010), Chong and Li (2017)
Trehalos lipids		*Rhodococcus fascians* BD8	Janek et al. (2018)
Sophorolipids		Stenotrophomonas sp. BAB-6435	Singh et al. (2019)
		Torulopsis bombicola	Mulligan et al. (2001), Siñeriz et al. (2001)
Mannosylerythritol lipids		*Pseudomonas aeruginosa* mutant	Raza et al. (2006)
		Listeria monocytogenes	Liu et al. (2020)
Surfactin	Lipopeptides	*Brevisbacillus brevis* BAB-6437	Singh et al. (2019)
		Bacillus subtilis	Mulligan et al. (2001)
		Bacillus licheniformis	Kong et al. (2010)
		Bacillus sp. LCF1	Jasim et al. (2016)
Iturine		*Bacillus amyloliquefaciens* LL3	Dang et al. (2019)
		Bacillus sp. LCF1	Jasim et al. (2016)
Fengicyne		*Bacillus* sp. LCF1	Jasim et al. (2016)
Viscosin		*Pseudomonas fluorescens* SBW25	Alsohim et al. (2014)
Lichenysin		*Bacillus licheniformis* WX-02	Qiu et al. (2014)
Serrawettin		*Serratia marcescens, Serratia surfactantfaciens*	Clements et al. (2019)
Phospholipids		*Klebsiella pneumoniae* IVN51	Nwaguma et al. (2016)
Gramicidin	Surface Active antibiotic	*Bacillus brevis*	Winnick et al. (1961), Conti et al. (1997)
		Aneurinibacillus migulanus	Berditsch et al. (2017)
Polymixin		*Paenibacillus polymyxa* E681	Kim et al. (2015)
Antibiotic TA			
Fatty acids/Natural lipids		*Acinetobacter* sp. HO1-N	Celiešiutė et al. (2009)
		Nocardia erythropolis	Pirog et al. (2014)
Corynomicolic acid		*Corynebacterium lepus*	Mulligan et al. (2001)

5.2 *Biosurfactants in the Petroleum Industry*

In the petroleum industry, biosurfactants are used for microbial enhanced oil recovery (MEOR). Both strategies viz., *ex situ* (production of biosurfactant in offsite fermenters and injection into the oil reservoir) and *in situ* (injected allochthonous microorganisms; and injection of nutrients to stimulate growth of indigenous bacteria) are used for MEOR (Perfumo et al. 2010). MEOR includes use of microorganisms and the exploitation of their metabolic processes to increase production of oil from marginally producing reservoirs. The mechanism responsible for oil release is acidification of the solid phase. Certain microorganisms, such as *Bacillus subtilis*, *Pseudomonas aeruginosa* and *Torulopsis bombicola* have been reported to utilize crude oil and hydrocarbons as sole carbon sources and can be used for oil spill clean-ups (Das and Mukherjee 2007).

6. Biosurfactant as a Tool for Oil Pollution Remediation

Biosurfactants are considered as successful soil remediation agent for industrial effluents and pollutants (Singh and Rathore 2019, Singh et al. 2020). The SWOT analysis of biosurfactants is presented in Figure 4, also shows that biosurfactants are environment friendly and non-toxic biologically originating compounds having high efficiency for bioremediation of oil contaminated sites. Increasing oil contamination has created a big market for biosurfactants and versatile byproducts have increased the opportunity to degrade the various components of petroleum hydrocarbons. The production of biosurfactants is costly in comparison to chemical surfactants, which restrict their entry into the market at higher rates.

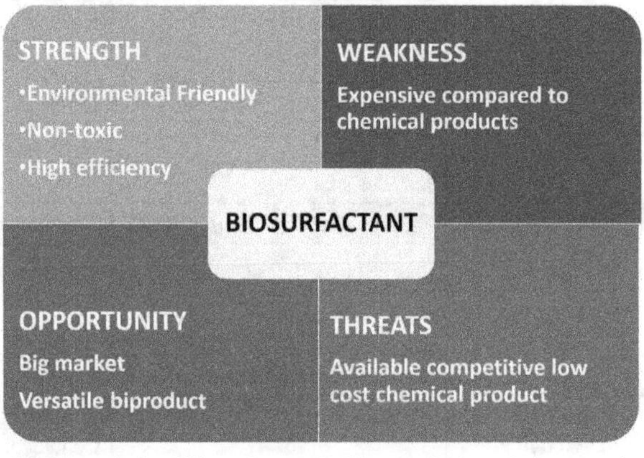

Figure 4. SWOT analysis of biosurfactant for bioremediation of oil contaminated sites.

6.1 *Bioremediation Mechanism of Biosurfactants*

After release of biosurfactants, micelles form a spherical structure in such a manner that the hydrophobic part is directed towards the center and the hydrophilic part is directed to the sphere surface, which makes an interface with water. This causes a reduction of surface tension between water and oil and results in micelle formation (Aparna et al. 2012) which facilitates hydrocarbon availability to bacteria favoring the biodegradation of PHC molecules (Soberón-Chávez and Maier 2010, Souza et al. 2014, Tondo et al. 2010). Critical micelle concentration (CMC) determines the efficiency of biosurfactants, and is the concentration point where micelles start to form (Soberón-Chávez and Maier 2010). Oil solubility increases with the start of the solubilization process at concentrations above CMC level, leading to the formation of micelles (Urum and Pekdemir 2004).

Microorganisms synthesize surfactants known as biosurfactants, which have both hydrophobic and hydrophilic parts. These parts are responsible for lowering the surface tension and the interfacial tension of the growth medium. Biosurfactants are synthesized during the bacterial growth period on water-immiscible substrates, providing an alternative to chemically prepared conventional surfactants. In the case of the bacterial degradation of hydrocarbons, microbial cells synthesize biosurfactants that solubilize oil droplets into the aqueous phase making the oil uptake by microbial cellseasier as shown in Figure 5 (Ganesh and Lin 2009).

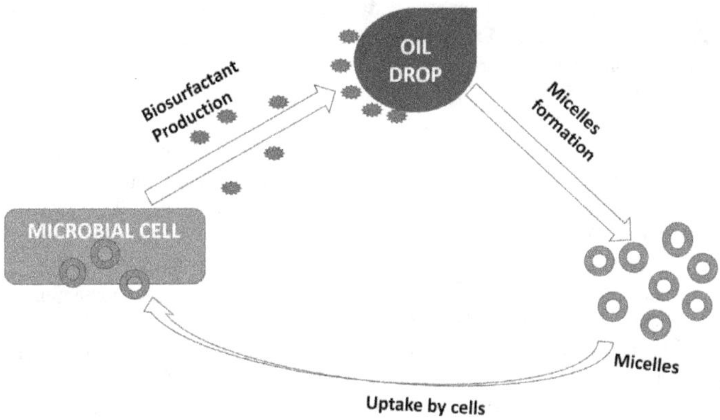

Figure 5. Bioremediation mechanism of biosurfactants.

6.2 *Economic Aspects of Biosurfactant Applications*

The economical production of biosurfactants depends on the selection of low-cost substrates. Biosurfactants must compete with petrochemical surfactants in terms of function, production capacity and cost (Banat et al. 2010, Coimbra et al. 2009). The high cost of biosurfactant production can be absorbed when these compounds are used in small amounts, such as in cosmetics, medications and food. However, the high production cost hinders the use of biosurfactants in applications such as oil recovery, which requires large amounts of surfactant (Cameotra and Makkar 1998). Rahman and Gapke (2008) suggest four factors for reducing the cost of biosurfactants: (1) microorganisms selected, adapted and cultivated for large-scale production; (2) adjustments to the production process to ensure low operational costs; (3) low-cost culture media; and (4) the processing of recycled products. Low-cost biosurfactant production is hampered when there is a need for extensive refining. Thus, biosurfactants should be capable of being recovered through the use of simple, inexpensive methods. Extraction with solvents (chloroform-methanol, dichloromethane-methanol, butanol, ethyl acetate, pentane, hexane, acetic acid) constitutes the most common biosurfactant recovery process (Sobrinho et al. 2014). Biosurfactant market revenue generation was over $1.8 Billion in 2018 and is expected to reach USD 2.6 Billion by 2025 (540 kilo tons by 2024) with the rhamnolipid market set to witness a gain of over 8% (Global Markets Insights 2018).

References

Abdel-Mawgoud, A.M., F. Lépine and E. Déziel. 2010. Rhamnolipids: Diversity of structures, microbial origins and roles. Appl. Microbiol. Biotechnol. 86: 1323–1336.

Abdel-Shafy, H.I. and M.S. Mansour. 2016. A review on polycyclic aromatic hydrocarbons: Source, environmental impact, effect on human health and remediation. Egypt. J. Petrol. 25: 107–123.

Alsohim, A.S., T.B. Taylor, G.A. Barrett, J. Gallie, X.X. Zhang et al. 2014. The biosurfactant viscosin produced by *Pseudomonas fluorescens* SBW25 aids spreading motility and plant growth promotion. Environ. Microbiol. 16(7): 2267–81. Doi: 10.1111/1462-2920.12469. Epub 2014 Apr 29. PMID: 24684210.

Aparna, A., G. Srinikethan and H. Smitha. 2012. Production and characterization of biosurfactant produced by a novel *Pseudomonas* sp. 2B. Colloids Surf. B Biointer. 95: 23–29.

Arpia, A.A., W.H. Chen, S.S. Lam, P. Rousset, M.D.G. de Luna et al. 2021. Sustainable biofuel and bioenergy production from biomass waste residues using microwave-assisted heating: A comprehensive review. Chem. Eng. J. 403: 126233.

Asia, I.O., I.B. Enweani and I.O. Eguavoen. 2006. Characterization and treatment of sludge from the petroleum industry. Afr. J. Biotechnol. 5(5): 461–466.

Banat, I.M., A. Franzetti, I. Gandolfi, G. Bestetti, M.G. Martinotti et al. 2010. Microbial biosurfactants production, applications and future potential. Appl. Microbiol. Biotechnol. 87: 427–444.

Berditsch, M., M. Trapp, S. Afonin et al. 2017. Antimicrobial peptide gramicidin S is accumulated in granules of producer cells for storage of bacterial phosphagens. Sci. Rep. 7: 44324. https://doi.org/10.1038/srep44324.

Bhattacharya, D., P.M. Sarma, S. Krishnan, S. Mishra, B. Lal et al. 2003. Evaluation of genetic diversity among *Pseudomonas citronellolis* strains isolated from oily sludge contaminated sites. Appl. Environ. Microbiol. 69: 1431–1441.

Broszeit, S., C. Hattam and N. Beaumont. 2016. Bioremediation of waste under ocean acidification: Reviewing the role of *Mytilus edulis.* Mar. Pollut. Bull. 103: 5–14.

Burnes, K.A. and R. Jones. 2016. Assessment of sediment hydrocarbon contamination from the Montara oil blow out in the Timor Sea. Environ. Pollut. 211: 214–225.

Cameotra, S.S. and R.S. Makkar. 2010. Biosurfactant-enhanced bioremediation of hydrophobic pollutants. Pure. Appl. Chem. 82(1): 97–116.

Cavalero, D.A. and D.G. Cooper. 2003. The effect of medium composition on the structure and physical state of sophorolipids produced by Candida bombicola ATCC 22214. J. Biotechnol. 103: 31–41.

Celiešiutė, R., S. Grigiškis and V. Čipinytė. 2009. Biological surface active compounds application possibilities and selection of strain with emulsifying activity. pp. 267–272. *In*: Institute of Electrical and Electronics Engineers (IEEE) (ed.). Proceedings of the 7th International Scientific and Practical Conference: Environment, Technology, Resources. IEEE, Rezekne.

Chakrabarti, S. 2012. Bacterial biosurfactant: Characterization, antimicrobial and metal remediation properties. Ph.D. Thesis, National Institute of Technology, India.

Chong, H. and Q. Li. 2017. Microbial production of rhamnolipids: Opportunities, challenges and strategies. Microb. Cell. Fact. 16: 137. Doi: 10.1186/s12934-017-0753-2.

Clark, R.B. 1999. Marine Pollution (fourth ed.), Oxford University Press.

Clements, T., T. Ndlovu and S. Khan. 2019. Biosurfactants produced by *Serratia* species: Classification, biosynthesis, production and application. Appl. Microbiol. Biotechnol. 103: 589–602. https://doi.org/10.1007/s00253-018-9520-5.

Coimbra, C.D., R.D. Rufino, J.M. Luna and L.A. Sarubbo. 2009. Studies of the cell surface properties of *Candida* species and relation with the production of biosurfactants for environmental applications. Current Microbiol. 58: 245–249.

Conti, E., T. Stachelhaus, M.A. Marahiel and P. Brick. 1997. Structural basis for the activation of phenylalanine in the non-ribosomal biosynthesis of gramicidin S. EMBO J. 16: 4174–4183. https://doi.org/10.1093/emboj/16.14.4174.

Cunha, C.D., M. DoRosario, A.S. Rosado and S.G.F. Leite. 2004. *Serratia* sp. SVGG 16: A promising bio-surfactant producer isolated from tropical soil during growth with ethanol-blended gasoline. Process Biochem. 39: 2277–2282.

Dabbs, C.W. 1999. Oil Production and Environmental Damage. http://www1.american.edu/ted/projects/tedcross/xoilpr15.htm.

Dadrasnia, A., N. Shahsavari and C.U. Emenike. 2013. Remediation of contaminated sites. pp. 66–82. *In*: Kutcherov, V. and A. Kolesnikov (eds.). Hydrocarbon. IntechOpen DOI: 10.5772/51591. Available from: https://www.intechopen.com/books/hydrocarbon/remediation-of-contaminated-sites.

Dang, Y., F. Zhao and X. Liu. 2019. Enhanced production of antifungal lipopeptide iturin A by *Bacillus amyloliquefaciens* LL3 through metabolic engineering and culture conditions optimization. Microb. Cell Fact. 18: 68. https://doi.org/10.1186/s12934-019-1121-1.

Das, K. and A.K. Mukherjee. 2007 Crude petroleum-oil biodegradation efficiency of *Bacillus subtilis* and *Pseudomonas aeruginosa* strains isolated from petroleum oil contaminated soil from North-East India. Bioresource Technol. 98: 1339–1345.

de Franca, I.W.L., A.P. Lima, J.A.M. Lemos, C.G.F. Lemos, V.M.M. Melo et al. 2015. Production of a biosurfactant by *Bacillus subtilis* ICA56 aiming bioremediation of impacted soils. Catal. Today 255: 10–15.

Deleu, M. and M. Paquot. 2004. From renewable vegetables resources to microorganisms: New trends in surfactants. Computers Rendus. Chimie. 7: 641–646.

Desai, J.D. and I.M. Banat. 1997. Microbial production of surfactants and their commercial potential. Microbiol. Mol. Biol. Rev. 61: 47–64.

Dhiman, R., K.R. Meena, A. Sharma and S.S. Kanwar. 2016. Biosurfactants and their screening methods. Res. J. Recent Sc. 5(10): 1–6.

Enzler, M.S. 2006. Environmental Disasters. Lenntech. http://www.lenntech.com/environmental-disasters.htm.

Esbaugh, A.J., E.M. Mager, J.D. Stieglitz, R. Hoenig, T.L. Brown et al. 2016. The effects of weathering and chemical dispersion on Deepwater Horizon crude oil toxicity to mahi-mahi (Coryphaenahippurus) early life stages. Sci. Total Environ. 543: 644–651.

Ferguson, A., H. Solo-Gabriele and K. Mena. 2020. Assessment for oil spill chemicals: Current knowledge, data gaps, and uncertainties addressing human physical health risk. Marine Pollut. Bullet. 150: 110746.

Ganesh, A. and J. Lin. 2009. Diesel degradation and biosurfactant production by gram positive isolates. Afri. J. Biotechnol. 8(21): 5847–5854.

Global Market Insights. 2018. Available at https://www.gminsights.com/industry-analysis/biosurfactants-market-report. Accessed February 20, 2020.

Gutzkow, K.B. 2015. Genetoxicity, Mutagenicity and Carcinogenicity and Reach. ICAW, Norwegian Institute of Public Health.

Han, Y., I.M. Nambi and T.P. Clement. 2018. Environmental impacts of the Chennai oils pill accident—A case study. Sci. Total Environ. 626: 795–806.

Hassanshahian, M. and S. Cappello. 2013. Crude oil biodegradation in the marine environments. *In*: Chamy, R. and F. Rosenkranz (eds.). Biodegradation—Engineering and Technology. IntechOpen, DOI: 10.5772/55554. Available from: https://www.intechopen.com/books/biodegradation-engineering-and-technology/crude-oil-biodegradation-in-the-marine-environments.

Hoffman, A.J. and P.D. Jennings. 2010. The BP oils pill as a cultural anomaly institutional context, conflict and change. Ross School of Business working paper, no. 1151. University of Michigan, USA.

Hošková, M., O. Schreiberová, R. Ježdík, J. Chudoba, J. Masák et al. 2013. Characterization of rhamnolipids produced by non-pathogenic *Acinetobacter* and *Enterobacter* bacteria. Bioresource Technol. 130: 510–516. https://doi.org/10.1016/j.biortech.2012.12.085.

Hu, G. 2016. Development of novel oil recovery methods for petroleum refinery oily sludge treatment. Ph.D. Dissertation. University of Northern British Columbia.

Hu, Y. and L.K. Ju. 2001. Purification of lactonicsophorolipids by crystallization. J. Biotechnol. 87: 263–272.

ITOPF Statistics. 2020. Avaialable at https://www.itopf.org/knowledge-resources/data-statistics. Access on November 22, 2020.

Janek, T., A. Krasowska, Ż. Czyżnikowska and M. Łukaszewicz. 2018. Trehalose lipid biosurfactant reduces adhesion of microbial pathogens to polystyrene and silicone surfaces: An experimental and computational approach. Front. Microbiol. 9: 2441. Doi: 10.3389/fmicb.2018.02441.

Jasim, B., K.S. Sreelakshmi, J. Mathew and E.K. Radhakrishnan. Surfactin, Iturin, and Fengycin biosynthesis by endophytic *Bacillus* sp. from *Bacopa monnieri*. Microb. Ecol. 72(1): 106–119. Doi: 10.1007/s00248-016-0753-5.

Johnson, O.A. and A.C. Affam. 2019. Petroleum sludge treatment and disposal: A review. Environmental Engineering Research 24(2): 191–201.

Joutey, N.T., W. Bahafid, H. Sayel and N. El-Ghachtouli. 2013. Biodegradation: Involved microorganisms and genetically engineered microorganisms. pp. 290–320. *In*: Chamy, R. and F. Rosenkranz (eds.). Biodegradation-Life Science, IntechOpen, DOI: 10.5772/56194. Available from: https://www.intechopen.com/books/biodegradation-life-of-science/biodegradation-involved-microorganisms-and-genetically-engineered-microorganisms.

Kankara, R.S., S. Arockiaraj and K. Prabhu. 2016. IEA Environmental sensitivity mapping and risk assessment for oil spill along the Chennai Coast in India. Mar. Pollut. Bull. 106: 95–103.

Kim, S.Y., S.Y. Park, S.K. Choi and S.H. Park. 2015. Biosynthesis of polymyxins B, E, and P using genetically engineered polymyxin synthetases in the surrogate host *Bacillus subtilis*. J. Microbiol. Biotechnol. 25(7): 1015–25. Doi: 10.4014/jmb.1505.05036.

Kong, H.G., J.C. Kim, G.J. Choi, K.Y. Lee, H.J. Kim et al. 2010. Production of surfactin and iturin by *Bacillus licheniformis* N1 responsible for plant disease control activity. Plant Pathol. J. 26: 170–177.

Kvenvolden, K.A. and C.K. Cooper. 2003. Natural seepage of crude oil into the marine environment. Geo-Marine Letters 23(3-4): 140–146.

Lawal, A.T. 2017. Polycyclic aromatic hydrocarbons. A review. Cogent Environ. Sci. 3: 1339841.

Li, J., M. Deng, Y. Wang and W. Chen. 2016. Production and characteristics of biosurfactant produced by *Bacillus pseudomycoides* BS6 utilizing soybean oil waste. Int. Biodeterior. Biodegrad. 112: 72–79.

Li, X., X. Wang, Z.L. Ren, Y. Zhang, N. Li et al. 2015. Sand amendment enhances bioelectrochemical remediation of petroleum hydrocarbon contaminated soil. Chemosphere 141: 62–70.

Ling, C.C. and M.H. Isa. 2006. Bioremediation of oily sludge contaminated soil by co-composting with sewage sludge. J. Sci. Ind. Res. 65: 364–369.

Liu, X., Q. Shu, Q. Chen, X. Pang, Y. Wu et al. 2020. Antibacterial efficacy and mechanism of mannosylerythritol lipids-A on *Listeria monocytogenes*. Molecul. 25: 4857. Doi: 10.3390/molecules25204857.

Liu, B., J. Liu, M. Ju, X. Li, Q. Yu et al. 2015. Purification and characterization of biosurfactant produced by *Bacillus licheniformis* Y-1 and its application in remediation of petroleum contaminated soil. Mar. Pollut. Bull. 107: 46–51.

Maletic, S.P., B.D. Dalmacija and S.D. Roncevic. 2013. Petroleum hydrocarbon biodegradability in soil—Implications for bioremediation. pp 3. *In*: Kutcherov, V. and A. Kolesnikov (eds.). Hydrocarbon. IntechOpen, DOI: 10.5772/50108. Available from: https://www.intechopen.com/books/hydrocarbon/petroleum-hydrocarbon-biodegradability-in-soil-implications-for-bioremediation.

Mandal, A.K., D. Manish, S. Abu, P.M. Sarma, B. Lal et al. 2007. Remediation of oily sludge at various installations of ONGC: A Biotechnological Approach, Proceedings of Petrotech. 7th International Oil & Gas Conference and Exhibition, New Delhi, India, Paper no. 753.

Martins, P.C. and V.G. Martins. 2018. Biosurfactant production from industrial wastes with potential remove of insoluble paint. Int. Biodeterior. Biodegrad. 127: 10–16.

Ministry of Environment and Forest (MoEF), Government of India. 2000. Hazardous Wastes (Management and Handling) Rules Amendment 2000.

Mishra, A.K. and G.S. Kumar. 2015. Weathering of oil spill: Modeling and analysis. Aquat. Procedia. 4: 435–442.

Moya, R.I., K. Tsaousi, M. Rudden, R. Marchant, E.J. Alameda et al. 2015. Rhamnolipid and surfactin production from olive oil mill waste as sole carbon source. Bioresour. Technol. 198: 231–236.

Mulligan, C.N., R.N. Yong and B.F. Gibbs. 2001. Surfactant-enhanced remediation of contaminated soil: A review. Eng. Geol. 60: 371–380.

Nwaguma, I.V., C.B. Chikere and G.C. Okpokwasili. 2016. Isolation, characterization, and application of biosurfactant by *Klebsiella pneumoniae* strain IVN51 isolated from hydrocarbon-polluted soil in Ogoniland, Nigeria. Bioresour. Bioprocess 3: 40. https://doi.org/10.1186/s40643-016-0118-4.

Oil tanker spill statistics. 2016. The International Tanker Owners Pollution Federation Limited (ITOPF).

Ossai, I.C., A. Ahmed, A. Hassan and F.S. Hamid. 2020. Remediation of soil and water contaminated with petroleum hydrocarbon: A review. Environ. Technol. Inno. 17: 100526.

Pacwa-Plociniczak, M., G.A. Plaza, Z. Piotrowska-Seget and S.S. Cameotra. 2011. Environmental applications of biosurfactants: Recent advances. Int. J. Mol. Sci. 13: 633–654.

Perfumo, A., I. Rancich and I.M. Banat. 2010. Possibilities and challenges for biosurfactants uses in petroleum industry. pp. 135–145. *In*: Sen, R. (ed.). Biosurfactants. Advances in Experimental Medicine and Biology. Austin: Landes Bioscience.

Peter, A.O. 2011. Enhanced bioremediation of soil contaminated with used lubricating oil (Ph.D. Thesis). Institute of Biological Sciences, University of Malaya.

Pirog, T.P., A.D. Konon, K.A. Beregovaya and M.A. Shulyakova. 2014. Antiadhesive properties of the surfactants of *Acinetobacter calcoaceticus* IMB B-7241, *Rhodococcus erythropolis* IMB Ac-5017, and *Nocardia vaccinii* IMB B-7405. Microbiol. 83: 732–739.

Prasad, S.J., N.T.M. Balakrishnan, P.A. Francis and T. Vijayalaksmi. 2014. Hindcasting and validation of mumbai oil spills using GNOME. Int. Res. J. Environ. Sci. 3(12): 18–27.

Qiu, Y., F. Xiao, X. Wei, Z. Wen, S. Chen et al. 2014. Improvement of lichenysin production in Bacillus licheniformis by replacement of native promoter of lichenysin biosynthesis operon and medium optimization. Appl. Microbiol. Biotechnol. 98(21): 8895–903. Doi: 10.1007/s00253-014-5978-y. Epub 2014 Aug 3. PMID: 25085615.

Rahman, P.K.M. and E. Gakpe. 2008. Production, characterization and applications of biosurfactants—Review. Biotechnol. 7(2): 360–370.

Raza, Z.A., M.S. Khan, Z.M. Khalid and A. Rehman. 2006. Production, kinetics and tensioactive characteristics of biosurfactant from a *Pseudomonas aeruginosa* mutant grown on waste frying oils. Biotechnol. Letters 28: 1623–1631. https://doi.org/10.1007/s10529-006-9134-3.

Repanich, J. 2010. The Deepwater Horizon Spill by the Numbers. Available at: http://www.popularmechanics.com/science/energy/coal-oil-gas/bp-oil-spill-statistics.

Ron, E.Z. and E. Rosenberg. 2014. Enhanced bioremediation of oil spills in the sea. Curr. Opin. Biotechnol. 27: 191–194.

Roy, A. 2017. Review on the biosurfactants: Properties, types and its applications. J. Fundam. Renewable Energy Appl. 8: 248. Doi: 10.4172/20904541.1000248.

Shekhar, S., A. Sundaramanickam and T. Balasubramanian. 2015. Biosurfactant producing microbes and their potential applications: A review. Crit. Rev. Environ. Sci. Technol. 45: 1522–1554.

Siñeriz, F., R.K. Hommel and H.P. Kleber. 2001. Production of biosurfactants. pp. 1–19. *In*: Doelle, H.W. (ed.). Biotechnology. Vol. V. EOLLS Publishers, Oxford.

Singh, A., J.D. VanHanne and O.P. Ward. 2007. Surfactants in microbiology and biotechnology: Part 2. Application aspects. Biotechnol. Adv. 25: 99–121.

Singh, R., B.R. Glick and D. Rathore. 2018. Biosurfactants as a biological tool to increase micronutrient availability in soil: A review. Pedosphere 28(2): 170–189.

Singh, R. and D. Rathore. 2019. Impact assessment of azulene and chromium on growth and metabolites of wheat and chilli cultivars under biosurfactant augmentation. Ecotoxicol. Environ. Saf. 186: 109789. https://doi.org/10.1016/j.ecoenv.2019.109789.

Singh, R., S.K. Singh and D. Rathore. 2019. Analysis of biosurfactants produced by bacteria growing on textile sludge and their toxicity evaluation for environmental application. J. Dispers. Sci. Technol. 1–13. https://doi.org/10.1080/01932691.2019.1592686.

Singh, R., B.R. Glick and D. Rathore. 2020. Role of textile effluent fertilization with biosurfactant to sustain soil quality and nutrient availability. J. Environ. Manag. 268: 110664. https://doi.org/10.1016/j.jenvman.2020.110664.

Soberón-Chávez, G. and R.M. Maier. 2010. Biosurfactants: A general overview. pp. 1–11. *In*: Soberón-Chávez, G. (ed.). Biosurfactants: From Genes to Applications. Springer, Münster, Germany.

Sobrinho, H.B.S., J.M. Luna, R.D. Rufino, A.L.F. Porto, L.A. Sarubbo et al. 2014. Biosurfactants: Classification, properties and environmental applications. *In*: Ahmad, M. and J.N. Govil (eds.). Biotechnology. Vol. 11. Bidegradation and Bioremediation, Studium Press LLC, USA.

Souza, E.C., T.C. Vessoni-Penna and R.P.D.S. Oliveira. 2014. Biosurfactant-enhanced hydrocarbon bioremediation: An overview. Int. Biodet. Biodeg. 89: 88–94.

Tondo, D.W., E.C. Leopoldino, B.S. Souza, G.A. Micke, A.C.O. Costa et al. 2010. Synthesis of a new zwitterionic surfactant containing an imidazolium ring. Evaluating the chameleon-like behavior of zwitterionic micelles. Langmuir 26(20): 15754–15760.

Urum, K. and T. Pekdemir. 2004. Evaluation of biosurfactants for crude oil contaminated soil washing. Chemosphere 57: 1139–1150.

Varjani, S.J. and V.N. Upasani. 2016. Carbon spectrum utilization by an indigenous strain of *Pseudomonas aeruginosa* NCIM 5514: Production, characterization and surface active properties of biosurfactant. Bioresour. Technol. 221: 510–516.

Varjani, S.J. 2017. Microbial degradation of petroleum hydrocarbons. Bioresour. Technol. 223: 277–286.

Velikonja, J. and N. Kosaric. 1993. Biosurfactants in food applications. pp. 419–448. *In*: Kosaric, N. and F.V. Sukan (eds.). Biosurfactants: Production, Properties, Applications. CRC Press: New York, NY, USA.

Winnick, R.E., H. Lis and T. Winnick. 1961. Biosynthesis of gramicidin S I. General characteristics of the process in growing cultures of *Bacillus brevis*. Biochimica et Biophysica Acta 49(3): 451–462. https://doi.org/10.1016/0006-3002(61)90242-6.

Yan, P., M. Lu, Q. Yang, H. Zhang, Z. Zhang et al. 2012. Oil recovery from refinery oily sludge using a rhamnolipid biosurfactant-producing Pseudomonas. Bioresour. Technol. 116: 24–28.

Zhu, X., D.V. Albert, T.S. Makram and L. Kenneth. 2001. U.S. Environmental Protection Agency (USA). Guidelines for the bioremediation of marine shorelines and freshwater wetlands.

14

Biosurfactants Current Trends and Applications

Nisha Dutta[1,*] and *Akansha Bhatnagar*[2]

1. Introduction

Biosurfactants are molecules which possess the qualities of a surfactant but are of a biological origin. These are synthesized mainly by the microbes such as bacteria, fungi or yeast; that is either a metabolite product (may be extracellular or not) or produced by the cell or the cell itself from the surface chemistry. Generally, these are amphiphilic molecules which comprise of two different moieties; the hydrophobic moiety and the hydrophilic moiety, both of which lie on different ends of the molecule. The hydrophilic or the polar moiety of these amphiphilic molecules generally consists of monosaccharides, oligosaccharides, polysaccharides, amino acids, peptides and proteins while the hydrophobic or the non-polar moiety consists of hydroxylated fatty acids, fatty alcohols, saturated and unsaturated fatty acids (Chavez et al. 2016, Nitschke and Silva 2018, Bustos et al. 2018, Kubicki et al. 2019).

Structurally they may be of different origins being produced by different organisms over different substances and thus possess varied chemical structures. These are also termed as surface active biomolecules owing to their unique functional property of acting on the surface and interfacial tension of the liquid in which they are added. According to Sourav et al. (2015) and Chavez et al. (2016) good biosurfactants are those which have the capability of reducing the surface tension of water to 35 mN/m from 72 mN/m and interfacial tension from 40 mN/m to 1 mN/m of water/nhexadecane.

According to Varjani and Upasani (2017) biosurfactants tend to reduce the surface and interfacial tension of immiscible liquids while bio emulsifiers are mostly used in oil in water emulsion stabilization. Though some scientists have tried to

[1] Department of Sports Nutrition, School of Sports Sciences, CURAJ, Kishangarh, Ajmer, Rajasthan.
[2] Department of Home Science, Swami Vivekanand Subharti University, Meerut.
* Corresponding author: nishaduttajsr@gmail.com

differentiate between the biosurfactants and bio-emulsions/bio-emulsifiers but mostly these terms are used interchangeably, this might be due to the fact that there is only a thin line differentiating the two and both are capable of the properties and activities possessed by the other one. The most desirable properties which are typical of any biosurfactant include reduction in the surface and interfacial tensions, critical micelle concentration (CMC), enhancement of solubility, wettability, emulsion formation and foaming capacity.

Microorganisms produce biosurfactants as secondary metabolites and these facilitate microbe growth. Rodrigues et al. (2006) are of the opinion that biosurfactants increase the availability of substrate and transport of nutrient for its producer. Studies have clearly portrayed that these biomolecules have been found to play a key role in cell signaling in bacteria, cellular differentiation, bioflim formation, bacterial pathogenesis and act as a biocidal agent whenever a microorganism comes across any interface (Lang 2002, Kitamoto et al. 2002, Cameotra and Makkar 2004, Van Hamme et al. 2006). These naturally occurring surface active compounds possess the potential for application in varied fields, such as agriculture, bioremediation, mining, food industry, bakery industry, cosmetics, synthetic medicine and pharmaceutical industry.

In spite of the huge potential to cater to varied industries they have not yet gained extensive commercialization owing to the relative high cost of production and recovery.

2. Classification

Biosurfactants are biomolecules which are obtained from varied microorganisms. As they are obtained from diverse microbes, they are generally structurally distinct molecules possessing surface active compounds. Unlike the synthetic surfactants that are generally categorized based on their ionic strength, biosurfactants are mostly either anionic or neutral compounds while few which contain amine groups are cationic in nature. The hydrophilic moiety of biosurfactants could be made up of carbohydrates, cyclic peptides, amino acids, phosphate carboxyl acids or alcohols whereas the hydrophobic moiety is composed of long chains of fatty acids.

According to Neu (1996) these compounds could be roughly categorized on the basis of their molecular weight (Figure 1) into 2 subgroups such as; (1) lower molecular weight compounds and (2) higher molecular weight compounds.

The compounds which possess lower molecular weight such as glycolipids, lipoproteins and proteins are called biosurfactants and the compounds that possess higher molecular weight such as polysaccharides, lipopolysaccharide proteins, or lipoproteins are known as bioemulsans or bioemulsifiers (Rosenberg and Ron 1997, Smyth et al. 2010). Based on the chemical structure that these biosurfactants possess they could be categorized into five groups namely glycolipids, lipoproteins or lipopeptides, phospholipids, fatty acids or natural lipids, polymeric biosurfactants and particulate biosurfactants (Santos et al. 2016, Varjani and Upasani 2017), of which glycolipids are the largest group of these compounds and the most widely and extensively studied and the hydrophobic moiety mostly consists of long chains of hydroxyl aliphatic acids linked via the ester group to the carbohydrate fraction which

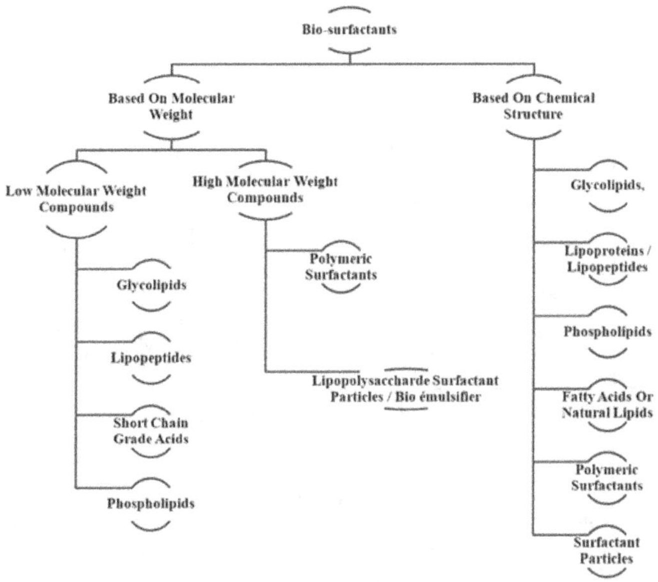

Figure 1. Classification of biosurfactants.

forms the hydrophilic moiety of the compound. Table 1 gives a vivid description of these compounds along with the names of few microorganisms that are capable of producing these biosurfactants and their importance.

3. Properties of Biosurfactants

Biosurfactants have been termed as multifunctional materials owing to the wide range of their application in diverse fields. These biomolecules have been observed to find new applications owing to the chemical structure and properties that they possess. Being termed as surface active substances they tend to possess the ability to lower the surface and interfacial tensions, which is one the most typical desirable properties they have. The general properties of biosurfactants have been discussed below with the idea that their understanding would clarify the proposed mechanism of their action as therapeutic agents.

3.1 Surface and Interfacial Activity

Biosurfactants are chemicals that accumulate at the interface of two immiscible liquids and help in decreasing the surface as well as interfacial tension thereby making two liquids mix together in the form of micelles. The concentration of the biosurfactant at which the micelle formation occurs is known as the critical micell concentration (CMC) (Figure 2). CMC is the parameter which is used to assess the activity of a surfactant and can be determined by measuring the surface tension of the surfactant at varied concentrations (Ohadi et al. 2018) and the initial concentration that is capable of maintaining a constant surface tension is known as CMC (Das et al. 2014).

Table 1. Biosurfactants types, microorganisms and their activities.

Class	Types	Microbes for Biosynthesis	Activity
Glycolipid	Rhamnolipids	*Pseudomonas aeruginosa, Pseudomonas* sp., *Burkholderia glumae, Burkholderia plantarii, Burkholderia thailandensis*	Antimicrobial, anti-adhesive activity against several bacterial and yeast strains, emulsification and removal of heavy metals
	Trehalose lipids	*Rhodococcus erythropolis, Nocardia erythropolis, Mycobacterium* sp., *Arthobacter* sp.	Enhances bioavailability of hydrocarbons, possess antiviral activity
	Sophorolipids	*Torulopsis bombicola, Torulopsis apicola, Torulopsis petrophilum*	Aids in removal of heavy metals and enhances oil recovery
	Mannosylerythritol lipid	*Candida antartica*	Possess antimicrobial, immunological and neurological properties
	Cellobiolipids	*Ustilago zeae, Ustilago maydis*	
Lipopeptides and lipoproteins	Surfactin/iturin/ fengycin	*Bacillus subtilis, Bacillus licheniformis*	Enhances biodegradation of hydrocarbons and chlorinated pesticides; facilitates removal of heavy metals; increases effectiveness of phytoextraction, possess antimicrobial, antifungal and antitumour, antiviral activities, inhibits clot formation, haemolysis and helps in formation of ion channels in lipid membranes, effective against human immunodeficiency virus 1 (HIV-1), and is an immunological adjuvant
	Viscosin	*Pseudomonas fluorescens*	Antimicrobial activity
	Lichenysin	*Bacillus licheniformis*	Enhances oil recovery, antibacterial activity chelating properties that helps the membrane-disrupting effect of lipopeptides
	Serrawettin	*Serratia marcescens*	Chemorepellent
	Subtilism	*Bacillus subtilis*	Antimicrobial activity
	Gramicidin	*Brevibacterium brevis*	Antibiotic, disease control
	Polymixin	*Bacillus polymyxa*	Bactericidal and fungicidal activity
	Antibiotic TA	*Myxococcus xanthus*	Bactericidal activity, chemotherapeutic applications

Table 1 Contd. ...

...Table 1 Contd.

Class	Types	Microbes for Biosynthesis	Activity
Fatty acids/ neutral lipids/ phospholipids	Corynomycolic acid	*Corynebacterium lepus*	Enhances recovery of bitumen
	Spiculisporic acid	*Penicillium spiculisporum*	Aids in sequestrants of heavy metals, disperses polar pigments, emulsion formation
	Phosphati-dylethanolamine	*Acinetobacter* sp., *Rhodococcus erythropolis, Mycococcus* sp.	Increases heavy metal resistance in a microbe
Polymeric surfactants	Emulsan	*Acinetobacter calcoaceticus*	Aids in emulsion formation
	Alasan	*Acinetobacter radioresistens*	Stabilizes hydrocarbon-in water emulsions
	Biodispersan	*Acinetobacter calcoaceticus* A2	Aids dispersion of limestone in water
	Polysaccharide protein complex	*Acinebacter calcoaceticus*	Bioemulsifier
	Liposan	*Candida lipolytica*	
	Mannoprotein	*Saccharomyces cerevisiae*	Stabilises hydrocarbon- in-water emulsions
	Protein PA	*Pseudomonas aeruginosa*	Bioemulsifier
Particulate biosurfactants	Vesicles	*Acinetobacter calcoaceticus, Pseudomonas marginalis*	
	Whole microbial cells	*Cyanobacteria*	Helps in degradation as well as removal of hydrocarbons

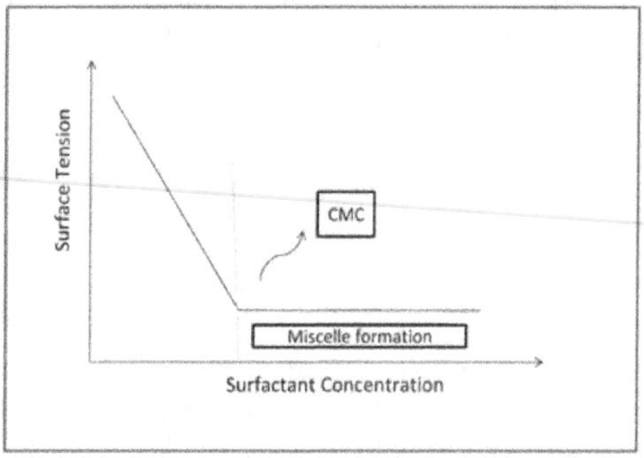

Figure 2. Graphical representation of CMC and micelle formation.

At the interface of the two immiscible liquids the biosurfactants intervene replacing the hydrophobic or the hydrophilic molecules with their corresponding moieties thus bringing down the inter-molecular forces between the solvent molecules and thus reduce the interfacial and surface tension forces (Jahan et al. 2020). Tayeb et al. (2019) suggest that by altering the structural composition of protein biosurfactants their interfacial properties can be improved which may enhance their applicability in industry. The zwitter ionic surfactant cocamidopropyl betaine (CAPB) in conjunction with rhamnolipid and sophorolipid biosurfactants can help in the reduction of surface tension (Zhou et al. 2019). Removal of hydrophobic organic compounds from land and aquatic bodies is a matter of environmental concern and as a remedy saponins aids in removal of hydrophobic organic compounds by forming saponin micelles (Mulligan 2009).

3.2 *Biocompatibility and Toxicity*

Biosurfactants are biocompatible in nature owing to multitudinous usage in the industry (Gudina et al. 2013). Rhamnolipids are also biocompatible and natural antimicrobial and antioxidant agents which are used for various industrial and biomedical applications (Haque et al. 2020, Maqsood and Seddiq-Shams 2014). The nano particles of iron oxide coated with biosurfactant polyethylene glycol is nontoxic but with surfactin and ramnolipid shows cytotoxicity for MRI (Magnetic resonance imaging) (Nguyen et al. 2010) while Lipopeptides and glycolipids are safe for drug delivery (Gudina et al. 2013).

There are various types of tests to assess the toxicity of biosurfactants and the most common being the germination index test and bioassays. A combination of measures of relative seed and root germination and their growth is termed as germination index. Singh and Rathore (2020) reported that the biosurfactants obtained from *Stenotrophomonas* sp. BAB-6435 and *Brevisbacillus brevis* BAB-6437 did not exhibit any toxic effects when tested for germination index. The researchers further documented that they were thus proved to be nontoxic and to possess antimicrobial properties. Polysaccharide bioemulsifiers produced by *V. paradoxus* 7bCT5 and *Candida lipolytica* UCP 0988 also showed no toxicity for crustaceans and had no effect on the germination index (Franzetti et al. 2012). The biosurfactants obtained from *Stenotrophomonas* sp. BAB-6435 and *Brevisbacillus brevis* BAB-6437 also proved non-toxic and have antimicrobial properties (Singh et al. 2020).

3.3 *Specificity*

Biosurfactants have specific functional groups owing to specific properties. Generally, they are neutral or anionic in nature but those biosurfactants which contain the amine group in their chemical structure are cationic in nature. Fatty acid chains form the hydrophobic moiety while the hydrophilic moiety generally consists of carbohydrates, amino acids, cyclic peptides, phosphate carboxyl acids or alcohols. Specific actions are of considerable interest in the pharmaceutical, food and cosmetic industries (Mulligan et al. 2014).

3.4 *Biodegradability*

Biosurfactants are biodegradable, i.e., they can be broken down into simpler components by the action of microorganisms while synthetic surfactants are non-biodegradable (Klosowska-Chomiczewska et al. 2011). The presence of long chain fatty acids may affect the biodegradability of lipopeptide biosurfactants (Cipinyte et al. 2009). Rodríguez-López et al. (2020) in his study on crude extracts of corn mills evaluated the biodegradability of biosurfactant present in them and concluded that without inoculums addition they are linked to environmental conditions (Rodriguez et al. 2020).

3.5 *Stability*

Biosurfactants can be used at varied pH levels and high temperatures with no change in their surface activity (Purwasena et al. 2019, Gudina et al. 2010, Pinto et al. 2018) along with stability in saline conditions (Campos et al. 2013, Luna et al. 2013, Santos et al. 2014). The biosurfactant produced by *Bacillus licheniformis* DS1 was found to be stable at 120°C temperature and pH 4 to 10 (Purwasena et al. 2019) and biosurfactant produced by *Lactobacillus paracasei* strain was stable and viable at a pH of 6 to 10 at 60°C for about 120 hours (Gudina et al. 2010). Pinto et al. (2018) noted that biosurfactant synthesized by *Candida bombicola* was stable at a wide temperature range of 4 to 120°C and pH (2 to 12).

3.6 *Wettability*

Biosurfactants can be a means to facilitate the wettability of the soil, which might be defined as the ability of liquid to maintain contact with the solid which can be assessed using the contact angle of fluid with the solid surface. This property of biosurfactants can be used in enhancing the mechanism of oil recovery from the soil (Al-Sulaimani et al. 2012, Ukwungwu et al. 2017). Some studies suggest the use of biosurfactants in the field of agriculture also, as they help in improving the availability of nutrients present in the soil by the method of solubilization and easy uptake by the plant roots, however the mobility of toxic elements is contraindicated (Singh et al. 2018). Advances in the usage of biosurfactants in the field of agriculture can improve the micronutrient malnutrition among the population which is of growing concern in the field of public health.

3.7 *Availability and Production*

Biosurfactants are sometimes referred to as microbial surfactants also, owing to the ability of these microbes to synthesize surfactant biomolecules using cheap raw materials adding to the sustainability aspect (Muthusamy et al. 2008, Marcelino et al. 2020). The genus *Pseudomonas* is the major and most active producer of biosurfactants (Renard et al. 2016). The biosurfactant produced from *Pseudomonas ceacia* CCT6659 on industrial waste have a low production cost and excellent applicability even under extreme environmental conditions (Rita et al. 2017).

4. Biosynthesis of Biosurfactants

Biosurfactants are synthesized from various microbes. According to Slydatk et al. (1987) linear and saturated n-alkanes are generally the major substrate used by the yeast and fungi while certain strains of bacteria also might utlizeisoalkanes, cycloalkanes and unsaturated aromatic compounds. Scientists have postulated diverse mechanisms that might lead to the precursor molecule synthesis which ultimately would result in the production of biosurfactant. Concisely its synthesis may be divided into three major steps. Since all the biosurfactants are amphiphilic in nature and structure, involving both hydrophilic and hydrophobic moieties, synthesis of both the moieties takes place using different metabolic *de novo* pathways and then synthesis of the biosurfactant from these two parts. Syldatk and Wagner (1987) suggested four different biosynthetic pathways on the basis of certain assumptions which might be involved in the synthesis mechanism (Figure 3). Its production depends not only on the microorganism and the substrate involved in its synthesis but also depends on several aspects such as the nature and type of the carbon and nitrogen source, carbon: nitrogen ratio, pH, temperature, nutritional availability, aeration, agitation, salt concentration and others.

Rhamnolipid, a type of glycolipid that is synthesized by *P. aeruginosa* and Surfactin, a type of lipopeptide produced from *B. subtilis* are the most studied and reported biosurfactants in the molecular biosynthetic regulation of biosurfactants (Das et al. 2008).

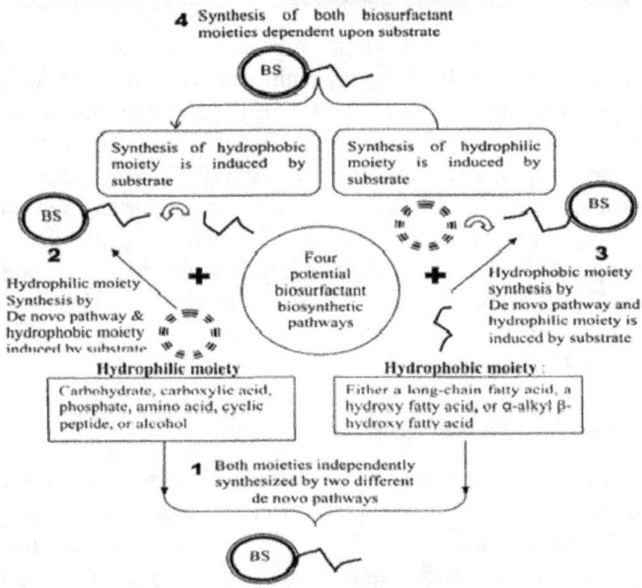

Figure 3. Biosynthetic pathways involved in synthesis of biosurfactants by various microorganisms. Source: Production of biosurfactants. pp. 89–120. *In*: Kosaric, N., W.L. Cairns and N.C.C. Gray (eds.). Biosurfactants and Biotechnology, Surfactant Science Series, Vol. 25. New York: Marcel Dekker Inc.

5. Applications of Biosurfactants

Biosurfactants are being studied in various aspects owing to the wide potential that they have demonstrated due to the varied properties that they possess. Biosurfactants exhibited an efficient and successful application in industries such as cosmetic, pharmaceutical, food, petroleum, agriculture, textile, and wastewater treatment.

Several scientists have emphasized on application in the biomedical field as they act as biologically active compounds. They have shown to possess antifungal, antibacterial, antiviral, antimicrobial activities against numerous pathogenic organisms. Healing of wounds (Piljac et al. 2008) is another promising application for biosurfactants.

5.1 *Bioremediation of Soil*

Soil hydrocarbon contamination is a major environmental issue which is a result of waste release from the petrochemical industries and to resolve this, bioremediation is required which involves removal of pollutants leading to soil detoxification. Biosurfactants can be used in bioremediation of soil effectively (Guo and Wen 2020). Lipopeptide biosurfactants produced by *Bacillus subtilis* SPB1 and *Acinetobacter radioresistens* RI7 strains, increase hydrocarbon mobility leading to biodegradation hence can be used for the bioremediation of hydrocarbon contaminated soil (Mnif et al. 2018). Methods including soil washing and bio stimulated treatment can be used in soil treatment to reduce crude oil contamination using the mixture of surfactin and rhamnolipid which is a sustainable and eco-friendly method for efficient treatment of soil contaminated with heavy oils (Fanaei et al. 2020).

In a study conducted by Patel and Patel 2020, concluded that biosurfactants produced by *Stenotrophomonas* sp. S1VKR-26 reduce the surface tension of various polycyclic aromatic hydrocarbons including naphthalene, phenanthrene, fluoranthene, and pyrene, total petroleum hydrocarbons and phenolic compounds of petroleum waste water leading to higher germination and seedling capacity as compared to untreated waste water. Therefore, the treatment of petroleum refinery wastewater with strain S1VKR-26 could be more effective in the sense of environmental safety and irrigation for crop production in agriculture (Patel and Patel 2020). However, thorough investigation of the efficacy and toxicity of biosurfactants is to be performed before implementing them in bioremediation. Biosurfactants can be a potential replacement to chemical surfactants only if they meet the large-scale production and cheap prices of synthetic surfactants. Use of inexpensive substrates, employing high yield strains, and developing cost-effective downstream processing are some of the approaches to reduce the cost of biosurfactants (Sajna and Gottumukkala 2019).

5.2 *Biosurfactants in Environmental Protection*

Increasing pollution is a global concern, since various pollutants such as polyaromatic hydrocarbons (PAHs), waxes, asphaltenes, monoaromatic hydrocarbons (BTEX), paraffin, and heavy metals due to their toxigenic effect on healthy microorganisms, plants, animals, and humans require remediation. Consequently, the need to remediate them becomes pertinent in different environmental media. This is where surfactants come into play in environmental remediation for which biosurfactants have a high potential.

Soil washing is a technique for removal of harmful pollutants from the soil and rhamnolipids have been confirmed as soil-washing agents for improved removal of hydrocarbons and metals. Rhamnolipid-enhanced soil washing targeted for hydrocarbons results from mobilization and solubilization to facilitate separation of the pollutants from the solid particles and increase the partition of the contaminants in the aqueous phase (Mao et al. 2015, Zhong et al. 2016). The presence of a biosurfactant and its property of mobilization, emulsification, and solubilization makes the hydrophobic organic matter, including hydrocarbons, soluble and bioavailable.

The oil spills during transportation and leaks from drilling rigs into aquatic bodies leads to disruption of the aquatic food chain, death of aquatic lives, poor penetration of sunlight and then remediation of the polluted marine water becomes necessary where biosurfactants could serve the purpose. The biosurfactant dispersant increases the surface area (formation of micelles due to emulsification) along with the solubility and mobility of hydrocarbon pollutants. Critical micelle concentrations in the range 1 to 1.5 CMC is effective in marine remediation (Patel et al. 2019, Whang et al. 2008).

References

Al-Sulaimani, H., Y. Al-Wahaibi, S. Al-Bahry, A. Elshafie, A. Al-Bemani et al. 2012. Residual-oil recovery through injection of biosurfactant, chemical surfactant, and mixtures of both under reservoir temperatures: induced-wettability and interfacial-tension effects. SPE Reservoir Evaluation & Engineering 1; 15(02): 210–7.

Bustos, G., U. Arcos, X. Vecino, J.M. Cruz, A.B. Moldes et al. 2018. Recycled Lactobacillus pentosus biomass can regenerate biosurfactants after various fermentative and extractive cycles. Biochemical Engineering Journal 132: 191–195.

Cameotra, S.S. and R.S. Makkar. 2004. Recent applications of biosurfactants as biological and immunological molecules. Current Opinion in Microbiology 7(3): 262–266.

Campos, J.M., T.L. Stamford and L.A. Sarubbo. 2014. Production of a bioemulsifier with potential application in the food industry. Applied Biochemistry and Biotechnology 1; 172(6): 3234–52.

Chávez González, J.D., M.A. Rodríguez Barrera, Y. Romero Ramírez, A. Ayala Sánchez, J.C. Ruvalcaba Ledezma et al. 2016. Pantoeaagglomerans isolated from producer surfactant pasture rhizosphere Tanzania and Llanero. Revista Mexicana de Ciencias Agrícolas 7(4): 961–968.

Čipinytė, V., S. Grigiškis and E. Baškys. 2009. Selection of fat-degrading microorganisms for the treatment of lipid-contaminated environment. Biologija 55(3): 84–92.

Das, P., S. Mukherjee and R. Sen. 2008. Antimicrobial potential of a lipopeptide biosurfactant derived from a marine *Bacillus circulans*. J. Appl. Microbiol. 104: 1675–1684. Doi: 10.1111/j.1365-2672.2007.03701.x.

Das, P., X.P. Yang and L.Z. Ma. 2014. Analysis of biosurfactants from industrially viable Pseudomonas strain isolated from crude oil suggests how rhamnolipids congeners affect emulsification property and antimicrobial activity. Frontiers in Microbiology 22; 5: 696.

Fanaei, F., G. Moussavi and S. Shekoohiyan. 2020. Enhanced treatment of the oil-contaminated soil using biosurfactant-assisted washing operation combined with H2O2-stimulated biotreatment of the effluent. Journal of Environmental Management Oct 1; 271: 110941.

Franzetti, A., I. Gandolfi, C. Raimondi, G. Bestetti, I.M. Banat et al. 2012. Environmental fate, toxicity, characteristics and potential applications of novel bioemulsifiers produced by Variovorax paradoxus 7bCT5. Bioresource Technology 108: 245–51.

Gudina, E.J., J.A. Teixeira and L.R. Rodrigues. 2010. Isolation and functional characterization of a biosurfactant produced by Lactobacillus paracasei. Colloids and Surfaces B: Biointerfaces 1; 76(1): 298–304.

Gudiña, E.J., V. Rangarajan, R. Sen and L.R. Rodrigues. 2013. Potential therapeutic applications of biosurfactants. Trends in Pharmacological Sciences 1; 34(12): 667–75.

Gudiña, E.J., J.A. Teixeira and L.R. Rodrigues. 2016. Biosurfactants produced by marine microorganisms with therapeutic applications. Mar. Drugs 14: E38. 10.3390/md14020038.

Haque, E., K. Kayalvizhi and S. Hassan. 2020. Biocompatibility, antioxidant and anti-infective effect of biosurfactant produced by marinobacterlitoralis MB15. International Journal of Pharmaceutical Investigation 8; 10(2): 172–7.

Jahan, R., A.M. Bodratti, M. Tsianou and P. Alexandridis. 2020. Biosurfactants, natural alternatives to synthetic surfactants: Physicochemical properties and applications. Advances in Colloid and Interface Science 1; 275: 102061.

Kitamoto, D., H. Isoda and T. Nakahara. 2002. Functions and potential applications of glycolipid biosurfactants—from energy-saving materials to gene delivery carriers. Journal of Bioscience and Bioengineering 94(3): 187–201.

Klosowska-Chomiczewska, I., K. Medrzycka and E. Karpenko. 2011. Biosurfactants–biodegradability, toxicity, efficiency in comparison with synthetic surfactants. Adv. Chem. Mech. Eng. 2: 1–9.

Kubicki, S., A. Bollinger, N. Katzke, K.E. Jaeger, A. Loeschcke et al. 2019. Marine biosurfactants: biosynthesis, structural diversity and biotechnological applications. Marine Drugs 17(7): 408.

Lang, S. 2002. Biological amphiphiles (microbial biosurfactants). Current Opinion in Colloid & Interface Science 7(1-2): 12–20.

Luna, J.M., R.D. Rufino, L.A. Sarubbo and G.M. Campos-Takaki. 2013. Characterisation, surface properties and biological activity of a biosurfactant produced from industrial waste by *Candida sphaerica* UCP0995 for application in the petroleum industry. Colloids and Surfaces B: Biointerfaces 1; 102: 202–9.

Mao, X., R. Jiang, W. Xiao and J. Yu. 2015. Use of surfactants for the remediation of contaminated soils: A review. J. Hazard. Mater. 285: 419–435.

Maqsood, M. and M. Seddiq-Shams. 2014. Rhamnolipids: well-characterized glycolipids with potential broad applicability as biosurfactants. Industrial Biotechnology 1; 10(4): 285–91.

Marcelino, P.R., F. Gonçalves, I.M. Jimenez, B.C. Carneiro, B.B. Santos et al. 2020. Sustainable production of biosurfactants and their applications. Lignocellulosic Biorefining Technologies 23: 159–83.

Mnif, I., R. Sahnoun, S. Ellouz-Chaabouni and D. Ghribi. 2017. Application of bacterial biosurfactants for enhanced removal and biodegradation of diesel oil in soil using a newly isolated consortium. Process Safety and Environmental Protection Jul 1; 109: 72–81.

Mulligan, C.N. 2009. Recent advances in the environmental applications of biosurfactants. Current Opinion in Colloid & Interface Science 1; 14(5): 372–8.

Mulligan, C.N., S.K. Sharma and A. Mudhoo. 2014. Biosurfactants. Biosurfactants: Research Trends and Applications, 10: 309.

Muthusamy, K., S. Gopalakrishnan, T.K. Ravi and P. Sivachidambaram. 2008. Biosurfactants: properties, commercial production and application. Current Science, 736–747.

Nguyen, T.T., A. Edelen, B. Neighbors and D.A. Sabatini. 2010. Biocompatible lecithin-based microemulsions with rhamnolipid and sophorolipid biosurfactants: Formulation and potential applications. Journal of Colloid and Interface Science 15; 348(2): 498–504.

Nitschke, M. and S.S.E. Silva. 2018. Recent food applications of microbial surfactants. Critical Reviews in Food Science and Nutrition 58(4): 631–638.

Ohadi, M., G. Dehghannoudeh, H. Forootanfar, M. Shakibaie, M. Rajaee et al. 2018. Investigation of the structural, physicochemical properties, and aggregation behavior of lipopeptide biosurfactant produced by Acinetobacter junii B6. International Journal of Biological Macromolecules 1; 112: 712–9.

Patel, K. and M. Patel. 2020. Improving bioremediation process of petroleum wastewater using biosurfactants producing Stenotrophomonas sp. S1VKR-26 and assessment of phytotoxicity. Bioresource Technology Nov 1; 315: 123861.

Patel, S., A. Homaei, S. Patil and A. Daverey. 2019. Microbial biosurfactants for oil spill remediation: Pitfalls and potentials. Appl. Microbiol. Biotechnol. 2103: 27–37.

Pinto, M.I., B. Ribeiro, J.M. Guerra, R. Rufino, L. Sarubbo et al. 2018. Production in bioreactor, toxicity and stability of a low-cost biosurfactant. Chemical Engineering Transactions 1; 64: 595–600.

Purwasena, I.A., D.I. Astuti, M. Syukron, M. Amaniyah, Y. Sugai et al. 2019. Stability test of biosurfactant produced by Bacillus licheniformis DS1 using experimental design and its application for MEOR. Journal of Petroleum Science and Engineering 1; 183: 106383.

Renard, P., I. Canet, M. Sancelme, N. Wirgot, L. Deguillaume and A.M. Delort. 2016. Screening of cloud microorganisms isolated at the puy de Dôme (France) station for the production of biosurfactants. Atmospheric Chemistry and Physics 16(18): 12347–12358.

Rita de Cássia, F., D.G. Almeida, H.M. Meira, E.J. Silva, C.B. Farias et al. 2017. Production and characterization of a new biosurfactant from Pseudomonas cepacia grown in low-cost fermentative medium and its application in the oil industry. Biocatalysis and Agricultural Biotechnology 1; 12: 206–15.

Rodríguez-López, L., M. Rincón-Fontán, X. Vecino, A.B. Moldes, J.M. Cruz et al. 2020. Biodegradability study of the biosurfactant contained in a crude extract from corn steep water. Journal of Surfactants and Detergents 23(1): 79–90.

Rosenberg, E. and E.Z. Ron. 1997. Bioemulsans: Microbial polymeric emulsifiers. Curr. Opin. Biotechnol. 8: 313–316.

Sajna, K.V. and L.D. Gottumukkala. 2019. Biosurfactants in bioremediation and soil health. In Microbes and Enzymes in Soil Health and Bioremediation. Springer, Singapore, pp. 353–378.

Santos, D.K., Y.B. Brandão, R.D. Rufino, J.M. Luna, A.A. Salgueiro et al. 2014. Optimization of cultural conditions for biosurfactant production from Candida lipolytica. Biocatalysis and Agricultural Biotechnology 1; 3(3): 48–57.

Santos, D.K.F., R.D. Rufino, J.M. Luna, V.A. Santos, L.A. Sarubbo et al. 2016. Biosurfactants: Multifunctional biomolecules of the 21st century. International Journal of Molecular Sciences 17(3): 401.

Singh, R., S.K. Singh and D. Rathore. 2020. Analysis of biosurfactants produced by bacteria growing on textile sludge and their toxicity evaluation for environmental application. Journal of Dispersion Science and Technology 41(4): 510–22.

Sourav, D., S. Malik, A. Ghosh, R. Saha, B. Saha et al. 2015. A review on natural surfactants. RSC Advances 5(81): 65757–65767.

Syldatk, C. and F. Wagner. 1987. Production of biosurfactants. pp. 89–120. *In*: Kosaric, N., W.L. Cairns and N.C.C. Gray (eds.). Biosurfactants and Biotechnology, Surfactant Science Series, Vol. 25. New York: Marcel Dekker Inc.

Syldatk, C., S. Lang, U. Matulovic and F. Wagner. 1987. Production of four interfacial active rhamnolipids from *n*-alkanes or glycerol by resting cells of *Pseudomonas* sp. DSM 2874. Zeitschrift für Naturforschung 40(c): 61–67.

Tayeb, H.H., M. Stienecker, A.P. Middelberg and F. Sainsbury. 2019. Impact of site-specific bioconjugation on the interfacial activity of a protein biosurfactant. Langmuir 26; 35(42): 13588–94.

Ukwungwu, S.V., A.J. Abbas, G.G. Nasr, H.E. Allison, S. Goodman et al. 2017. Wettability effects on Bandera Gray sandstone using biosurfactants. Journal of Engineering Technology 2017.

Van Hamme, J.D., A. Singh and O.P. Ward. 2006. Physiological aspects: Part 1 in a series of papers devoted to surfactants in microbiology and biotechnology. Biotechnology Advances 24(6): 604–620.

Varjani, S.J. and V.N. Upasani. 2017. Critical review on biosurfactant analysis, purification and characterization using rhamnolipid as a model biosurfactant. Bioresource Technology 232: 389–397.

Whang, L.M., P.W.G. Liu, C.C. Ma, S.S. Cheng et al. 2008. Application of biosurfactants, rhamnolipid, and surfactin, for enhanced biodegradation of diesel-contaminated water and soil. J. Hazard. Mater. 151: 155–163.

Zhong, H., X. Yang, F. Tan, M.L. Brusseau, L. Yang et al. 2016. Aggregate-based sub-CMC solubilization of *n*-alkanes by monorhamnolipid biosurfactant. New J. Chem. 40: 2028–2035.

Zhou, Y., S. Harne and S. Amin. 2019. Optimization of the surface activity of biosurfactant-surfactant mixtures. Journal of Cosmetic Science 70(3): 127–36.

Index

www.ingramcontent.com/pod-product-compliance
Lightning Source LLC
LaVergne TN
LVHW010627300926
849162LV00001B/87
* 9 7 8 1 0 3 2 1 9 6 3 5 0 *